AF745991

The Cell Biology of Stem Cells

ADVANCES IN EXPERIMENTAL MEDICINE AND BIOLOGY

Editorial Board:
NATHAN BACK, *State University of New York at Buffalo*
IRUN R. COHEN, *The Weizmann Institute of Science*
ABEL LAJTHA, *N.S. Kline Institute for Psychiatric Research*
JOHN D. LAMBRIS, *University of Pennsylvania*
RODOLFO PAOLETTI, *University of Milan*

Recent Volumes in this Series

Volume 687
BCL-2 PROTEIN FAMILY: ESSENTIAL REGULATORS OF CELL DEATH
Edited by Claudio Hetz

Volume 688
SPHINGOLIPIDS AS SIGNALING AND REGULATORY MOLECULES
Edited by Charles Chalfant and Maurizio Del Poeta

Volume 689
HOX GENES: STUDIES FROM THE 20TH TO THE 21ST CENTURY
Edited by Jean S. Deutsch

Volume 690
THE RENIN-ANGIOTENSIN SYSTEM: CURRENT RESEARCH PROGRESS IN THE PANCREAS
Edited by Po Sing Leung

Volume 691
ADVANCES IN TNF FAMILY RESEARCH
Edited by David Wallach

Volume 692
NEUROPEPTIDE SYSTEMS AS TARGETS FOR PARASITE AND PEST CONTROL
Edited by Timothy G. Geary and Aaron G. Maule

Volume 693
POST-TRANSCRIPTIONAL REGULATION BY STAR PROTEINS
Edited by Talila Volk and Karen Artzt

Volume 694
PROTEIN METABOLISM AND HOMEOSTASIS IN AGING
Edited by Nektarios Tavernarakis

Volume 695
THE CELL BIOLOGY OF STEM CELLS
Edited by Eran Meshorer and Kathrin Plath

A Continuation Order Plan is available for this series. A continuation order will bring delivery of each new volume immediately upon publication. Volumes are billed only upon actual shipment. For further information please contact the publisher.

The Cell Biology of Stem Cells

Edited by
Eran Meshorer, PhD
Department of Genetics, Institute of Life Sciences, The Hebrew University of Jerusalem, Jerusalem, Israel

Kathrin Plath, PhD
UCLA School of Medicine, Department of Biological Chemistry, Los Angeles, California, USA

Springer Science+Business Media, LLC

Landes Bioscience

LOYOLA UNIVERSITY MEDICAL CENTER LIBRARY

QU
325
C392
2010

Springer Science+Business Media, LLC
Landes Bioscience

Copyright ©2010 Landes Bioscience and Springer Science+Business Media, LLC

All rights reserved.
No part of this book may be reproduced or transmitted in any form or by any means, electronic or mechanical, including photocopy, recording, or any information storage and retrieval system, without permission in writing from the publisher, with the exception of any material supplied specifically for the purpose of being entered and executed on a computer system; for exclusive use by the Purchaser of the work.

Printed in the USA.

Springer Science+Business Media, LLC, 233 Spring Street, New York, New York 10013, USA
http://www.springer.com

Please address all inquiries to the publishers:
Landes Bioscience, 1002 West Avenue, Austin, Texas 78701, USA
Phone: 512/ 637 6050; FAX: 512/ 637 6079
http://www.landesbioscience.com

The chapters in this book are available in the Madame Curie Bioscience Database.
http://www.landesbioscience.com/curie

The Cell Biology of Stem Cells, edited by Eran Meshorer and Kathrin Plath. Landes Bioscience / Springer Science+Business Media, LLC dual imprint / Springer series: Advances in Experimental Medicine and Biology.

ISBN: 978-1-4419-7036-7

While the authors, editors and publisher believe that drug selection and dosage and the specifications and usage of equipment and devices, as set forth in this book, are in accord with current recommendations and practice at the time of publication, they make no warranty, expressed or implied, with respect to material described in this book. In view of the ongoing research, equipment development, changes in governmental regulations and the rapid accumulation of information relating to the biomedical sciences, the reader is urged to carefully review and evaluate the information provided herein.

Library of Congress Cataloging-in-Publication Data

The cell biology of stem cells / edited by Eran Meshorer, Kathrin Plath.
p. ; cm. -- (Advances in experimental medicine and biology ; v. 695)
Includes bibliographical references and index.
ISBN 978-1-4419-7036-7
1. Stem cells. I. Meshorer, Eran. II. Plath, Kathrin. III. Series: Advances in experimental medicine and biology ; v. 695.
[DNLM: 1. Stem Cells--physiology. W1 AD559 v.695 2010 / QU 325 C3925 2010]
QH588.S83.C443 2010
616'.02774--dc22

2010024633

PREFACE

Stem cells have been gaining a lot of attention in recent years. Their unique potential to self-renew and differentiate has turned them into an attractive model for the study of basic biological questions such as cell division, replication, transcription, cell fate decisions, and more. With embryonic stem (ES) cells that can generate each cell type in the mammalian body and adult stem cells that are able to give rise to the cells within a given lineage, basic questions at different developmental stages can be addressed. Importantly, both adult and embryonic stem cells provide an excellent tool for cell therapy, making stem cell research ever more pertinent to regenerative medicine.

As the title *The Cell Biology of Stem Cells* suggests, our book deals with multiple aspects of stem cell biology, ranging from their basic molecular characteristics to the in vivo stem cell trafficking of adult stem cells and the adult stem-cell niche, and ends with a visit to regeneration and cell fate reprogramming. In the first chapter, "Early embryonic cell fate decisions in the mouse", Amy Ralson and Yojiro Yamanaka describe the mechanisms that support early developmental decisions in the mouse pre-implantation embryo and the current understanding of the source of the most immature stem cell types, which includes ES cells, trophoblast stem (TS) cells and extraembryonic endoderm stem (XEN) cells. From the derivation of these stem cell types, we turn to examining the nuclear architecture and genome organization of pluripotent ES cells in the second chapter "Nuclear architecture in stem cells" by Kelly Morris, Mita Chotalia and Ana Pombo. The chapter addresses the structure and function of the three-dimensional space of the nucleus in ES cells, emphasising the unique properties of chromatin, nuclear bodies and gene positioning in these cells. ES cell epigenetics is analyzed in more depth in the third chapter "Epigenetic regulation of pluripotency" by Eleni Tomazou and Alexander Meissner. The authors describe the epigenetic profiles of key chromatin modifications, including DNA methylation and histone modifications, and discuss functional aspects of these epigenetic marks. Remaining at the DNA level, the fourth chapter, "Autosomal lyonization of replication domains during early mammalian development", by Ichiro Hiratani and David Gilbert, illustrates the dynamics and regulation of DNA replication in ES cells by taking us through 50 years of research history of this exciting field, reviving the old concept of 'autosomal lyonization' to explain the process of heterochromatinization.

Genomic DNA, the fundamental unit of life, is constantly being damaged and repaired. Peter Stambrook and Elisia Tichy discuss mutation rates, signaling pathways and the mechanisms of DNA damage and repair in ES cells in their chapter, "Preservation of genomic integrity in mouse embryonic stem cells". Having talked about DNA packaging, replication and damage, the book now turns to focus on RNA with the sixth chapter, "Transcriptional regulation in embryonic stem cells", by Jian-Chien Dominic Heng and Huck-Hui Ng. This chapter discusses the transcriptional networks that are at the heart of the pluripotent state and describes the recent technological advances that allow a systemic look at transcriptional regulation in ES cells and during their differentiation. From transcriptional control, we continue to RNA splicing. David Nelles and Gene Yeo authored the seventh chapter entitled "Alternative splicing in stem cell self-renewal and differentiation", in which they review the recent literature on splicing, highlighting several key examples of alternatively spliced genes in ES cells, and address novel genome-wide approaches to analyze splicing and alternative splicing patterns at a global scale. Chapter eight, "MicroRNA regulation of embryonic stem cell self-renewal and differentiation" by Collin Melton and Robert Blelloch, elucidates microRNA regulation in ES cells, emphasizing several prominent examples of microRNAs, including Let-7, Lin-28, miR-134, miR-296 and others, that regulate self-renewal and/or pluripotency of ES cells. Chapter 9, "Telomeres and telomerase in adult stem cells & pluripotent embryonic stem cells" by Rosa Marión and Maria Blasco gives an overview of telomere biology and telomerase regulation in multipotent and pluripotent cells, discussing the potential mechanisms enabling the remodeling of telomeric chromatin during nuclear reprogramming from somatic cells to pluripotency. In the mouse, nuclear reprogramming to pluripotency also entails the reactivation of the somatically silenced X chromosome in female cells. The next chapter, "X chromosome inactivation and embryonic stem cells" by Tahsin Stefan Barakat and Joost Gribnau discusses the regulation of X chromosome inactivation (XCI) as female ES cells are induced to differentiate and explains the cis-and trans-acting mechanisms that act in concert to precisely orchestrate this transcriptional silencing of an entire chromosome, while presenting hypotheses for why this intriguing process occurs in female cells only.

Having covered the molecular biology in the nucleus of pluripotent ES cells, the next three chapters deal with somatic or adult stem cells. While pluripotent cells only exist during a brief phase in early embryonic development, adult stem cell populations are maintained throughout the entire lifespan of the organism until they are required for tissue homeostasis and/or repair. The signals that keep adult stem cells in check and regulate their differentiation versus self-renewal are thought to be controlled by interactions with the cells and extracellular matrix that constitute the stem cell niche. In Chapter 11, "Adult stem cells and their niches", Francesca Ferraro, Cristina Celso and David Scadden explain the niche concept, discuss the signaling pathways that operate at different mammalian niches, and link the current understanding of niche biology to carcinogenesis and aging. In Chapter 12 "Adult stem cell differentiation and trafficking and their implications in disease", Ying Zhuge, Zhao-Jun Liu and Omaida Velazquez present trafficking of hematopoietic stem cells (HSCs), mesenchymal stem cells (MSCs), and endothelial progenitor cells (EPCs) and discuss the mechanisms that control their regulated movement in mammals. Zhuge et al. also explain how understanding these fundamental processes may translate into therapeutic applications.

In the next chapter, "Vertebrates that regenerate as models for guiding stem cells", Christopher Antos and Elly Tanaka focus on the mechanisms of regeneration in several vertebrate animal models such as frog, fish and salamander. They describe the extensive cellular plasticity involved in the regeneration of several structures: the eye, heart, nervous system and appendages, and they summarize some of the molecules that underlie transdifferentiation and dedifferentiation in select tissues. The final chapter of the book "Reprogramming of somatic cells to pluripotency" by Masato Nakagawa and Shinya Yamanaka comes to the most recent exciting development in stem cell biology: cellular reprogramming to pluripotency. The authors give a brief history of somatic cell nuclear transfer experiments conducted in frog oocytes in the '50s and '60s, discuss cell fusion experiments leading to reprogrammed cells, albeit tetraploid, and describe their lab's own seminal contribution to the reprogramming field—the generation of pluripotent cells from somatic cells upon expression of a specific set of transcription factors—leading to the new thriving field of induced pluripotent stem (iPS) cells.

Human adult and embryonic stem cells, and now induced pluripotent stem cells, could be used for the generation of cells and tissues for cell-based therapies. With iPS cells, one is now able to generate patient-specific pluripotent cells with tremendous potential for disease studies and drug screenings. To be able to take full advantage of the huge capacity of stem cells, our knowledge of the underlying biology still needs to grow. In its 14 chapters, *The Cell Biology of Stem Cells* provides much of the current understanding of the cell biology of stem cells and discusses many of the open questions that remain to be answered.

Eran Meshorer, PhD
Department of Genetics, Institute of Life Sciences, The Hebrew University of Jerusalem, Jerusalem, Israel

Kathrin Plath, PhD
UCLA School of Medicine, Department of Biological Chemistry, Los Angeles, California, USA

ABOUT THE EDITORS...

ERAN MESHORER, PhD, studies chromatin plasticity in embryonic and neuronal stem cells at the Department of Genetics at the Hebrew University of Jerusalem. He received his PhD in Molecular Neuroscience from the Hebrew University and conducted his post-doctoral studies at the National Cancer Institute, NIH. His lab focuses on understanding pluripotency, differentiation and reprogramming from a chromatin perspective, taking both genome-wide and single cell approaches. He is a member of the International Society for Stem Cell Research and holds the Joseph H. and Belle R. Braun Senior Lectureship in Life Sciences.

ABOUT THE EDITORS...

KATHRIN PLATH, PhD, is an Assistant Professor in the Department of Biological Chemistry at the University of California Los Angeles since 2004. After she received her PhD from the Humboldt University of Berlin, Germany, she was at the University of California, San Francisco and the Whitehead Institute in Cambridge, MA for her postdoctoral studies. Dr. Plath's main research interest is to understand how developmental cues induce changes in chromatin structure at the molecular level and how these changes regulate cell fate decisions and gene expression in mammalian development. She is a member of the International Society for Stem Cell Research and of the editorial board of several stem cell journals.

PARTICIPANTS

Christopher L. Antos
DFG-Center for Regenerative Therapies
Dresden Technische Universität
Dresden
Germany

Tahsin Stefan Barakat
Department of Reproduction
and Development
Erasmus MC
University Medical Center
Rotterdam
The Netherlands

Maria A. Blasco
Telomeres and Telomerase Group
Molecular Oncology Program
Spanish National Cancer Centre (CNIO)
Madrid
Spain

Robert Blelloch
The Eli and Edythe Broad Center
of Regeneration Medicine
and Stem Cell Research
Center for Reproductive Sciences
Program in Biomedical Science
Department of Urology
University of California San Francisco
San Francisco, California
USA

Cristina Lo Celso
Division of Cell and Molecular Biology
Imperial College London
London, England
UK

Mita Chotalia
Genome Function Group
MRC Clinical Sciences Centre
Imperial College School of Medicine
Hammersmith Hospital Campus
London, England
UK

Francesca Ferraro
Center for Regenerative Medicine
Massachusetts General Hospital and
Harvard Medical
Department of Stem Cell
and Regenerative Biology
Harvard University
and
Harvard Stem Cell Institute
Cambridge, Massachusetts
USA

David M. Gilbert
Department of Biological Science
Florida State University
Tallahassee, Florida
USA

Joost Gribnau
Department of Reproduction and Development
Erasmus MC
University Medical Center
Rotterdam
The Netherlands

Jian-Chien Dominic Heng
Gene Regulation Laboratory
Genome Institute of Singapore
and
NUS Graduate School for Integrative Sciences and Engineering
Singapore

Ichiro Hiratani
Department of Biological Science
Florida State University
Tallahassee, Florida
USA

Zhao-Jun Liu
Division of Vascular Surgery
DeWitt Daughtry Family Department of Surgery
and
Sylvester Comprehensive Cancer Center
Miller School of Medicine
University of Miami
Miami, Florida
USA

Rosa M. Marión
Telomeres and Telomerase Group
Molecular Oncology Program
Spanish National Cancer Centre (CNIO)
Madrid
Spain

Alexander Meissner
Department of Stem Cell and Regenerative Biology
Harvard University
and
Harvard Stem Cell Institute
and
Broad Institute
Cambridge, Massachusetts
USA

Collin Melton
The Eli and Edythe Broad Center of Regeneration Medicine
Stem Cell Research Center for Reproductive Sciences
Program in Biomedical Science and Department of Urology
University of California San Francisco
San Francisco, California
USA

Eran Meshorer
Department of Genetics
Institute of Life Sciences
The Hebrew University of Jerusalem, Jerusalem
Israel

Kelly J. Morris
Genome Function Group
MRC Clinical Sciences Centre
Imperial College School of Medicine
Hammersmith Hospital Campus
London, England
UK

Masato Nakagawa
Center for iPS Research and Application
Institute for Integrated Cell-Material Sciences
Kyoto University
Kyoto
Japan

David A. Nelles
Department of Cellular and Molecular Medicine
Stem Cell Program
University of California, San Diego
La Jolla, California
USA

Huck-Hui Ng
Gene Regulation Laboratory
Genome Institute of Singapore
and
NUS Graduate School for Integrative Sciences and Engineering
Singapore

Kathrin Plath
UCLA School of Medicine
Department of Biological Chemistry
Los Angeles, California
USA

Ana Pombo
Genome Function Group
MRC Clinical Sciences Centre
Imperial College School of Medicine
Hammersmith Hospital Campus
London, England
UK

Amy Ralston
Department of Molecular, Cell, and Developmental Biology
University of California Santa Cruz
Santa Cruz, California
USA

David Scadden
Department of Stem Cell and Regenerative Biology
Harvard University
and
Harvard Stem Cell Institute
Cambridge, Massachusetts
and
Cancer Center, Massachusetts General Hospital
Boston, Massachusetts
USA

Peter J. Stambrook
Department of Molecular Genetics, Biochemistry and Microbiology
University of Cincinnati College of Medicine
Cincinnati, Ohio
USA

Elly M. Tanaka
Max-Planck Institute for Molecular Cell Biology and Genetics
Dresden
Germany

Elisia D. Tichy
Department of Molecular Genetics, Biochemistry and Microbiology
University of Cincinnati
College of Medicine
Cincinnati, Ohio
USA

Eleni M. Tomazou
Department of Stem Cell Regenerative Biology
Harvard University
Cambridge, Massachusetts
USA

Omaida C. Velazquez
Division of Vascular Surgery
DeWitt Daughtry Family Department of Surgery
and
Sylvester Comprehensive Cancer Center
Miller School of Medicine
University of Miami
Miami, Florida
USA

Shinya Yamanaka
Center for iPS Research and Application
Institute for Integrated Cell-Material Sciences
Kyoto University
Kyoto
Japan

Yojiro Yamanaka
Goodman Cancer Center
Department of Human Genetics
McGill University
Montreal, Quebec
Canada

Gene W. Yeo
Department of Cellular and Molecular Medicine
Stem Cell Program
University of California, San Diego
La Jolla, California
USA

Ying Zhuge
Division of Vascular Surgery
DeWitt Daughtry Family Department of Surgery
Miller School of Medicine
University of Miami
Miami, Florida
USA

CONTENTS

CHAPTER 1

EARLY EMBRYONIC CELL FATE DECISIONS IN THE MOUSE

Yojiro Yamanaka* and Amy Ralston*

Abstract: During development, initially totipotent cells of the embryo specialize to form discrete tissue lineages. The first lineages to form in the mouse are the extraembryonic tissues. Meanwhile, cells that do not become extraembryonic retain a pluripotent fate since they can give rise to all the germ layers of the fetus. Pluripotent stem cell lines have been derived from the fetal lineage at several stages of development. Interestingly, multipotent stem cell lines have been derived from the extraembryonic lineages around the same time. Examining the regulation of early embryonic cell fate decisions is therefore a rare opportunity to examine establishment of stem cell progenitors. Classical studies have provided considerable insight into specification of the first three lineages and use of modern molecular and imaging techniques has advanced this field further. Here we describe current understanding of the diverse molecular mechanisms that lead to establishment and maintenance of the first three lineages during mouse development.

INTRODUCTION

During the earliest days of mouse development, initially totipotent cells become restricted in their developmental potential to give rise to the first lineages of the mouse. While in nonmammalian species the first lineage decisions might involve specification of the major body axes, mammals have an altogether different first priority: implantation. Thus discrimination between fetal and extraembryonic tissue lineages comprises the first two lineage decisions (Fig. 1) and precedes establishment of the germ layers (ectoderm, mesoderm, endoderm) and the germline by several days. This uniquely mammalian developmental strategy involves unique cell types that can be isolated and expanded

*Corresponding Authors: Yojiro Yamanaka—Goodman Cancer Center, Department of Human Genetics, Faculty of Medicine, McGill University, Montreal, QC H3A1A3. Email: yojiro.yamanaka@mcgill.ca. Amy Ralston—Department of Molecular, Cell and Developmental Biology, University of California Santa Cruz, California, USA. Email: ralston@biology.ucsc.edu

The Cell Biology of Stem Cells, edited by Eran Meshorer and Kathrin Plath.
©2010 Landes Bioscience and Springer Science+Business Media.

in culture as stable stem cell lines. Understanding the origins of the extraembryonic tissues therefore illuminates our understanding of establishment and differentiation of stem cells. Classical studies provided considerable insight into specification of the first three lineages and use of modern molecular and imaging techniques has advanced this field further.

Three days after fertilization, the mouse embryo, or blastocyst contains three tissue lineages: epiblast (EPI), trophectoderm (TE) and primitive endoderm (PE). Isolation and study of stem cell lines from these lineages has reinforced and extended our understanding of early embryonic cell fate decisions. Three types of stem cell lines have been derived from the blastocyst: embryonic, trophoblast and extraembryonic endoderm stem cells (ES, TS and XEN cells). Each of these exhibits stem cell properties, such as the ability to either self-renew or to differentiate into multiple mature cell types. Yet each stem cell line exhibits features of the lineage from which it derives, including tissue-specific developmental potential, morphology, transcription factor expression and growth factor requirements.[1] These stem cell lines not only provide an expandable source of pure cell populations for studies requiring large amounts of starting material, but they provide an opportunity to understand where stem cells come from.

Studies performed in ES cells have enabled deeper molecular analysis of the role of genes in cell fate selection. Manipulation of levels of certain lineage-regulating genes causes corresponding changes in stem cell fate. For example, the trophoblast transcription factor Cdx2 is sufficient to convert ES cells to TS-like cells.[2] These kinds of observations demonstrate the remarkable plasticity of ES cells, as well as the central role of genes such as *Cdx2* as lineage-determining factors. ES cells also provide an opportunity to examine molecular interactions between lineage-determining genes and thus serve as a model for understanding cell fate selection in the embryo. However, examination of the role of lineage-determining genes in the embryo has revealed that lineage-determining genes play a relatively late role in lineage specification, raising the question as to how the first three lineages are initially specified.

A variety of mechanisms are probably involved in specifying the first lineages, including cell position, shape, polarization, signaling and division plane. A new paradigm is emerging, in which an early pre-stem cell program specifies the tissue lineages as the blastocyst forms. Later, around the time of implantation and thereafter, cell fates are maintained by a program that is active in stem cell lines (Fig. 2).

LINEAGE ESTABLISHMENT AND THE PRE-STEM CELL PROGRAM: FORMATION OF THE BLASTOCYST

Here we will consider the first phase of lineage specification: establishment of the TE and inner cell mass (ICM) as the blastocyst forms. The TE will give rise to placenta, while the ICM contains a mixture of fetal and primitive endoderm progenitors. In the blastocyst, the TE surrounds the ICM and hollow blastocoel and lineage-tracing experiments have shown that TE and ICM populations begin as the outside and inside cell populations of the embryo.[3] That is, as cell cleavage partitions the zygote into two, four, eight and sixteen cells, a small number of cells become enclosed by outside cells. Continued cleavages increase numbers of inside and outside cells, the TE epithelializes and the blastocoel expands, forming the blastocyst structure. The mechanism by which topology becomes linked to cell fate has been elusive. Several models have been put

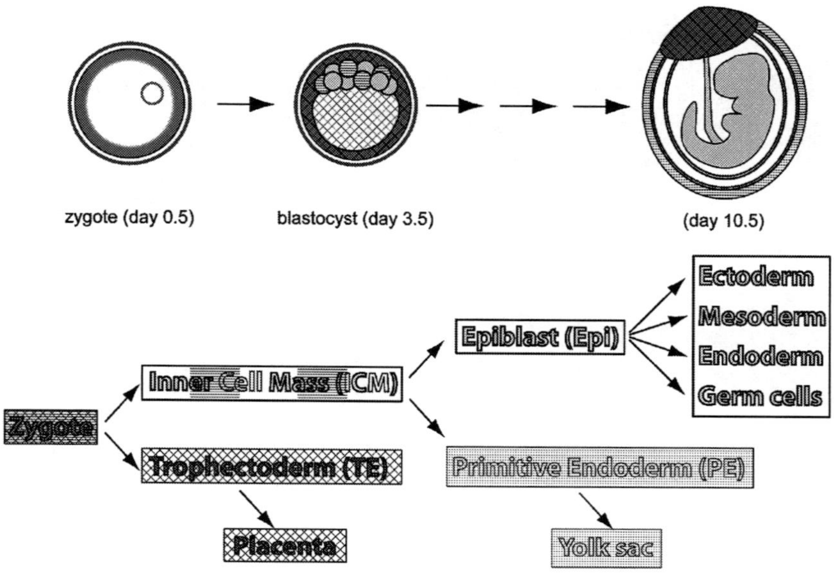

Figure 1. Overview of the first two lineages decisions during mouse development. The initially totipotent zygote develops to the blastocyst, which contains three lineages: EPI (blue), TE (red, crosshatched) and PE (yellow, lined). These lineages will give rise to the fetus, the placenta and a portion of the yolk sac at later stages of development. A color version of this figure is available at www.landesbioscience.com/curie.

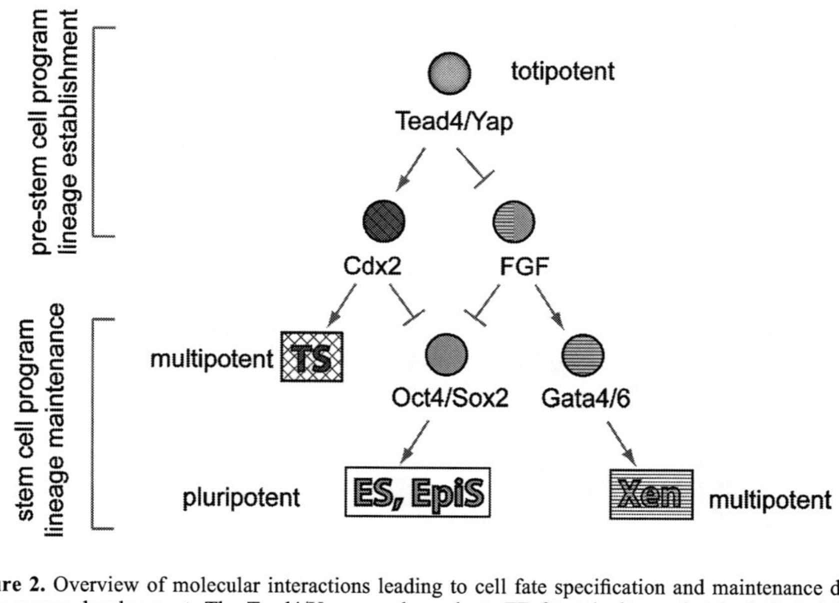

Figure 2. Overview of molecular interactions leading to cell fate specification and maintenance during early mouse development. The Tead4/Yap complex selects TE fates (red, crosshatched) from initially totipotent cells (grey). Cells that do not become TE, then adopt a mixture of EPI (blue) and PE (yellow, lined) fates. Signaling within this lineage facilitates the sorting out of EPI and PE fates. Lineage-specific transcription factors participate in maturation of each lineage. A color version of this figure is available at www.landesbioscience.com/curie.

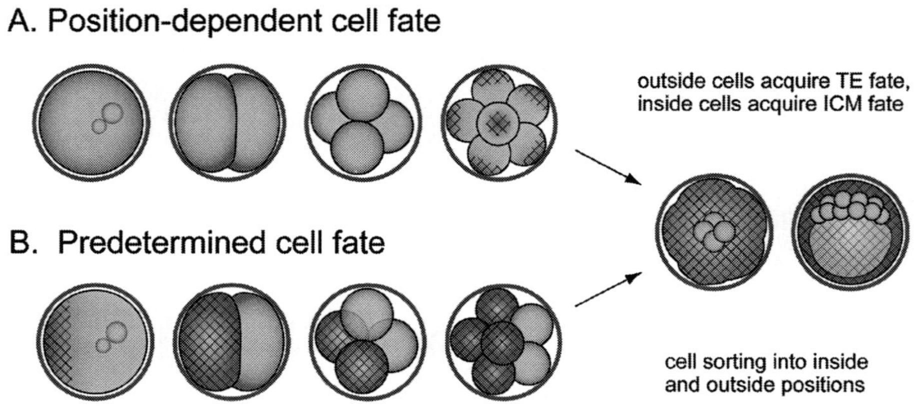

Figure 3. Two possible models of TE specification. A) Cell position dictates cell fate, as outer cells, or outer portions of cells, adopt TE cell fate (red, crosshatched). B) TE fate is predetermined and a specific subset of cells inherits TE fate-determining molecules. A color version of this figure is available at www.landesbioscience.com/curie.

forward. For example, cell fate could be a consequence of cell position (Fig. 3A). Alternatively, predetermined cell fates could drive cells into appropriate topological positions (Fig. 3B). This latter mechanism predicts that pre-inside and pre-outside cells would be detectable prior to formation of overt inside and outside cell populations. In spite of extensive effort in the field, however, there is currently no support for this predetermination mechanism.

Two main strategies have been used to look for evidence of predetermination among cells prior to the blastocyst stage: lineage tracing and molecular analysis. In terms of lineage tracing, reports of biased developmental potential among cells at the two-cell stage[4-13] are not relevant to the TE/ICM lineage decision since these studies demonstrate contribution of both cells to the TE and ICM. Likewise, all cells of four and eight-cell embryos can also contribute to both TE and ICM lineages.[14,15] Although one group reported restricted lineage potential from the four-cell stage,[7] extraembryonic lineages were incompletely scored. Thus there is no evidence from lineage tracing experiments to suggest that cells are predetermined to make TE or ICM prior to formation of inside and outside groups. In terms of molecular analyses, no protein has been detected within a subset of cells prior to the 16-cell stage that instructs the TE/ICM lineage decision. The level of one type of histone methylation is reported to exhibit uneven distribution among blastomeres at the 4-cell stage and correlates with reduced potential to contribute to viable mice in chimeras.[16] The functional importance of these observations in TE/ICM lineage specification needs to be clarified. Therefore, no molecular evidence supports the existence of pre-TE or pre-ICM cells prior to formation of inside and outside cell populations. Rather, inside and outside cells could acquire fates once they have acquired their positions within the embryo.

If cell position acts upstream of cell fate, mechanisms must exist for cells to sense their position within the embryo. Longstanding evidence that cells polarize around the 8-cell stage[17] supports the claim that there are differences along the inside/outside axis at the cellular level. Polarization by conserved polarity proteins such as atypical PKC (aPKC),

Par3 and Par6 is required for maintaining cell position[8] and cell contact has been shown to be required for cell polarization.[17] However the link between position, polarization and cell fate has not been examined at the molecular level. This area is challenging to study using conventional knockout techniques. Many of the proteins involved in cell position and cell contact, such as aPKC, are members of large gene families, suggesting that genetic redundancy may mask their requirements in single gene knockout studies. In addition, this early developmental stage may be regulated in part by maternally supplied protein, requiring germline gene deletion to detect a phenotype. Finally, many of these proteins are involved in basic cellular processes, such as cell division, making it difficult to study their effects during development. On the other hand, overexpression of dominant-negative or siRNA constructs leads to only short-term or partial loss of function, which can also impede phenotype resolution.

Ultimately, to convert inside/outside differences into changes in gene expression, a differentially localized transcription factor is needed. Several strategies have led to the identification of transcription factors involved in early lineage development. Candidates have been identified by microarray analysis of transcripts expressed in pre-implantation development, followed by in situ hybridization to screen for those with restricted expression in the blastocyst.[18] Alternatively, candidates have been identified by microarray comparison of blastocyst-derived stem cell lines.[19] Advances have also come from fortuitous discovery of an unexpectedly early lethal phenotype in knockouts,[20-22] which led to identification of *Cdx2* and *Tead4*.

While required for TE development, *Cdx2* probably does not play an instructive role in TE formation.[23,24] Nevertheless, *Cdx2* mRNA,[25] but not protein,[24,26] has been reported to localize to the outside surface of cells at the 8-cell stage. Since *Cdx2* is not required for specification of TE at either morphological[23,24] or molecular levels, evidenced by the continued expression of the TE marker Gata3 in *Cdx2* null embryos,[19] it is difficult to imagine that localized *Cdx2* mRNA plays an instructive role in lineage establishment. Recently, a new pathway, involving Tead4 and cofactors, has been shown to play an instructive role in the first lineage decision. The transcriptional coactivator Yap and related protein Taz, exhibit cell position-sensitive changes in activation of Cdx2 expression.[27] Prior to the blastocyst stage, Yap/Taz localize to nuclei of outside cells and cytoplasms of inside cells. This localization is regulated by phosphorylation by the Hippo signaling pathway members Lats1/2. In addition, manipulation of cell position led to corresponding changes in Yap localization: outside cells embedded inside an aggregate of cells lost nuclear Yap, while inside cells stripped of surrounding outer cells acquired nuclear Yap. Yap/Taz interact directly with Tead4 a DNA binding protein required for expression of Cdx2[21,22] and other trophectoderm markers.[19] The identity or nature of Yap/Taz-regulating signals that can sense cell position are unknown, but probably involve the Hippo signaling pathway and possibly proteins involved in cell contact such as cadherins. This will undoubtedly be an exciting area of research to follow in the future.

Besides what is working upstream of Yap/Tead4, it is not entirely clear what is working downstream. *Tead4* is required for Cdx2 expression, but *Tead4* null embryos die prior to blastocyst formation, while *Cdx2* null embryos die after blastocyst formation. *Tead4* is not required in the ICM,[21,22] so additional genes must operate in parallel to *Cdx2* in the TE. Some of these, such as *Gata3* are beginning to be identified.[19] It will be important to identify Tead4 targets that participate in promoting outside cell proliferation and construction of the blastocyst.

LINEAGE MAINTENANCE AND THE STEM CELL PROGRAM: BEYOND THE BLASTOCYST

In the blastocyst, interactions between lineage-determining transcription factors reinforce TE and ICM fates established at earlier stages. Central players at this stage are *Oct4* (*Pou5f1*) and *Cdx2*. *Oct4* is required for maturation of the ICM,[28] while *Cdx2* is required for maturation of the TE.[23] Mutual antagonism between these two factors was initially speculated to cause the first lineage decision. *Cdx2* is required for repression of *Oct4* and other ICM genes in the TE of the blastocyst.[23] But the TE still forms in *Cdx2* null embryos and other TE markers are still expressed.[19] Similarly, Oct4 represses Cdx2 in the ICM, but not until implantation, a full day after blastocyst formation.[19] Thus lineage specification is initially normal in the absence of either *Oct4* or *Cdx2*, but embryos fail to maintain correct expression of lineage genes. Nevertheless, in spite of adoption of ICM gene expression, *Cdx2* null TE does not fully adopt ICM fate. The TE marker Gata3 is still expressed in the TE of *Cdx2* null embryos[19] and *Cdx2* null embryos exhibit higher levels of apoptosis in the TE than do wild type embryos.[23] Cdx2 must therefore enable survival and/or proliferation of cells that are already committed to being TE. This is consistent with its continued expression in the proliferative region of the trophoblast at later stages.[29] The reason for the lethality of *Oct4* null embryos is currently unclear.

The antagonistic relationship between Oct4 and Cdx2 is borne out by stem cells from the blastocyst. ES cells cannot be derived from *Oct4* null embryos and TS cells cannot be derived from *Cdx2* null embryos.[23,28] Loss of *Oct4* from existing ES cell lines leads to upregulation of *Cdx2* and formation of TS-like cells in the presence of TS cell culture medium.[30] Similarly, overexpression of *Cdx2* in ES cells leads to repression of *Oct4* and formation of TS-like cells.[2] Other trophoblast factors, such as Eomes and Gata3 can also induce trophoblast gene expression in ES cells[2,19] and these also play relatively late roles in trophoblast maturation rather than allocation.[23,31,32] Maintenance of the TE/ICM lineage restriction in stem cells therefore appears to use genetic programs that become active once the blastocyst has formed. This makes sense given that stem cell derivation requires culture beyond the blastocyst stage. Understanding the further development of the ICM, however, requires a look at the second lineage decision in development, discussed next.

THE SECOND LINEAGE DECISION: SUBDIVIDING THE ICM

Three days after fertilization, the ICM of the blastocyst contains two cell types: the epiblast (EPI) and the primitive endoderm (PE). Only the EPI gives rise to the fetus, whereas the PE is an extraembryonic lineage, which contributes to the yolk sac (Fig. 1).[33-36] The PE lineage plays two important roles just after implantation. The first is that it provides nutrients to the embryo and the second is that it serves as a signaling center that helps confer anterior-posterior polarity upon the gastrulating embryo.[37] As for the TE lineage, a special stem cell line can be derived from the PE lineage. Multipotent stem cell lines, called XEN cells, have been derived from the PE lineage (Fig. 2).[38] In addition, PE-like cells can be induced from ES cells by overexpression of PE transcription factors, such as *Gata4* and *Gata6*.[39] Yet *Gata4/6* act relatively late in PE development,[40,41] suggesting that, as for the TE lineage, the PE is specified by a mechanism acting upstream of the stem cell genes. Insight into specification of the PE lineage has revealed a unique cell signaling-based strategy.

Heterogeneity and Progenitor Sorting

Four days after fertilization, the blastocyst implants. At this stage, the PE appears as a distinct monolayer on the blastocoel surface of the ICM. For this reason, the PE was originally assumed to arise from ICM cells directly facing the blastocoel around the time of implantation. Microenvironmental differences between blastocoel-facing and deeper cells were postulated to participate in lineage specification at this stage. However, recent studies have shown that EPI and PE progenitors can be detected in the blastocyst one full day before implantation.[36,42,43] At this stage, the ICM appears as a mixed population of EPI and PE progenitors, expressing lineage-specific transcription factors. Prior to this stage, Nanog and Gata6 are coexpressed in all cells of the ICM and expression gradually becomes mutually exclusive to specify the two progenitors in a position-independent manner during blastocyst expansion.[36,44] Notably, there is no stereotyped pattern of distribution of the two progenitors within the ICM. Rather, they are sprinkled randomly throughout the ICM like salt and pepper.

These results suggest that the two randomly distributed lineage progenitors sort out to form two morphologically distinct layers by implantation. Indeed, support for this model has been provided by live imaging of blastocyst expansion in transgenic mice expressing fluorescent lineage markers. In the *Pdgfra*$^{H2B\text{-}GFP}$ mouse line histone H2B-GFP is expressed in the PE and revealed that separation of the two lineages involves both apoptosis and cell migration.[36] Cells within the growing ICM appear to rearrange constantly,[36,45] but once PE progenitors come to the ICM surface they stay there. Consistent with this, the maturation of the PE takes place progressively and this is correlated with position within the ICM.[46] One outstanding question is whether PE cells sort out by directional cell movement or a combination of random movement and position recognition.

Several mutants exhibit a defect in formation of a cohesive PE layer.[47-51] In these mutants, Gata4-expressing, presumptive PE cells, are found clustered within the middle of the ICM, suggesting that PE progenitors are specified but fail to form a morphologically distinct surface layer. This contrasts with the TE, in which lineage allocation (position) precedes lineage specification. For the PE, lineage specification precedes allocation. Understanding how PE fates are selected from within the ICM is therefore key to understanding PE/EPI lineage choice.

CELL SIGNALING REGULATES PE/EPI SPECIFICATION

Early heterogeneity in the ICM suggests that position-independent mechanisms regulate specification of PE and EPI lineages. FGF signaling has been shown to be necessary for PE formation in vivo and in vitro.[52-54] How extracellular signaling pathways, such as the FGF signaling pathway, could participate in the generation of a salt and pepper distribution of PE and EPI within the ICM is not clear. For example, certain pre-PE cells within the embryo could be predisposed to respond to signals, or cells could randomly receive signals and thereby become PE progenitors.

These possibilities are summarized in two models: the origin-dependent model and the signaling-dependent model (Fig. 4A,B).[17,55] The origin-dependent model relies on understanding the process of inner cell generation during the cleavage stages.[56] Inner cells of the morula, which will become the ICM of the blastocyst, are generated from two rounds of asymmetric divisions at 8-16 and 16-32 cell stages.[20] According to the

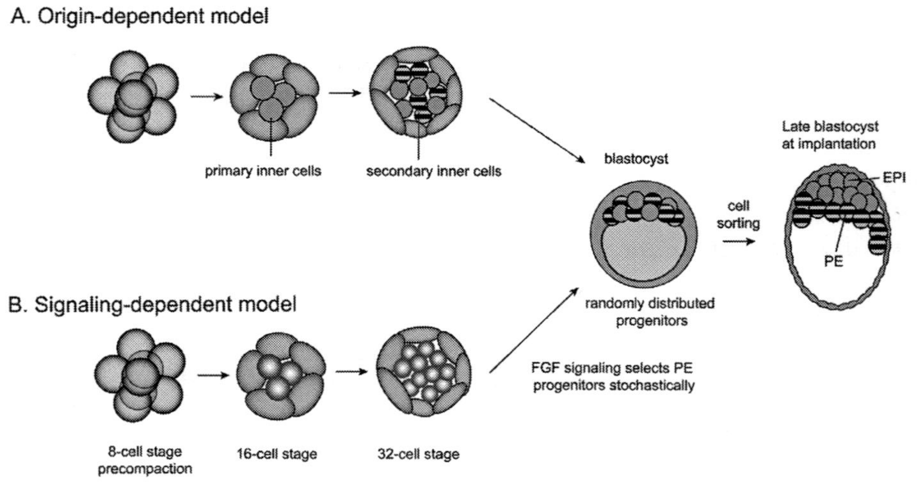

Figure 4. Two models of PE/EPI formation in the mouse embryo. A) Origin-dependent model in which the developmental origin of ICM cells regulates EPI/PE specification. ICM cells are generated from two rounds of asymmetric divisions after the 8-cell stage. Primary inner cells (blue) give rise to the EPI lineage and secondary inner cells (yellow, lined) to the PE lineage. B) Signaling-dependent model in which no difference in lineage potential exists between primary and secondary inner cells. Each inner cell is stochastically capable of responding to FGF signaling. Responding cells become the PE lineage and nonresponding cells become the EPI lineage. After the PE/EPI lineage decision, EPI and PE progenitors express lineage-specific transcription factors, Nanog or Gata6 and are distributed randomly in the ICM of the blastocyst. These two progenitors then sort out to form the two distinct layers of EPI and PE by day 4.5 at implantation. A color version of this figure is available at www.landesbioscience.com/curie.

origin-dependent model, the developmental origins of individual ICM cells determine their fate. That is, inner cells generated in the first round of divisions (primary inner cells) would preferentially adopt the EPI fate, whereas cells generated in the second round (secondary inner cells) would preferentially become PE (Fig. 4A).[42,57] Secondary inner cells would be predisposed to become extraembryonic due to their prolonged external position since TE cells are also external.[17] To test the origin-dependent model, generation of inner cells was first directly observed in live embryos and then the contribution of their progeny to EPI and PE lineages was analyzed at later stages.[44] No difference in lineage potential was detected between primary and secondary inner cells since both primary and secondary inner cell progeny contributed to EPI and PE lineages without an obvious bias. These observations therefore suggest that the origin-dependent model is unlikely.

The second model is a signaling-dependent model, in which individual ICM cells stochastically respond to certain levels of FGF signaling to choose EPI or PE fates (Fig. 4B). As described above, FGF signaling is necessary for PE formation in the embryo.[52-54] When FGF signaling is blocked, using either chemical inhibitors or by gene knockouts, all ICM cells adopt EPI fates.[42,58] Interestingly, high doses of exogenous FGF4 can induce the converse phenotype: all ICM cells adopt PE fates.[44] This suggests that all early ICM cells have the potential to respond to FGF signaling and become PE. During normal development, however, limited amounts of endogenous FGFs would restrict the proportion of FGF-responding ICM cells (Fig. 5). Whether or not individual ICM cells

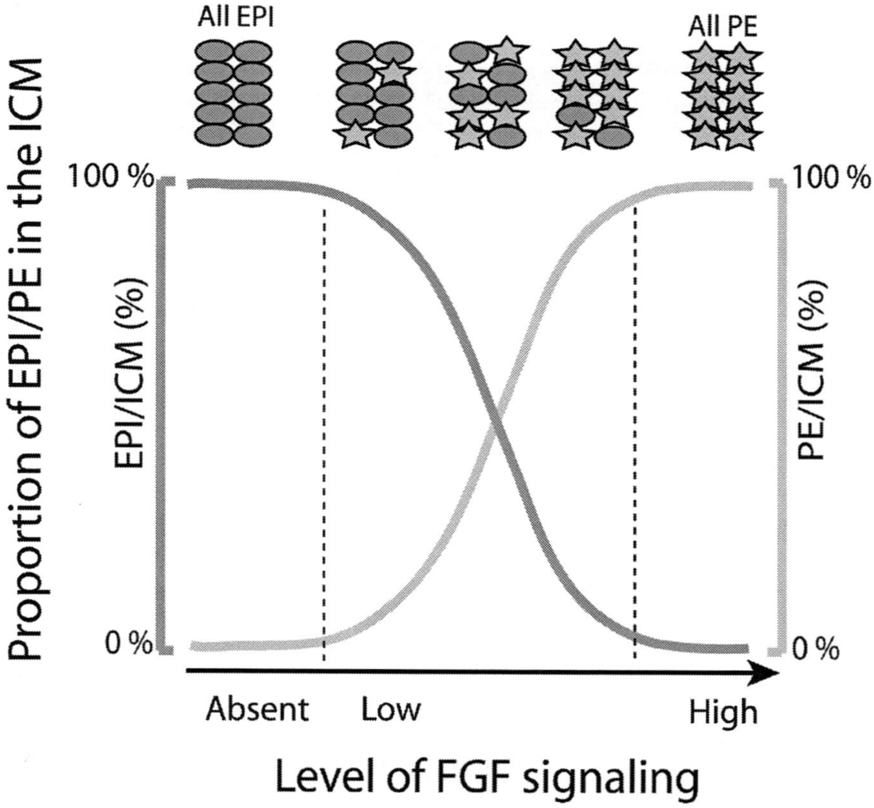

Figure 5. Schematic model of FGF signaling-dependent specification of PE and EPI lineages. The X-axis indicates the proposed activation level of the FGF signaling. The Y-axis indicates the proportion of the EPI (blue) and PE (yellow, stars) in the ICM. When signaling is below threshold, all ICM cells adopt the EPI fate. However, when the signal is high, all ICM cells adopt the PE fate. At intermediate levels of activation, individual ICM cells stochastically respond to FGF signaling. In this model, the level of FGF signaling controls the proportion of the two lineages in the ICM, but not the distribution. A color version of this figure is available at www.landesbioscience.com/curie.

respond to the limited amount of FGFs could be stochastically determined by cell-to-cell variation in sensitivity determined by cell-autonomous or non-autonomous mechanisms.[59] Endogenous levels of FGFs, governed by developmental genetic programs, would thereby generate roughly equal proportions of EPI/PE lineages reproducibly, without need for deterministic developmental mechanisms.

ESTABLISHMENT AND MODULATION OF PLURIPOTENCY IN THE EPI LINEAGE

After two rounds of lineage specification, first the TE and then the PE, the EPI is established as a pluripotent lineage. While pluripotency is generically defined as the ability to form tissues from all three embryonic germ layers, recent identification of multiple pluripotent stem cell lines makes apparent that pluripotency is not a single state. Rather, pluripotency may comprise a range of states with developmental equivalence in the embryo.[60] There are at least two states of pluripotency in the mouse embryo, represented by two types of pluripotent stem cells: ES cells and epiblast-derived stem cells (EpiSCs).[58] These cell lines are derived from the EPI lineage, but represent two distinct embryonic stages: ES cells are equivalent to the EPI cells of the implanting embryo,[19,58] while EpiSCs are equivalent to EPI cells of the embryo just after implantation and prior to gastrulation.[61,62] Although EpiSCs cannot contribute to embryos when they are injected into blastocysts, probably due to failure to integrate into host ICMs, they can generate all three germ layers in teratomas. Pluripotency genes such as *Oct4* and *Sox2* are both expressed during early and late stages, but several features differ between EPI cells over the course of implantation. For example, cell morphology changes from an unorganized cell mass to an epithelial monolayer. In addition, expression of some genes change dramatically during the transition, such as *Rex1*, which is downregulated and *Fgf5*, which is upregulated.[63] After the transition, late EPI cells are competent to receive inductive signals to generate three germ layers. Understanding how the pluripotent state is safeguarded during establishment of first, second and subsequent lineages is an active area of research.

One gene potentially involved in safeguarding the pluripotent state is *Nanog*. Nanog was originally identified as a transcription factor essential for maintaining pluripotency in ES cells,[64,65] but subsequent studies have revealed that it acts rather as a gate-keeper instead. That is, *Nanog* levels fluctuate in ES cells in an FGF-signaling dependent manner[66,67] and ES cells are more prone to differentiate when *Nanog* levels are low. Downregulation of *Nanog* does not initiate differentiation but permits it. This is consistent with the endogenous *Nanog* expression which is transiently downregulated during implantation. *Nanog* null blastocysts are morphologically normal but ICM cells degenerate soon after the blastocyst stage. *Nanog* null ICM cells appear to be trapped in a prepluripotent state, specified as neither EPI nor PE, but as a nonviable indeterminate state.[66]

Not much is known regarding mechanisms regulating the transition from early to late EPI tissues in the embryo. In vitro stem cell studies have provided some insight. ES cells have been found to readily become EpiSCs, when cultured in EpiSC culture conditions, including FGF2 and activin A.[68] Interestingly, *Fgf4* is required for ES cell differentiation.[69] These observations suggest that the quality or amount of FGF signaling may participate in the transition from ES cells to EpiSCs and possibly ICM to EPI fates.

Interestingly, it is also possible to reverse the transition between the two pluripotent states. That is, EpiSCs can become ES-like following overexpression of KLF4, one of the original reprogramming factors,[70] although reversal occurs with very low frequency.[68] However, when EpiSCs or epiblast cells from gastrula embryos are cultured in conventional ES cell culture conditions, reprogrammed ES-cell-like cells (rES cells) emerge after 10-20 days.[71] Although the reversion takes more time than progression from ES cells to EpiSCs, rES cells have fully reestablished the early pluripotent state.

CONCLUSION

Here, we have described lineage specification in the mouse, currently the most extensively analyzed mammalian embryo. One of the most interesting lessons from the mouse lays in the observation that multiple pluripotent states exist, evidenced by the existence of ES cells and EpiSCs. Thus far, rat ES cells are the only other mammalian ES cell line similar to mouse ES cells. In contrast, human ES (hES) cells are more similar to mouse EpiSCs than they are to mouse ES cells in morphology, gene expression, and growth factor dependency.[61,62,72] At this point, it is not known whether the human embryo also has multiple pluripotent states. Interestingly, adult human and mouse cells reprogrammed by identical factors resemble ES cells of their respective species. That is, human iPS cells resemble hES cells,[73,74] while mouse iPS cells resemble mouse ES cells[70] and not mouse EpiSCs. Perhaps there are species-specific differences in the stability of pluripotent states. Even though all mammals develop using a blastocyst, the developmental timing of implantation and morphology of early postimplantation embryos are highly varied. Further analysis of lineage specification and stem cells from other mammalian species should provide exciting insight into these issues.

REFERENCES

1. Ralston A, Rossant J. Genetic regulation of stem cell origins in the mouse embryo. Clin Genet 2005; 68(2):106-112.
2. Niwa H, Toyooka T, Shimosato D et al. Interaction between Oct3/4 and Cdx2 determines trophectoderm differentiation. Cell 2005; 123(5):917-929.
3. Fleming T. A quantitative analysis of cell allocation to trophectoderm and inner cell mass in the mouse blastocyst. Dev Biol 1987; 119(2):520-531.
4. Gardner R. The early blastocyst is bilaterally symmetrical and its axis of symmetry is aligned with the animal-vegetal axis of the zygote in the mouse. Development 1997; 124(2):289-301.
5. Gardner R. Specification of embryonic axes begins before cleavage in normal mouse development. Development 2001; 128(6):839-847.
6. Piotrowska K, Wianny F, Pedersen R et al. Blastomeres arising from the first cleavage division have distinguishable fates in normal mouse development. Development 2001; 128(19):3739-3748.
7. Fujimori T, Kurotaki Y, Miyazaki J et al. Analysis of cell lineage in two- and four-cell mouse embryos. Development 2003; 130(21):5113-5122.
8. Plusa B, Frankenberg S, Chalmers A et al. Downregulation of Par3 and aPKC function directs cells towards the ICM in the preimplantation mouse embryo. J Cell Sci 2005; 118(Pt 3):505-515.
9. Kurotaki Y, Hatta K, Nakao K et al. Blastocyst axis is specified independently of early cell lineage but aligns with the ZP shape. Science 2007; 316(5825):719-723.
10. Alarcón V, Marikawa Y. Deviation of the blastocyst axis from the first cleavage plane does not affect the quality of mouse postimplantation development. Biol Reprod 2003; 69(4):1208-1212.
11. Chróścicka A, Komorowski S, Maleszewski M. Both blastomeres of the mouse 2-cell embryo contribute to the embryonic portion of the blastocyst. Mol Reprod Dev 2004; 68(3):308-312.
12. Motosugi N, Bauer T, Polanski Z et al. Polarity of the mouse embryo is established at blastocyst and is not prepatterned. Genes Dev 2005; 19(9):1081-1092.
13. Bischoff M, Parfitt D, Zernicka-Goetz M. Formation of the embryonic-abembryonic axis of the mouse blastocyst: relationships between orientation of early cleavage divisions and pattern of symmetric/asymmetric divisions. Development 2008; 135(5):953-962.
14. Piotrowska-Nitsche K, Perea-Gomez A, Haraguchi S et al. Four-cell stage mouse blastomeres have different developmental properties. Development 2005; 132(3):479-490.
15. Johnson M, McConnell J. Lineage allocation and cell polarity during mouse embryogenesis. Semin Cell Dev Biol 2004; 15(5):583-597.
16. Torres-Padilla M, Parfitt D, Kouzarides T et al. Histone arginine methylation regulates pluripotency in the early mouse embryo. Nature 2007; 445(7124):214-218.
17. Yamanaka Y, Ralston A, Stephenson R et al. Cell and molecular regulation of the mouse blastocyst. Dev Dyn 2006; 235(9):2301-2314.
18. Yoshikawa T, Piao Y, Zhong J et al. High-throughput screen for genes predominantly expressed in the ICM of mouse blastocysts by whole mount in situ hybridization. Gene Expr Patterns 2006; 6(2):213-224.
19. Ralston A, Cox B, Nishioka N et al. Gata3 regulates trophoblast development downstream of Tead4 and in parallel to Cdx2. Development 2010; 137(3):395-403.
20. Chawengsaksophak K, James R, Hammond V et al. Homeosis and intestinal tumours in Cdx2 mutant mice. Nature 1997; 386(6620):84-87.

21. Nishioka N, Yamamoto S, Kiyonari H et al. Tead4 is required for specification of trophectoderm in pre-implantation mouse embryos. Mech Dev 125(3-4):270-283.
22. Yagi R, Kohn M, Karavanova I et al. Transcription factor TEAD4 specifies the trophectoderm lineage at the beginning of mammalian development. Development 2007; 134(21):3827-3836.
23. Strumpf D, Mao CA, Yamanaka Y et al. Cdx2 is required for correct cell fate specification and differentiation of trophectoderm in the mouse blastocyst. Development 2005; 132(9):2093-2102.
24. Ralston A, Rossant J. Cdx2 acts downstream of cell polarization to cell-autonomously promote trophectoderm fate in the early mouse embryo. Dev Biol 2008; 313(2):614-629.
25. Jedrusik A, Parfitt D, Guo G et al. Role of Cdx2 and cell polarity in cell allocation and specification of trophectoderm and inner cell mass in the mouse embryo. Genes Dev 2008; 22(19):2692-2706.
26. Dietrich J, Hiiragi T. Stochastic patterning in the mouse pre-implantation embryo. Development 2007; 134(23):4219-4231.
27. Nishioka N, Inoue K, Adachi K et al. The Hippo signaling pathway components Lats and Yap pattern Tead4 activity to distinguish mouse trophectoderm from inner cell mass. Dev Cell 2009; 16(3):398-410.
28. Nichols J, Zevnik B, Anastassiadis K et al. Formation of pluripotent stem cells in the mammalian embryo depends on the POU transcription factor Oct4. Cell 1998; 95(3):379-391.
29. Beck F, Erler T, Russell A et al. Expression of Cdx-2 in the mouse embryo and placenta: possible role in patterning of the extra-embryonic membranes. Dev Dyn 1995; 204(3):219-227.
30. Niwa H, Miyazaki J, Smith A. Quantitative expression of Oct-3/4 defines differentiation, dedifferentiation or self-renewal of ES cells. Nat Genet 2000; 24(4):372-376.
31. Russ A, Wattler S, Colledge W et al. Eomesodermin is required for mouse trophoblast development and mesoderm formation. Nature 2000; 404(6773):95-99.
32. Ma G, Roth M, Groskopf J et al. GATA-2 and GATA-3 regulate trophoblast-specific gene expression in vivo. Development 1997; 124(4):907-914.
33. Gardner RL, Rossant J. Investigation of the fate of 4-5 day postcoitum mouse inner cell mass cells by blastocyst injection. J Embryol Exp Morphol 1979; 52:141-152.
34. Gardner RL. Investigation of cell lineage and differentiation in the extraembryonic endoderm of the mouse embryo. J Embryol Exp Morphol 1982; 68:175-198.
35. Gardner RL. An in situ cell marker for clonal analysis of development of the extraembryonic endoderm in the mouse. J Embryol Exp Morphol 1984; 80:251-288.
36. Plusa B, Piliszek A, Frankenberg S et al. Distinct sequential cell behaviours direct primitive endoderm formation in the mouse blastocyst. Development 2008; 135(18):3081-3091.
37. Rossant J, Tam PP. Blastocyst lineage formation, early embryonic asymmetries and axis patterning in the mouse. Development 2009; 136(5):701-713.
38. Kunath T, Arnaud D, Uy G et al. Imprinted X-inactivation in extra-embryonic endoderm cell lines from mouse blastocysts. Development 2005; 132(7):1649-1661.
39. Fujikura J, Yamato E, Yonemura S et al. Differentiation of embryonic stem cells is induced by GATA factors. Genes Dev 2002; 16(7):784-789.
40. Morrisey E, Ip H, Lu M et al. GATA-6: a zinc finger transcription factor that is expressed in multiple cell lineages derived from lateral mesoderm. Dev Biol 1996; 177(1):309-322.
41. Koutsourakis M, Keijzer R, Visser P et al. Branching and differentiation defects in pulmonary epithelium with elevated Gata6 expression. Mech Dev 2001; 105(1-2):105-114.
42. Chazaud C, Yamanaka Y, Pawson T et al. Early lineage segregation between epiblast and primitive endoderm in mouse blastocysts through the Grb2-MAPK pathway. Dev Cell 2006; 10(5):615-624.
43. Kurimoto K, Yabuta Y, Ohinata Y et al. An improved single-cell cDNA amplification method for efficient high-density oligonucleotide microarray analysis. Nucleic Acids Res 2006; 34(5):e42.
44. Yamanaka Y, Lanner F, Rossant J. FGF signal-dependent segregation of primitive endoderm and epiblast in the mouse blastocyst. Development 2010; 137(5):715-724.
45. Meilhac S, Adams R, Morris S et al. Active cell movements coupled to positional induction are involved in lineage segregation in the mouse blastocyst. Dev Biol 2009; 331(2):210-221.
46. Gerbe F, Cox B, Rossant J et al. Dynamic expression of Lrp2 pathway members reveals progressive epithelial differentiation of primitive endoderm in mouse blastocyst. Dev Biol 2008; 313(2):594-602.
47. Morris S, Tallquist M, Rock C et al. Dual roles for the Dab2 adaptor protein in embryonic development and kidney transport. EMBO J 2002; 21(7):1555-1564.
48. Yang D, Smith E, Roland I et al. Disabled-2 is essential for endodermal cell positioning and structure formation during mouse embryogenesis. Dev Biol 2002; 251(1):27-44.
49. Gao F, Shi H, Daughty C et al. Maspin plays an essential role in early embryonic development. Development 2004; 131(7):1479-1489.
50. Smyth N, Vatansever H, Murray P et al. Absence of basement membranes after targeting the LAMC1 gene results in embryonic lethality due to failure of endoderm differentiation. J Cell Biol 1999; 144(1):151-160.

51. Stephens L, Sutherland A, Klimanskaya I et al. Deletion of beta 1 integrins in mice results in inner cell mass failure and peri-implantation lethality. Genes Dev 1995; 9(15):1883-1895.
52. Arman E, Haffner-Krausz R, Chen Y et al. Targeted disruption of fibroblast growth factor (FGF) receptor 2 suggests a role for FGF signaling in pregastrulation mammalian development. Proc Natl Acad Sci USA 1998; 95(9):5082-5087.
53. Feldman B, Poueymirou W, Papaioannou V et al. Requirement of FGF-4 for postimplantation mouse development. Science 1995; 267(5195):246-249.
54. Cheng A, Saxton T, Sakai R et al. Mammalian Grb2 regulates multiple steps in embryonic development and malignant transformation. Cell 1998; 95(6):793-803.
55. Rossant J, Chazaud C, Yamanaka Y. Lineage allocation and asymmetries in the early mouse embryo. Philos Trans R Soc Lond B Biol Sci 2003; 358(1436):1341-1348; discussion 1349.
56. Johnson MH, Chisholm JC, Fleming TP et al. A role for cytoplasmic determinants in the development of the mouse early embryo? J Embryol Exp Morphol 1986; 97 Suppl:97-121.
57. Chisholm JC, Houliston E. Cytokeratin filament assembly in the preimplantation mouse embryo. Development 1987; 101(3):565-582.
58. Nichols J, Silva J, Roode M et al. Suppression of Erk signalling promotes ground state pluripotency in the mouse embryo. Development 2009; 136(19):3215-3222.
59. Huang S. Non-genetic heterogeneity of cells in development: more than just noise. Development 2009; 136(23):3853-3862.
60. Rossant J. Stem cells and early lineage development. Cell 2008; 132(4):527-531.
61. Brons I, Smithers L, Trotter M et al. Derivation of pluripotent epiblast stem cells from mammalian embryos. Nature 2007; 448(7150):191-195.
62. Tesar P, Chenoweth J, Brook F et al. New cell lines from mouse epiblast share defining features with human embryonic stem cells. Nature 2007; 448(7150):196-199.
63. Pelton T, Sharma S, Schulz T et al. Transient pluripotent cell populations during primitive ectoderm formation: correlation of in vivo and in vitro pluripotent cell development. J Cell Sci 2002; 115(Pt 2):329-339.
64. Mitsui K, Tokuzawa Y, Itoh H et al. The homeoprotein Nanog is required for maintenance of pluripotency in mouse epiblast and ES cells. Cell 2003; 113(5):631-642.
65. Chambers I, Colby D, Robertson M et al. Functional expression cloning of Nanog, a pluripotency sustaining factor in embryonic stem cells. Cell 2003; 113(5):643-655.
66. Silva J, Nichols J, Theunissen T et al. Nanog is the gateway to the pluripotent ground state. Cell 2009; 138(4):722-737.
67. Lanner F, Lee K, Sohl M et al. Heparan Sulfation Dependent FGF Signalling Maintains Embryonic Stem Cells Primed for Differentiation in a Heterogeneous State. Stem Cells 2009.
68. Guo G, Yang J, Nichols J et al. Klf4 reverts developmentally programmed restriction of ground state pluripotency. Development 2009; 136(7):1063-1069.
69. Kunath T, Saba-El-Leil M, Almousailleakh M et al. FGF stimulation of the Erk1/2 signalling cascade triggers transition of pluripotent embryonic stem cells from self-renewal to lineage commitment. Development 2007; 134(16):2895-2902.
70. Takahashi K, Yamanaka S. Induction of pluripotent stem cells from mouse embryonic and adult fibroblast cultures by defined factors. Cell 2006; 126(4):663-676.
71. Bao S, Tang F, Li X et al. Epigenetic reversion of post-implantation epiblast to pluripotent embryonic stem cells. Nature 2009; 461(7268):1292-1295.
72. Thomson J, Itskovitz-Eldor J, Shapiro S et al. Embryonic stem cell lines derived from human blastocysts. Science 1998; 282(5391):1145-1147.
73. Takahashi K, Tanabe K, Ohnuki M et al. Induction of pluripotent stem cells from adult human fibroblasts by defined factors. Cell 2007; 131(5):861-872.
74. Yu J, Vodyanik M, Smuga-Otto K et al. Induced pluripotent stem cell lines derived from human somatic cells. Science 2007; 318(5858):1917-1920.

CHAPTER 2

NUCLEAR ARCHITECTURE IN STEM CELLS

Kelly J. Morris[§], Mita Chotalia[§] and Ana Pombo*

Abstract: Fundamental features of genome regulation depend on the linear DNA sequence, cell type specific modification of DNA and chromatin-associated proteins, which locally control the expression of single genes. Architectural features of genome organization within the three-dimensional (3D) nuclear space establish preferential positioning of genes relative to nuclear subcompartments associated with specific biochemical activities, thereby influencing states of expression. The structural and temporal organization of the genome within the nucleus of stem cells, together with specific features of epigenetic and transcriptional regulation are emerging as key players that influence pluripotency and differentiation.[1,2]

INTRODUCTION

Stem cells have a remarkable potential to develop into many different cell types during early life and growth. Research has primarily focused on two kinds of stem cells from both rodents and humans: pluripotent embryonic stem (ES) cells and multipotent tissue-specific stem cells. ES cells, derived from the inner cell mass of pre-implantation embryos at the blastocyst stage, are characterized by their ability to self-renew indefinitely in culture, while showing great genomic plasticity and potential to differentiate into all derivatives of the three primary germ layers (ectoderm, endoderm and mesoderm).

The ability of ES cells to self-renew in a pluripotent state and commit to any cell lineage is associated with the existence of opposing regulatory constraints at genes important for development, which are repressed but in an epigenetic state compatible

[§]Both authors contributed equally to this work.

*Corresponding Author: Ana Pombo—Genome Function Group, MRC Clinical Sciences Centre, Imperial College School of Medicine, Hammersmith Hospital Campus, Du Cane Road, London W12 0NN, UK. Email: ana.pombo@csc.mrc.ac.uk

The Cell Biology of Stem Cells, edited by Eran Meshorer and Kathrin Plath.
©2010 Landes Bioscience and Springer Science+Business Media.

with efficient activation during cellular differentiation. The process of self-renewal requires the ES cell genome to divide without disrupting its pluripotent state, whereas differentiation programs instruct complex changes in gene expression that allow ES cells to enter specific cell commitment pathways. Understanding the molecular mechanisms underlying ES cell plasticity is essential if we are to comprehend gene regulation during pluripotency and development, with implications in cell therapy.

FUNCTIONAL COMPARTMENTALIZATION OF THE ES CELL NUCLEUS

The eukaryotic cell nucleus contains chromosomes, each consisting of a single DNA polymer complexed with histones and other proteins forming chromatin. In humans, 22 pairs of autosomes and two sex chromosomes are compacted within the nuclear volume. They are separated from the cytoplasm by a nuclear envelope, interspersed with nuclear pores responsible for trafficking proteins and RNA in and out of the nucleus. The inner envelope provides a functional landmark involved in organizing chromosomes inside the nucleus and is characterized by the presence of structural proteins that constitute the nuclear lamina. Within the nuclear interior, further structural and functional compartmentalization is observed through establishment of higher-order complexes that aggregate to form large nuclear domains, such as the nucleolus and splicing speckles, or smaller nuclear bodies such as Cajal and Polycomb bodies. Chromatin interactions within the nuclear interior mediated by local interactions or associations with functional domains are thought to be both important for and dependent on programs of gene expression.

Organization of Chromosomes and Single Genes within the Nuclear Space

During the formation of the interphase cell nucleus, upon exit from mitosis, chromosomes partially decondense to occupy discrete nuclear territories that intermingle at their boundaries[3] and establish long-range interchromosomal interactions.[4] Positioning of chromosome territories (CTs) within the nuclear space, as well as positioning of genes relative to their CTs have implications in gene expression and genome stability.[5]

Studies in human ES (hES) cells show that chromosomes also form discrete CTs, with similar distributions to those observed in somatic cells.[6,7] For example, the nuclear position of gene-rich human chromosome 19 and gene-poor chromosome 18, determined using fluorescence in-situ hybridization (FISH) of whole chromosome probes, is strikingly similar between hES and lymphoblastoid cells (LCLs) occupying a more interior and peripheral position, respectively.[6] Similar correlations have been shown for a range of other chromosomes (10, 12, 15, 17 and 19) before and during hES early differentiation, collectively suggesting a level of conserved architecture that is independent of cell lineage commitment and already present early in development.[7]

Analysis of specific CT regions that contain pluripotency genes in hES cells including chromosome arm 12p harboring the *Nanog* gene, embedded in a region surrounded by other pluripotency associated genes, revealed a more central nuclear position in hES compared to LCLs.[6] Central positioning is a feature of transcriptional activity in many types of somatic cells with interesting exceptions.[8] In the case of *Oct4,* which maps to chromosome arm 6p among an active amjor histocompatibility complex (MHC) gene cluster, a more internal position is observed in both cell types, as expected due to the local active chromatin neighborhood. Further exploration of the positioning of particular gene loci relative to their CT demonstrates that the *Oct4* locus is decondensed and is preferentially positioned outside its CT in hES, but remains within the interior in LCLs.[6]

Chromatin 'looping' reflective of transcriptional activity is also observed in ES cells after induction of differentiation and *Hox* gene expression with retinoic acid.[9] *Hoxd* and *b* loci contain clusters of developmentally regulated *Hox* genes crucial for anterior/posterior patterning in the developing embryo, which are silent in ES cells but become expressed upon in vitro differentiation. In ES cells, *Hox* gene clusters are preferentially located within the CT, whereas upon differentiation, *Hoxb* decondenses and repositions, along with flanking genes, to the CT exterior.[9,10] The *Hoxb* cluster repositions towards the CT exterior in a manner that corresponds to temporal gene expression with active alleles colocalizing with the elongating form of RNA polymerase II (RNAPII).[10] Decondensation and nuclear localization of *Hox* loci to the exterior of its CT, upon differentiation, is accompanied by looping out of flanking genes but in this instance does not reflect an increase in expression of the latter.[10] Although, transcriptional activity of active flanking genes adjacent to the *Hoxb* cluster is slightly more abundant at the CT exterior, nascent transcripts are also detected at the CT interior.

Active histone modifications are observed at *Hoxb* loci before expression can be detected, suggesting a two step model where the locus is epigenetically altered, then it decondenses allowing a permissive state, followed by locus looping and temporal expression. Importantly, increasing histone acetylation levels in vitro is not sufficient to induce locus remodeling, suggesting that higher-order chromatin structure is not simply a reflection of the underlying histone modifications in this system. Re-organization of chromatin at *Hox* clusters is likely to result in increased chromatin mobility thereby increasing accessibility and opportunity to associate with transcriptional machinery inside or outside of CTs.[9]

Higher-order structural changes are also observed during the process of X chromosome inactivation (XCI), which initiates upon differentiation in female ES cells. XCI is a phenomenon where one X chromosome in female mammalian cells is silenced to equal the dosage of X-linked genes in male and female cells. Initiation of XCI is controlled by the X-inactivation centre (Xic), a specific genomic region that contains a multitude of noncoding genes including *Xist*, *Tsix* and *Xite*, which have been implicated in regulating XCI.[11,12] At the onset of XCI, *Xist* expression is up-regulated on the future inactive X chromosome. *Xist* is a large, noncoding RNA molecule that initiates XCI by coating the chromosome *in cis* and triggers transcriptional silencing via the modulation of chromatin structure.[11] The mechanisms of random XCI are not completely elucidated, but recent analyses suggest that the positioning of Xics during XCI in early differentiation has important roles.

Prior to the initiation of XCI, Xics transiently colocalize within the nuclear space[13-15] and this interaction is mediated by the recruitment of the chromatin insulator CTCF and is also dependent on the cotranscriptional activity of *Tsix* and *Xite*.[16] The pairing of homologous chromosomes may be an architectural feature unique to X chromosomes in the context of ES cells, as typically chromosome pairings are only favored between heterologous chromosomes in the interphase nucleus.[17] Computer simulations of polymer folding in the presence of binding factors suggests that the close spatial proximity between Xics in undifferentiated cells may be dependent on the binding affinity and concentration of a limiting factor, such as CTCF,[16] which would become preferentially associated with one of the Xic.[18]

Further analyses at the onset of XCI have shown chromatin decondensation and long-range chromatin interactions between XCI regulatory domains in a developmentally regulated and sex-specific manner.[19] 3D morphological analysis of human X chromosomes in female cells showed that the active X chromosome is elongated compared to its inactive X chromosome counterpart, which is more spherical in shape. Analysis of the morphology of human chromosome 7, with similar DNA content, shows that it is similar in shape and volume to the active X chromosome suggesting that active chromosomes display a flat, elongated morphology,[20] which may favor more contact points with chromosome neighbors and enhance transcriptional activity.[17] Further aspects of chromosome organization are related with how the X chromosome, coated with Xist, is maintained in a silent transcriptional state. The accumulation of Xist on the inactive X chromosome has been associated with the formation of a repressive nuclear compartment depleted of transcription factors and RNAPII. Interestingly, genes that escape silencing are preferentially positioned at the exterior of the inactive X CT, whereas silent genes display a more internal location without access to transcription machinery.[21]

In addition to differences in chromosome and gene positioning of pluripotency markers and developmental regulator genes, other distinct higher-order features of ES cell chromatin structures have been described. Centromere clusters form in ES cells but tend to be localized more towards the nuclear interior in ES than in differentiated cells, a possible reflection of the rapid ES cell cycle.[6,7,22] Upon differentiation of hES cells, centromeres stained with markers of centric heterochromatin (α-satellite/CENP-A) undergo higher-order re-organization and reposition towards the nuclear periphery.[7,22] The origin of these drastic changes likely implicates the large scale chromatin remodeling that occurs during cell commitment, in concert with changes in gene expression, although the mechanistic details are yet to be elucidated.

Lamina and the Nuclear Periphery

Lamins form the major structural component of the inner nuclear membrane and play broad roles in many nuclear processes including nuclear assembly after mitosis, chromatin organization and gene expression. The lamina is composed of a meshwork of intermediate filaments interspersed with lamina associated proteins that include the B-type lamins B1 and B2, which are essential for survival and ubiquitously expressed in all cell types. A-type lamins are derived from splice variants of a single gene (*LMNA*) and show expression in a tissue-specific pattern. The nuclear lamina is thought to be important for association between the nuclear envelope, heterochromatin and histones as its knockdown in mouse fibroblasts leads to a depletion of heterochromatin from the nuclear periphery.[23] Chromatin associations with the nuclear lamina have been implicated in the down-regulation of gene expression, for some, but not all associated genes; it is also an area where facultative heterochromatin typically accumulates in differentiating mammalian cells.[24-30]

Both human and mouse ES cells lack lamin A/C at the nuclear periphery (Fig. 1)[22,31,32] but express lamin B1 and B2. Lamin A/C appears early at the onset of differentiation[22,31] prior to the down regulation of the pluripotency marker *Oct4*.[31] The lack of lamin A/C in ES cells has implications in nuclear rigidity[32] and morphology.[22] Absence of lamin A/C is likely to contribute to the plasticity and chromatin mobility observed in the ES cell nucleus,[32] whereas its expression during differentiation may be an important factor in the process and/or maintenance of a differentiated phenoytype.[31]

Quantitative, cytological analysis of epigenetic status and nuclear distribution of genes in ES cells relative to the nuclear periphery has revealed an enrichment in H3

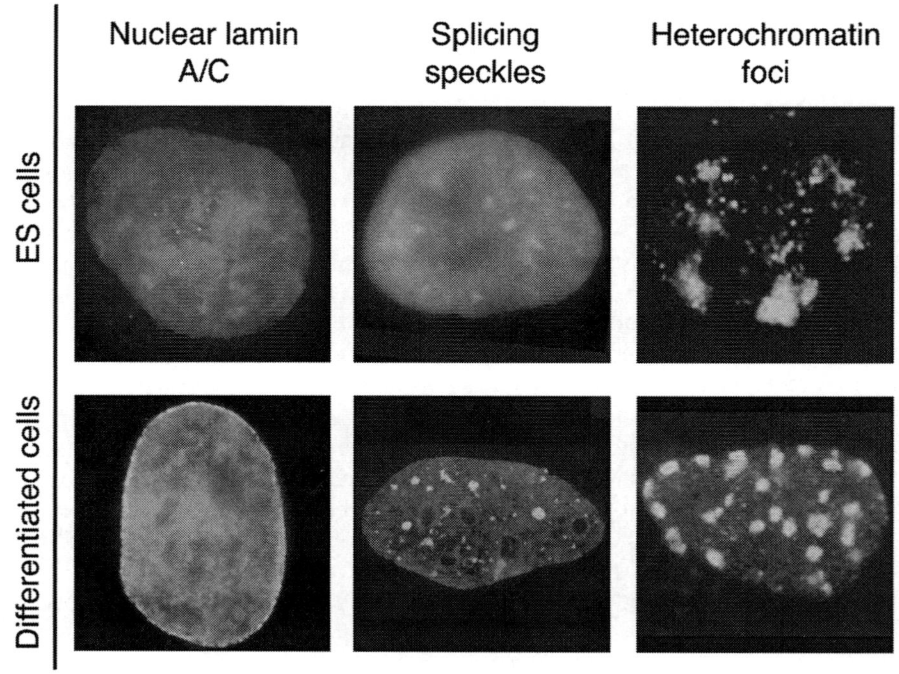

Figure 1. Distinguishing features of nuclear architecture in ES cells. Nucleoplasmic subcompartments are shown in embryonic stem (ES) cells (top panel) and in various differentiated cell types (bottom panel). *Nuclear Lamina.* Lamin A/C is absent in undifferentiated hES cells and expression is induced upon differentiation. Images were kindly provided by Butler JT and Lawrence JB. *Splicing speckles.* SC35 domain formation occurs as cells differentiate. In undifferentiated hES cells, diffuse SC35 patterns are observed, the initiation of differentiation correlates with SC35 domain formation. Images were kindly provided by Butler JT and Lawrence JB. *Heterochromatin foci.* Staining for heterochromatin binding protein HP1α is confined to fewer and larger foci in ES cells compared with neural progenitor cells (NPCs).[51] Reprinted from Meshorer et al. Hyperdynamic plasticity of chromatin proteins in pluripotent embryonic stem cells. Dev Cell 2006; 10(1):105-16. ©2006 with permission from Elsevier. A color version of this figure is available at www.landesbioscience.com/curie.

trimethylation on lysine 27 (H3K27me3; an epigenetic marker localized at repressed gene promoters) in ES cells relative to differentiated cells (embryo-derived NIH-3T3 fibroblasts and neural progenitor cells; NPCs). In addition, 10% of actively (H3 trimethylation on lysine4; H3K4me3) marked chromatin and hundreds of transcription sites localize to the nuclear periphery of ES and differentiated cells with similar distributions.[33]

Changes in gene positioning relative to the nuclear periphery were observed for the developmental gene *Mash1* (*Ascl1*), a transcription factor essential in neural precursor differentiation and specific neural commitment, that correlate with differentiation and gene expression alterations.[28] In ES cells, the *Mash1* promoter is associated with H3K27me3 and is positioned near the nuclear periphery. Interestingly, its position is not affected by knockdown of histone methyl transferase (HMT) Ezh2/Eed. On induction to the neural lineage, *Mash1* transcription increases >100 fold in parallel with repositioning of the locus to the interior, reduction in H3K27me3 and an increase in H3 acetylation on lysine 9 (H3K9ac; an active epigenetic mark) at the *Mash1* gene but not adjacent genes.

Interior positioning of the locus did not dictate expression of flanking genes suggesting that they are regulated by additional regulatory mechanisms that maintain expression state in a gene-specific way during changes in nuclear environment. These findings raise an intriguing possibility that the nuclear periphery may contribute to the epigenetic state of the pluripotent ES cell genome, although the molecular mechanisms that would achieve this state have yet to be elucidated.

STEM CELL FEATURES OF OTHER NUCLEOPLASMIC SUBCOMPARTMENTS

Dynamic compartmentalization of factors within the nuclear space involved in nuclear processes such as transcription, RNA processing, DNA replication and repair form macromolecular complexes proposed to be self-organizing structures that create functional sites.[34,35] Recent studies suggest that many of these structures differ between ES and somatic cells.

Splicing Speckles and Cajal Bodies

Splicing speckles are major nuclear domains enriched for components of the splicing machinery such as SC35, polyA$^+$ RNA and numerous mRNA metabolic factors.[36-38] Immunofluorescence of hES nuclei has revealed that SC35 marked speckles are distributed diffusely throughout the nucleoplasm, but upon differentiation become more concentrated into discrete foci (Fig. 1).[22] Similar findings have also been described for Cajal bodies, nuclear structures involved in the assembly or modification of transcription and splicing complexes.[39] hES nuclei do not contain Cajal bodies and coilin, a marker of Cajal bodies, is observed as a weak, diffuse signal distributed throughout the nucleoplasm. Cajal bodies are more defined and localize to the stem cell colony edge upon differentiation.[22] Gene associations with splicing speckles and Cajal bodies have been shown to correlate with their expression in some cell systems, and are likely to be related with specific aspects of transcript processing.[38,39]

Promyelocytic Leukemia Bodies

Promyelocytic leukemia (PML) bodies comprise several regulatory proteins and have been implicated in a range of cellular processes including chromatin organization, DNA replication and transcriptional regulation. In somatic cells PML bodies appear as uniform, spherical structures. In hES cell cultures, a small proportion of cells contain only a few spherical PML bodies while a substantial fraction of the population contain PML-defined structures with an elaborate morphology categorized into two types: long linear PML "rods" or large (>2 μm) "rosettes".[22] PML structures in hES cells not only differ in morphology but also protein composition, as they are not enriched for markers that define somatic PML bodies including SUMO, Sp100 and Daxx. These unique PML structures arise transiently in early stages of cell commitment and are absent after differentiation is established. They are also frequently associated with the nuclear envelope, a feature that is not observed in somatic cells.[22]

Polycomb Bodies

Polycomb proteins are vital for gene repression in many cell types.[40] In ES cells, they have been shown to be important for repressing developmental regulator genes.[41] Polycomb repressor complex 1 (PRC1) contains a subunit that is composed of CBX family proteins, which can bind H3K27me3 in vitro.[42] Fluorescence imaging revealed that different CBX family members exhibit distinct subnuclear distributions in undifferentiated murine ES cells, most CBX proteins are enriched in foci, known as Polycomb bodies, whereas upon differentiation the foci disappear as CBX proteins disperse.[43]

CHROMATIN FEATURES CHARACTERISTIC OF ES CELL NUCLEI

Structure and organization of the eukaryotic genome is multi-layered. At the local level DNA is wrapped around histone octamers to form nucleosomes and condenses to different degrees of chromatin compaction depending on the length of linker DNA and histone H1. Hetero-(closed) or eu-(open) chromatin states display more peripheral or central nuclear positions respectively, in most cell tissues, conferring regulatory potential by controlling accessibility of proteins to chromatin.[44] The highly dynamic associations of chromatin factors, with the exception of core histone proteins that bind to chromatin stably, characterize heterochromatin and euchromatin states in the mammalian cell nucleus.[45,46] Dynamic chromatin association of the basal transcription/repair factor IIH (TFIIH) during cellular differentiation of some cell types suggests that the ES cell genome is primed for lineage specific programs, which involves a cohort of regulatory factors.[47]

Hypermobility of Architectural Chromatin Proteins and Heterochromatin Formation

Architectural chromatin proteins including heterochromatin protein 1 group (HP1) and the linker histone H1 maintain condensed, often repressed heterochromatin domains.[48-50] Surprisingly, live cell imaging of HP1, H1 and core histones revealed that ES cells have a unique chromatin state compared to lineage restricted and differentiated cells.[51] In ES cells, several architectural chromatin proteins exist in a hyperdynamic, loosely bound or soluble fraction. Interestingly, a reduction in the protein mobility was associated with a loss of pluripotency rather than with ES cell differentiation per se. Hypermobility of chromatin proteins in ES cells was proposed to not only contribute to the maintenance of the pluripotent status of ES cells, but also be essential for reshaping the global architecture of the ES cell genome, particularly the organization of heterochromatin.[51]

Comparisons of heterochromatin domains in ES cells and differentiating cells have shown that heterochromatin undergoes drastic spatial rearrangements during the early stages of differentiation.[51-53] Visualization of heterochromatin directly (using a DNA probe against the major satellite repeat) and examination of heterochromatin binding proteins (HP1α) in ES cells revealed partial decondensation of heterochromatin when compared to NPCs (Fig. 1). In ES cells, heterochromatin is observed as a diffuse structure, whereas in NPCs heterochromatin domains display compaction and concentration in to distinct foci, a pattern more similar to what is seen in other somatic cell types.[51] Similar differences in heterochromatin foci have been confirmed in several ES cell lines.[54,55] Lamin A/C and lamin-associated proteins can interact directly with chromatin and histone proteins, thereby restricting chromatin structural dynamics and affecting the global organization of chromatin within the nucleus.[23] The absence of lamin A/C in ES cells may together with decondensed chromatin contribute to the nuclear plasticity observed in ES cells.

Formation of distinct heterochromatin foci during differentiation is accompanied by an increase in H3 trimethylation on lysine 9 (H3K9me3; an epigenetic marker of heterochromatin), a decrease in acetylation of H3 and H4 (epigenetic markers of transcriptionally active euchromatin) and an increase in DNA methylation. These changes in epigenetic marks are agreeable with the notion that ES cell chromatin assumes a more open conformation compared to lineage restricted and differentiated cell types.[51,56] One of the molecular players involved in maintaining this open chromatin state is the chromatin-remodeling factor Chd1, as murine ES cells deficient in Chd1 exhibit heterochromatin that resembles foci observed in differentiated cell types. In addition, Chd1 deficient cells loose their pluripotency implicating a role for Chd1 and possibly for heterochromatin formation, in establishing or maintaining the pluripotent characteristic of ES cells.[57]

Hypertranscription and DNA Replication in ES Cells

Genome-wide analyses of gene expression and labeling of sites of transcription after Br-UTP (5-bromouridine 5′-triphosphate) incorporation suggest that the ES cell genome may be more transcriptionally active than its differentiated progeny.[51,55,58] ES cell chromatin is characterized by increased levels of active histone marks (e.g., H3 dimethylation on lysine 36; H3K36me2;[55] associated with the elongation-termination form of RNAPII). Genome-wide tiling transcription arrays have revealed hypertranscription of both protein-coding and noncoding regions in murine ES cells, whereas the transcription landscape becomes more defined as cells differentiate down the neuronal pathway. Half of all annotated genes were transcriptionally elevated in ES cells in relation to NPCs. The observed hypertranscription may simply be a by-product of the unusual open chromatin structure,[51] as the hyperdynamic chromatin would be more accessible to transcription factors and the transcriptional machinery.[55] Closer inspection of the over-expressed genes revealed that the majority of genes encode for general transcription factors and chromatin remodeling proteins. This raises the possibility that elevated transcription of these chromatin remodeling factors is important for establishing or maintaining the unusual open chromatin conformation of ES cells. Despite the relationship between hypertranscription and open chromatin structure being unclear, the default hyperactive transcription in ES cells is suggested to be a key mechanism in maintaining pluripotency in ES cells, which may contribute to the plasticity of genome.

The timing of DNA replication is another indicator of the global chromatin state.[59-62] The interrelationship between chromatin structure and DNA replication stems from pertinent studies of the *β-globin* locus in erythroid cells, XCI in differentiating female ES cells[63] and ES associated genes in differentiating ES cells.[60,61] These studies suggest that changes in replication timing are reflective of alterations in chromatin structure and the degree of condensation. High resolution profiles of replication timing in murine ES cells revealed that despite the general trend for replication timing and transcription to change coordinately, in a comparison between ES cells and NPCs, the high number of exceptions suggest that a direct relationship between the two is unlikely. The temporal reorganization of replication domains is reflective of the spatial reorganization of chromosomes during differentiation down the neural pathway.[61]

Silencing Mechanisms at Developmental Regulator Genes

A distinguishing feature of the ES cell genome is that ~2,500 silent, developmentally regulated genes harbor bivalent histone modifications that confer active (; H3K4me3) and repressive (H3K27me3; and H2A monoubiquitination of lysine 119; H2Aub1) chromatin states,[64-66] the latter mediated by polycomb repressor complexes, PRC1 and PRC2. Surprisingly, developmental regulator genes are also associated with an unusual form of RNAPII, which is able to transcribe through coding regions in the absence of active Serine2-phosphorylation at its C-terminal domain.[66] Genome-wide comparisons of these bivalent histone marks in ES cells and differentiated cell types have shown that bivalent domains at developmental regulator genes are characteristic of pluripotent cells and tend to resolve upon ES differentiation for gene activation or repression, in accordance with associated changes in gene expression.[67]

Bivalent domains are not restricted to ES cells and have been identified in lineage-restricted multipotent stem cells,[68] differentiated cell types including progenitor and terminal neurons[69] and colorectal cancer cells.[70,71] Interestingly, analyses in lymphocytes has identified a subgroup of inducible primary response genes that harbor bivalent chromatin and respond rapidly to appropriate extracellular signals, suggesting that this permissive chromatin state has important implications for regulating gene expression in a broad range of cell types.[72,73]

CONCLUSION

From the intricate details of local chromatin folding to the global organization of chromosomes within the nucleus, it is clear that genome architecture is intimately linked with genome function. Recent studies of the unique and highly plastic genome of ES cells have provided the first insights into nuclear architecture and chromatin organization characteristic of the pluripotent state, as well as changes that ensue during the establishment of cell lineage-specific programs. Despite the limited analyses of nuclear subcompartments in ES cells, a common theme is emerging that the nuclear organization of ES cells is less structured than differentiated cells. In some cases, although the components that assemble to form functional subcompartments are already present in undifferentiated cells, they are dispersed in a seemingly disorganized fashion that is resolved as differentiation is established. In other cases, such as for lamin A/C, induced expression is sufficient to promote differentiation. The small number of studies that have probed the functional organization of the ES cell nucleus already associate pluripotency with a highly dynamic genome that is reflected in a unique nuclear architecture.

ACKNOWLEDGEMENTS

We thank J.T. Butler and J.B. Lawrence for the kind gift of unpublished images and the Medical Research Council (UK) for support.

REFERENCES

1. Keenen B, de la Serna IL. Chromatin remodeling in embryonic stem cells: regulating the balance between pluripotency and differentiation. J Cell Physiol 2009; 219(1):1-7.
2. Meshorer E, Misteli T. Chromatin in pluripotent embryonic stem cells and differentiation. Nat Rev Mol Cell Biol 2006; 7(7):540-546.
3. Branco MR, Pombo A. Intermingling of chromosome territories in interphase suggests role in translocations and transcription-dependent associations. PLoS Biol 2006; 4(5):e138.
4. Lieberman-Aiden E, van Berkum NL, Williams L et al. Comprehensive mapping of long-range interactions reveals folding principles of the human genome. Science 2009; 326(5950):289-293.
5. Parada L, Misteli T. Chromosome positioning in the interphase nucleus. Trends Cell Biol 2002; 12(9):425-432.
6. Wiblin AE, Cui W, Clark AJ et al. Distinctive nuclear organisation of centromeres and regions involved in pluripotency in human embryonic stem cells. J Cell Sci 2005; 118(Pt 17):3861-3868.
7. Bartova E, Galiova G, Krejci J et al. Epigenome and chromatin structure in human embryonic stem cells undergoing differentiation. Dev Dyn 2008; 237(12):3690-3702.
8. Solovei I, Kreysing M, Lanctot C et al. Nuclear architecture of rod photoreceptor cells adapts to vision in mammalian evolution. Cell 2009; 137(2):356-368.
9. Chambeyron S, Bickmore WA. Chromatin decondensation and nuclear reorganization of the HoxB locus upon induction of transcription. Genes Dev 2004; 18(10):1119-1130.
10. Morey C, Kress C, Bickmore WA. Lack of bystander activation shows that localization exterior to chromosome territories is not sufficient to up-regulate gene expression. Genome Res 2009; 19(7):1184-1194.
11. Senner CE, Brockdorff N. Xist gene regulation at the onset of X inactivation. Curr Opin Genet Dev 2009; 19(2):122-126.
12. Lee JT. Regulation of X-chromosome counting by Tsix and Xite sequences. Science 2005; 309(5735):768-771.
13. Xu N, Tsai CL, Lee JT. Transient homologous chromosome pairing marks the onset of X inactivation. Science 2006; 311(5764):1149-1152.
14. Bacher CP, Guggiari M, Brors B et al. Transient colocalization of X-inactivation centres accompanies the initiation of X inactivation. Nat Cell Biol 2006; 8(3):293-299.
15. Augui S, Filion GJ, Huart S et al. Sensing X chromosome pairs before X inactivation via a novel X-pairing region of the Xic. Science 2007; 318(5856):1632-1636.
16. Xu N, Donohoe ME, Silva SS et al. Evidence that homologous X-chromosome pairing requires transcription and Ctcf protein. Nat Genet 2007; 39(11):1390-1396.
17. Khalil A, Grant JL, Caddle LB et al. Chromosome territories have a highly nonspherical morphology and nonrandom positioning. Chromosome Res 2007; 15(7):899-916.
18. Scialdone A, Nicodemi M. Mechanics and dynamics of X-chromosome pairing at X inactivation. PLoS Comput Biol 2008; 4(12):e1000244.
19. Tsai CL, Rowntree RK, Cohen DE et al. Higher order chromatin structure at the X-inactivation center via looping DNA. Dev Biol 2008; 319(2):416-425.
20. Eils R, Dietzel S, Bertin E et al. Three-dimensional reconstruction of painted human interphase chromosomes: active and inactive X chromosome territories have similar volumes but differ in shape and surface structure. J Cell Biol 1996; 135(6 Pt 1):1427-1440.
21. Chaumeil J, Le Baccon P, Wutz A et al. A novel role for Xist RNA in the formation of a repressive nuclear compartment into which genes are recruited when silenced. Genes Dev 2006; 20(16):2223-2237.
22. Butler JT, Hall LL, Smith KP et al. Changing nuclear landscape and unique PML structures during early epigenetic transitions of human embryonic stem cells. J Cell Biochem 2009; 107(4):609-621.
23. Gruenbaum Y, Margalit A, Goldman RD et al. The nuclear lamina comes of age. Nat Rev Mol Cell Biol 2005; 6(1):21-31.
24. Francastel C, Schubeler D, Martin DI et al. Nuclear compartmentalization and gene activity. Nat Rev Mol Cell Biol 2000; 1(2):137-143.
25. Akhtar A, Gasser SM. The nuclear envelope and transcriptional control. Nat Rev Genet 2007; 8(7):507-517.
26. Finlan LE, Sproul D, Thomson I et al. Recruitment to the nuclear periphery can alter expression of genes in human cells. PLoS Genet 2008; 4(3):e1000039.
27. Ragoczy T, Bender MA, Telling A et al. The locus control region is required for association of the murine beta-globin locus with engaged transcription factories during erythroid maturation. Genes Dev 2006; 20(11):1447-1457.
28. Williams RR, Azuara V, Perry P et al. Neural induction promotes large-scale chromatin reorganisation of the Mash1 locus. J Cell Sci 2006; 119(Pt 1):132-140.

29. Kosak ST, Skok JA, Medina KL et al. Subnuclear compartmentalization of immunoglobulin loci during lymphocyte development. Science 2002; 296(5565):158-162.
30. Hewitt SL, High FA, Reiner SL et al. Nuclear repositioning marks the selective exclusion of lineage-inappropriate transcription factor loci during T helper cell differentiation. Eur J Immunol 2004; 34(12):3604-3613.
31. Constantinescu D, Gray HL, Sammak PJ et al. Lamin A/C expression is a marker of mouse and human embryonic stem cell differentiation. Stem Cells 2006; 24(1):177-185.
32. Pajerowski JD, Dahl KN, Zhong FL et al. Physical plasticity of the nucleus in stem cell differentiation. Proc Natl Acad Sci USA 2007; 104(40):15619-15624.
33. Luo L, Gassman KL, Petell LM et al. The nuclear periphery of embryonic stem cells is a transcriptionally permissive and repressive compartment. J Cell Sci 2009; 122(Pt 20):3729-3737.
34. Kaiser TE, Intine RV, Dundr M. De novo formation of a subnuclear body. Science 2008; 322(5908):1713-1717.
35. Misteli T. Self-organization in the genome. Proc Natl Acad Sci USA 2009; 106(17):6885-6886.
36. Hall LL, Smith KP, Byron M et al. Molecular anatomy of a speckle. Anat Rec A Discov Mol Cell Evol Biol 2006; 288(7):664-675.
37. Xie SQ, Martin S, Guillot PV et al. Splicing speckles are not reservoirs of RNA polymerase II, but contain an inactive form, phosphorylated on serine2 residues of the C-terminal domain. Mol Biol Cell 2006; 17(4):1723-1733.
38. Lawrence JB, Clemson CM. Gene associations: true romance or chance meeting in a nuclear neighborhood? J Cell Biol 2008; 182(6):1035-1038.
39. Morris GE. The Cajal body. Biochim Biophys Acta 2008; 1783(11):2108-2115.
40. Schwartz YB, Pirrotta V. Polycomb complexes and epigenetic states. Curr Opin Cell Biol 2008; 20(3):266-273.
41. Brookes E, Pombo A. Modifications of RNA polymerase II are pivotal in regulating gene expression states. EMBO Rep 2009; 10(11):1213-1219.
42. Bernstein E, Duncan EM, Masui O et al. Mouse polycomb proteins bind differentially to methylated histone H3 and RNA and are enriched in facultative heterochromatin. Mol Cell Biol 2006; 26(7):2560-2569.
43. Ren X, Vincenz C, Kerppola TK. Changes in the distributions and dynamics of polycomb repressive complexes during embryonic stem cell differentiation. Mol Cell Biol 2008; 28(9):2884-2895.
44. Bancaud A, Huet S, Daigle N et al. Molecular crowding affects diffusion and binding of nuclear proteins in heterochromatin and reveals the fractal organization of chromatin. EMBO J 2009; 28(24):3785-3798.
45. Kimura H, Cook PR. Kinetics of core histones in living human cells: little exchange of H3 and H4 and some rapid exchange of H2B. J Cell Biol 2001; 153(7):1341-1353.
46. Phair RD, Scaffidi P, Elbi C et al. Global nature of dynamic protein-chromatin interactions in vivo: three-dimensional genome scanning and dynamic interaction networks of chromatin proteins. Mol Cell Biol 2004; 24(14):6393-6402.
47. Giglia-Mari G, Theil AF, Mari PO et al. Differentiation driven changes in the dynamic organization of Basal transcription initiation. PLoS Biol 2009; 7(10):e1000220.
48. Cheutin T, McNairn AJ, Jenuwein T et al. Maintenance of stable heterochromatin domains by dynamic HP1 binding. Science 2003; 299(5607):721-725.
49. Festenstein R, Pagakis SN, Hiragami K et al. Modulation of heterochromatin protein 1 dynamics in primary Mammalian cells. Science 2003; 299(5607):719-721.
50. Misteli T, Gunjan A, Hock R et al. Dynamic binding of histone H1 to chromatin in living cells. Nature 2000; 408(6814):877-881.
51. Meshorer E, Yellajoshula D, George E et al. Hyperdynamic plasticity of chromatin proteins in pluripotent embryonic stem cells. Dev Cell 2006; 10(1):105-116.
52. Aoto T, Saitoh N, Ichimura T et al. Nuclear and chromatin reorganization in the MHC-Oct3/4 locus at developmental phases of embryonic stem cell differentiation. Dev Biol 2006; 298(2):354-367.
53. Kobayakawa S, Miike K, Nakao M et al. Dynamic changes in the epigenomic state and nuclear organization of differentiating mouse embryonic stem cells. Genes Cells 2007; 12(4):447-460.
54. Cammas F, Oulad-Abdelghani M, Vonesch JL et al. Cell differentiation induces TIF1beta association with centromeric heterochromatin via an HP1 interaction. J Cell Sci 2002; 115(Pt 17):3439-3448.
55. Efroni S, Duttagupta R, Cheng J et al. Global transcription in pluripotent embryonic stem cells. Cell Stem Cell 2008; 2(5):437-447.
56. Bibikova M, Chudin E, Wu B et al. Human embryonic stem cells have a unique epigenetic signature. Genome Res 2006; 16(9):1075-1083.
57. Gaspar-Maia A, Alajem A, Polesso F et al. Chd1 regulates open chromatin and pluripotency of embryonic stem cells. Nature 2009; 460(7257):863-868.
58. Faro-Trindade I, Cook PR. A conserved organization of transcription during embryonic stem cell differentiation and in cells with high C value. Mol Biol Cell 2006; 17(7):2910-2920.

59. Donaldson AD. Shaping time: chromatin structure and the DNA replication programme. Trends Genet 2005; 21(8):444-449.
60. Perry P, Sauer S, Billon N et al. A dynamic switch in the replication timing of key regulator genes in embryonic stem cells upon neural induction. Cell Cycle 2004; 3(12):1645-1650.
61. Hiratani I, Ryba T, Itoh M et al. Global reorganization of replication domains during embryonic stem cell differentiation. PLoS Biol 2008; 6(10):e245.
62. Hiratani I, Leskovar A, Gilbert DM. Differentiation-induced replication-timing changes are restricted to AT-rich/long interspersed nuclear element (LINE)-rich isochores. Proc Natl Acad Sci USA 2004; 101(48):16861-16866.
63. Goren A, Cedar H. Replicating by the clock. Nat Rev Mol Cell Biol 2003; 4(1):25-32.
64. Azuara V, Perry P, Sauer S et al. Chromatin signatures of pluripotent cell lines. Nat Cell Biol 2006; 8(5):532-538.
65. Bernstein BE, Mikkelsen TS, Xie X et al. A bivalent chromatin structure marks key developmental genes in embryonic stem cells. Cell 2006; 125(2):315-326.
66. Stock JK, Giadrossi S, Casanova M et al. Ring1-mediated ubiquitination of H2A restrains poised RNA polymerase II at bivalent genes in mouse ES cells. Nat Cell Biol 2007; 9(12):1428-1435.
67. Mikkelsen TS, Ku M, Jaffe DB et al. Genome-wide maps of chromatin state in pluripotent and lineage-committed cells. Nature 2007; 448(7153):553-560.
68. Cui K, Zang C, Roh TY et al. Chromatin signatures in multipotent human hematopoietic stem cells indicate the fate of bivalent genes during differentiation. Cell Stem Cell 2009; 4(1):80-93.
69. Mohn F, Weber M, Rebhan M et al. Lineage-specific polycomb targets and de novo DNA methylation define restriction and potential of neuronal progenitors. Mol Cell 2008; 30(6):755-766.
70. Rodriguez J, Munoz M, Vives L et al. Bivalent domains enforce transcriptional memory of DNA methylated genes in cancer cells. Proc Natl Acad Sci USA 2008; 105(50):19809-19814.
71. McGarvey KM, Van Neste L, Cope L et al. Defining a chromatin pattern that characterizes DNA-hypermethylated genes in colon cancer cells. Cancer Res 2008; 68(14):5753-5759.
72. Lim PS, Hardy K, Bunting KL et al. Defining the chromatin signature of inducible genes in T-cells. Genome Biol 2009; 10(10):R107.
73. Araki Y, Wang Z, Zang C et al. Genome-wide analysis of histone methylation reveals chromatin state-based regulation of gene transcription and function of memory CD8+ T-cells. Immunity 2009; 30(6):912-925.

CHAPTER 3

EPIGENETIC REGULATION OF PLURIPOTENCY

Eleni M. Tomazou and Alexander Meissner*

Abstract: Epigenetic regulation refers to the mechanisms that alter gene expression patterns in the absence of changes in the nucleotide sequence of the DNA molecule. The best understood epigenetic marks include posttranslational modifications of the histone tails and DNA methylation. Both play central roles in normal development and in diseases. Pluripotent stem cells have great promise for regenerative medicine and recent efforts have focused on identifying molecular networks that govern pluripotency. This chapter provides an overview of epigenetic regulation in embryonic stem cells. We present a brief introduction into epigenetic mechanisms and focus on their role in pluripotent cells.

INTRODUCTION

The term epigenetics was first introduced by Conrad Waddington in 1942.[1] It was used to describe the interactions of genes with their environment "to bring a phenotype into being". Today epigenetics refers to mitotically and, in some cases, meiotically heritable states of gene expression that are not due to changes in the DNA sequence.[2] The Greek prefix 'epi-' implies features that are "in addition" to genetics and this is reflected by the current definition. Epigenetic modifications can alter the functions of associated genes by modulating DNA accessibility, protein recruitment and chromatin structure. As such they have been implicated in playing vital roles in defining, maintaining and propagating cellular states.[3] Epigenetics is concerned not only with the current state of the cell, but also with its potential to acquire new states,[4,5] which establishes the relevance of epigenetic regulation in developmental plasticity that is found in embryonic stem (ES) cells.[6]

*Corresponding Author: Alexander Meissner—Department of Stem Cell and Regenerative Biology, Harvard University, 7 Divinity Avenue, Cambridge, Massachusettes 02138, USA. Email: alexander_meissner@harvard.edu

The Cell Biology of Stem Cells, edited by Eran Meshorer and Kathrin Plath.
©2010 Landes Bioscience and Springer Science+Business Media.

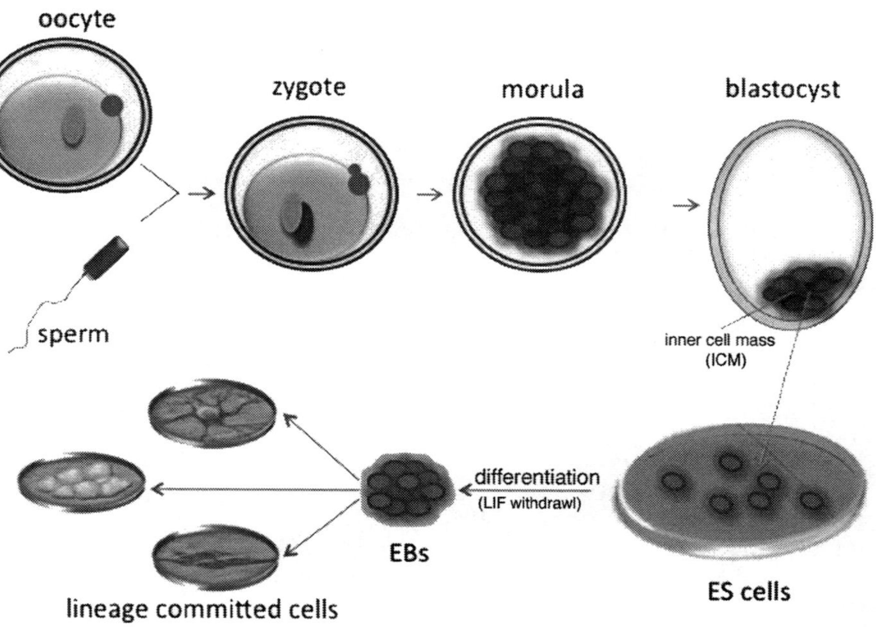

Figure 1. Mouse ES cell derivation and differentiation. mES cells originate from the ICM of the blastocyst-staged embryo during early mouse development (embryonic day 3.5). mES cells are cultivated in vitro in the presence of Leukemia Inhibitory Factor (LIF) and can be kept in culture indefinitely without going into senescence. Upon LIF withdrawal mES cells can form embryoid bodies (EBs) when placed into non-adherent culture dishes. The cells of EBs can be differentiated to all three primary germ layers (ectoderm, endoderm, mesoderm) using the appropriate culture conditions.

ES cells are derived from the inner cell mass (ICM) of pre-implantation embryos[7-10] (Fig. 1) and possess two main characteristics: self-renewal and pluripotency. "Self-renewal" describes the ability of ES cells to generate an unlimited number of identical successors while "pluripotency" refers to their capacity to respond to various developmental cues to generate multiple somatic cell lineages and germline cells in vitro and in vivo.[11] ES cells therefore offer the possibility to establish new models of mammalian development and to create new sources of cells for regenerative medicine.[12] Over the past decade, following the initial derivation of five human ES cell (hES cells) lines in 1998,[10] hundreds of additional lines have been derived.[13,14] Notably, using appropriate culture conditions pluripotent cells can also be derived from later stages of development. Examples include epiblast stem cells (EpiSCs), reverted embryonic stem cells (rES cells)[15] and embryonic germ (EG) cells.[16] The recent derivation of induced pluripotent stem (iPS) cells also enables the generation of ES-like cells from most somatic cell types through ectopic expression of four transcription factors.[17]

Understanding the molecular mechanisms governing the ES cell state and the pluripotent state in general is of immense biological interest. Previous work has identified a transcriptional network driven by specific transcription factors (TFs) such as OCT4, NANOG and SOX2, that is essential to maintain the undifferentiated state of pluripotent stem cells.[18] This network sustains the active state of genes that are required for stem cell survival and proliferation while suppressing genes required for differentiation.[19]

Interactions of the TFs with epigenetic modifiers are likely to contribute to the stability of the network regulating pluripotency.[20] In the following sections we will discuss the role of epigenetics in pluripotent cells in more detail.

EPIGENETIC MODIFICATIONS

In eukaryotes, DNA is packed into chromatin (Fig. 2). The basic element of chromatin is the nucleosome, which consists of about 147 bp of DNA wrapped twice around an octamer containing two copies of each core histone heterodimers (H3/H4 and H2A/

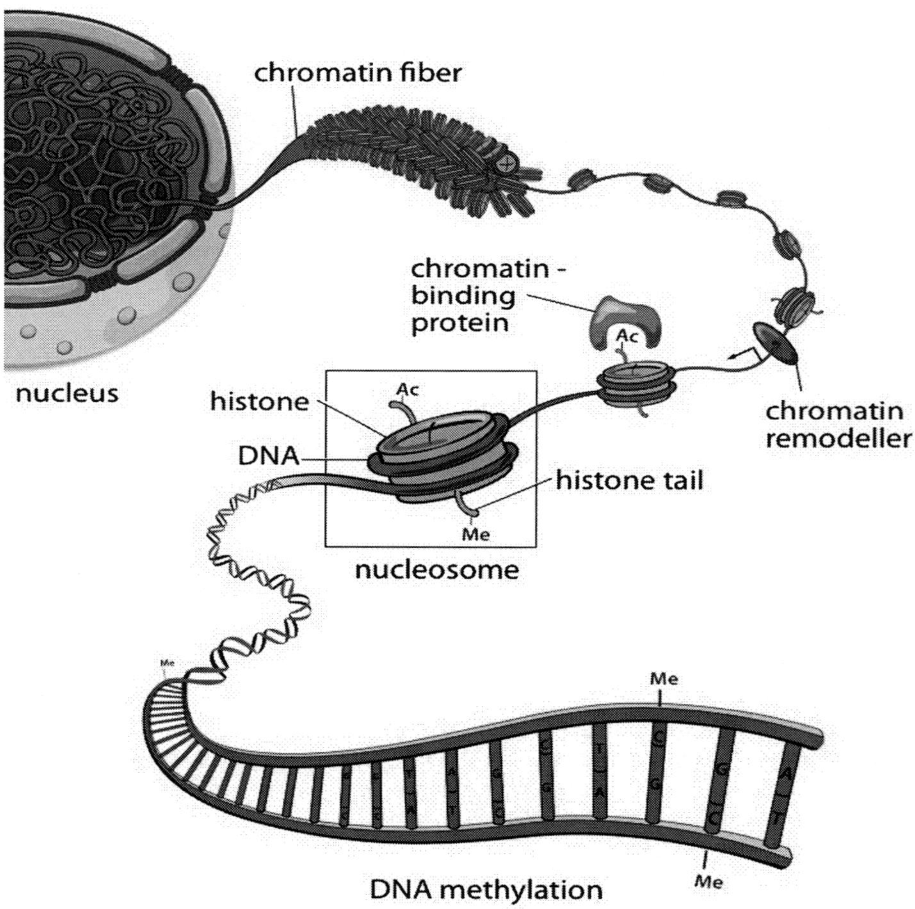

Figure 2. Epigenetic regulation. In the nucleus, the coding and structural information of DNA is organized into chromatin. The basic unit of chromatin is the nucleosome, which consists of 147 bp of DNA wrapped around a core of histone proteins. Nucleosomes can be organized into higher order structures and the level of packaging can have profound consequences on all DNA-mediated processes including gene regulation. A wide range of processes, including posttranslational modifications of histone tails, DNA methylation and binding of chromatin remodeling proteins, can influence nucleosome positioning and chromatin compaction.

H2B).[21] Adjacent nucleosomes are connected via linker DNA bound by the linker histone H1. Chromatin structure can influence gene expression by controlling the accessibility of regulatory elements and by modulating the affinity of transcription factors to their target regions.[22] Key effectors of chromatin structure are posttranslational histone modifications and methylation of genomic DNA,[23,24] while ATP-dependent chromatin remodeling factors can modulate chromatin states[25] (Fig. 2).

Modulators of Chromatin Structure and Dynamics

ATP-dependent chromatin remodeling factors are a class of enzymes that utilize the hydrolysis of ATP to disrupt contacts between histone proteins and DNA. This leads to changes in nucleosome conformation and positioning and subsequently results in altered higher order chromatin structure.[26,27] These factors increase DNA accessibility, allowing gene regulators and transcription factors to bind to their target region and can be grouped into four families: SWI/SNF, ISWI, NuRDMi-2/CHD and INO80. Each family is involved in diverse biological processes including DNA repair, checkpoint regulation, DNA replication, telomere maintenance and chromosome segregation, indicating that they are vital components of pathways maintaining genomic integrity.[25]

Histone Modifications

Core histones are the major protein components of nucleosomes (Fig. 2) and are evolutionary conserved. Their amino-terminal tails protrude outward from the nucleosome, while the globular carboxyl-terminal domain makes up the nucleosome scaffold. The N-terminal histone tails function as acceptors for a variety of posttranslational modifications, including acetylation, methylation and ubiquitinations of lysine (K) residues, phosphorylation of serine (S) and threonine (T) residues and methylation of arginine (R) residues. A subset of these modifications has been linked to the activation or repression of genes.[28] Key marks for active transcription and open chromatin include acetylation of histone 3 K9 and K14 (H3K9ac, H3K14ac) and H4K16 (H4K16ac), the mono-, di- and tri-methylation of H3K4 (H3K4me1, H3K4me2, H3K4me3) and the tri-methylation of H3K36 (H3K36me3). Histone modifications associated with transcriptional repression include tri-methylated H3K27 (H3K27me3) and tri-methylated H3K9 (H3K9me3). It has been suggested that a combination of multiple histone modifications located at one gene reflects not only its current expression state but also provides the cell with the ability to activate this gene at a later stage in development or in its daughter cells.[4] The concept of combinatorial pattern giving rise to distinct biological outcomes has been aptly termed the "histone code" hypothesis.[5]

A particular combination of relevance for understanding pluripotency and developmental potential is the co-occurrence of the repressive H3K27me3 with the active H3K4me3 modification. This specific combination was termed bivalent domain.[29] The current hypothesis is that bivalent domains maintain genes in a poised state for subsequent activation. A large proportion of them, which includes many master regulators of early development, are activated upon differentiation and concomitant loss of the repressive H3K27me3 mark. However, bivalent domains can be considered as bidirectional switches that not only allow their targets to be activated, but also to be switched off completely if no longer required after commitment to alternative lineages. After such cell fate decisions, non-induced bivalent genes tend to lose the active H3K4me3 mark, whereas the repressive

H3K27me3 mark is retained.[30,31] Although bivalent domains were initially thought to be exclusive to pluripotent cells, recent studies have identified such domains also in cells with more restricted developmental potential.[30]

In addition to the core histones, several histone variants of H2A and H3 are found in the DNA and understanding of their role and importance has advanced in recent years. These variants are non-allelic forms of the core histones and are incorporated into chromatin in a cell-cycle independent manner. Replacements of core histones by the variants can induce localized changes in chromatin structure resulting in altered transcriptional states.[32]

Many of the enzymes responsible for establishing and maintaining histone modifications are known. These include enzymes for acetylation,[33] methylation,[34] phosphorylation,[35] ubiquitination,[36] sumoylation,[37] ADP-ribosylation,[38] deimination,[39] and proline isomerization.[40] Most modifications have been found to be reversible, as enzymes that remove such modifications, including histone demethylases and deacetylaces, have also been identified confirming a dynamic role of histone modifications in transcriptional regulation.

The first identified and best-studied mediators include the Polycomb group (PcG) proteins.[41] These developmental regulators form multi-protein complexes called Polycomb-Repressive Complexes (PRCs). PcG proteins catalyze two distinct histone modifications: tri-methylation of lysine 27 of histone 3 (H3K27me3) and mono-ubiquitination of H2A (H2Aub1). H3K27 is tri-methylated by the enhancer of zeste (Ezh2), which together with embryonic ectoderm development (Eed) and suppressor of Zeste 12 (Suz12) are components of PRC2. H3K27me3 can recruit PRC1 (maintenance complex) mediating chromatin compaction. PRC1 mono-ubiquitinates H2A through Ring1b; this activity is stimulated by the presence of Bmi1 and Mel18.[42] In mammalian cells PRC2 components are tightly colocalized with H3K27me3 whereas PRC1 shows lower levels of localization indicating that PRC1 may provide additional stability to the silencing progress mediated by PcGs.[43]

DNA Methylation

DNA methylation is among the best-studied epigenetic marks and the only process that acts directly on the DNA molecule. On the biochemical level, it is a covalent modification of the 5-carbon position of cytosine: it replaces the hydrogen atom with a methyl group (Fig. 3). In mammals DNA methylation occurs predominantly in the context of cytidine-guanosine (CpG) dinucleotides[44] but non-CpG methylation has also been reported in certain cell types (Fig. 3).[45-49] Transient DNA demethylation occurs in the germline and during pre-implantation stages of embryonic development.[50] Gain of DNA methylation is linked to local spatial re-arrangements of chromatin (more compact chromatin structure) and alters transcription factor binding in promoter regions, resulting in highly stable gene silencing.[51]

Two important classes of DNA methyltransferases catalyze DNA methylation. DNA methyltransferase 1 (Dnmt1) has been described as a maintenance methyltransferase because it methylates hemi-methylated CpG dinucleotides in the nascent strand of DNA after DNA replication, thus maintaining DNA methylation patterns over multiple cell divisions.[52,53] The two de novo Dnmts, Dnmt3a and Dnmt3b are responsible for establishing new methylation patterns during embryonic development and differentiation.[54] They are highly expressed in ES cells and down regulated in

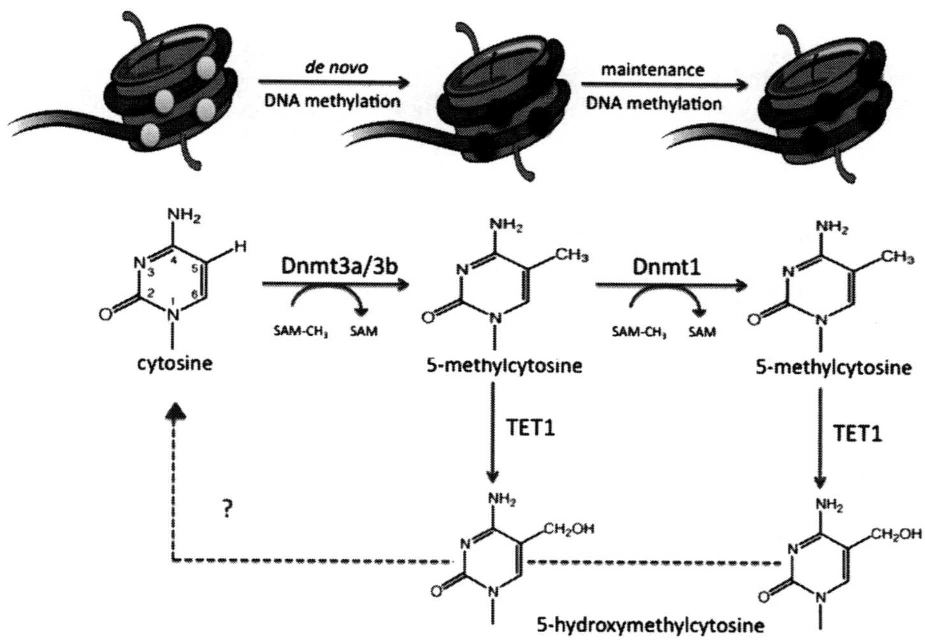

Figure 3. Mechanism of DNA methylation. 5-Methylcytosine is produced by the action of the DNA methyltransferases (DNMT 1, 3a or 3b), which catalyze the transfer of a methyl group (CH_3) from *S*-adenosylmethionine (SAM) to the 5-carbon position of cytosine. Recently it has been reported that TET1 converts 5-methylcytosine into 5-hydroxymethylcytosine. The latter has been detected in mES cells.[105] It has been speculated that 5-hydroxymethylcytosine is an intermediate during the process of active demethylation but further studies are required to confirm this argument.

differentiated cells. Dnmt3l is a close homolog of Dnmt3a and Dnmt3b,[55] but it lacks active methyltransferase activity and functions as a coregulator of both Dnmt3a and Dnmt3b. It has recently been shown to interact with the N-terminal tail of histone H3 when it lacks methylation at lysine 4.[56] The latter provides a possible mechanism by which Dnmt3l facilitates de novo DNA methylation to nucleosomal DNA in the germline.

THE EPIGENOME OF ES CELLS

Chromatin Structure and Dynamics

Murine ES cells possess an unusually open chromatin structure, which is rich in noncompact euchromatin[57-59] and diffuse heterochromatin.[60] This "open" chromatin structure allows easy access to transcription factors and the transcriptional machinery resulting in global "hyper-transcription" possibly contributing to ES cell plasticity; lineage commitment is accompanied by accumulation of highly condensed, transcriptionally inactive heterochromatin regions.[61]

Consistent with their chromatin state, ES cells show elevated levels of several ATP-dependent chromatin-remodeling factors. For instance, Brg1 (Smarca4), the ATPase subunit of the mammalian SWI/SNF complex has been implicated in regulating ES cell pluripotency through interactions with Nanog.[62] In a similar manner the NuRD (nucleosome remodeling and histone deacetylation) component Mbd3 is essential for differentiation[63] whereas Chd1-deficient cells are no longer pluripotent. Chd1 contains an ATPase SNF2-like helicase domain and associates with genes of active promoters. Interestingly, down-regulation of Chd1 leads to heterochromatin formation, which supports the notion that highly accessible chromatin is essential to the unique properties of stem cells.[64]

The elevated levels of chromatin remodelers that maintain ES cell chromatin in a "relaxed" state may also contribute to the cellular plasticity of ES cells.[57] Chromatin binding proteins are hyperdynamic in the sense that they are loosely bound to the chromatin in ES cells. Upon differentiation, proteins become immobilized on chromatin, which abolishes the pluripotent state. ES cells lacking the nucleosome assembly factor HirA exhibit elevated levels of unbound histones and formation of embryoid bodies (EBs) is accelerated. In contrast, ES cells with restricted H1 dynamic exchange display differentiation arrest.[60] This suggests that "hyperdynamic" chromatin is an important hallmark of pluripotency.

Histone Modifications

Histone Marks Associated with Active Genomic Regions

In addition to ES cells' open chromatin structure, active chromatin domains are widespread in ES cells. These domains are typically marked by histone modifications such as acetylated H3 and H4 and di- and tri-methylated H3K4 as well as H3K9,14ac.[30,65] Enzymes known to "read and write" these modifications have started to emerge as regulators of pluripotency. For instance, Rbp2 (Jarid1a) which specifically catalyses H3K4me2/3 demethylation, is bound on promoters of differentiation-related genes in ES cells, indicating that Rbp2 may control their epigenetic states.[66] p300, a histone acetyltransferase that associates with the elongating form of Pol II, is recruited to sites bound by Nanog, Oct4 and Sox2[67] whereas the histone acetyltransferase and chromatin remodeling complex Tip60-p400 aides in maintaining the pluripotent state of ES cells by regulating Nanog-bound loci.[68] Active promoters also possess H3K4me2, while H3K36me3 and H3K79me2 are located along actively transcribed genomic regions[65] (Fig. 4).

In addition to histone lysine modification, histone H3 arginine methylation at R17 and R26 has recently emerged as a novel epigenetic mechanism regulating pluripotency.[69,70] The promoter regions of the genes Oct4 and Sox2 possess H3R16 and H3R26 methylation and display detectable levels of co-activator-associated-protein-arginine methyltransferase 1 (CARM1). In ES cells CARM1 depletion leads to differentiation and its over-expression to H3 arginine methylation of the Nanog promoter resulting in delayed differentiation upon induction.[70]

Enhancer regions are enriched in H3K4me1 and H3K27ac in hES cells.[71] Enhancers are DNA sequences that activate cognate promoters and regulate gene expression in an orientation- and position-independent manner.[72]

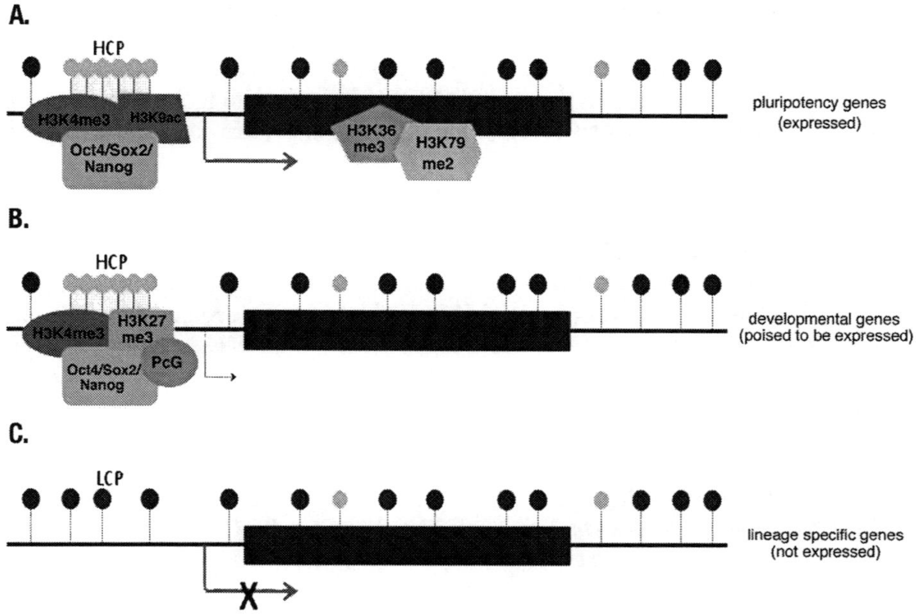

Figure 4. A simplified overview of epigenetic regulation in ES cells. Epigenetic patterns are sorted into three subclasses of genes: A) pluripotency genes that are expressed; B) developmental genes that are poised to be expressed; and C) genes that are expressed upon induction of differentiation (lineage specific genes). Gene promoters are further annotated as HCP and LCP according to different density of CpGs. CpGs are represented as lollipops. Open circles represent unmethylated and filled circles represent methylated CpGs. CpGs within the gene bodies and LCPs are mostly methylated whereas CpGs within HCPs are unmethylated. Pluripotency genes possess histone marks associated with active transcription (H3K4me3, H3K9ac) on their promoters and H3K36me3 and H3K79me2 within their gene bodies. Bivalent domains characterize the promoters of developmental associated genes. The interaction of polycomb proteins (PcGs) with such promoters is also shown. In some case these show transcription initiation activity (represented here with a small, faint arrow).

Histone Marks Associated within Promoters Poised for Expression

Genome wide studies have shown that in hES cells more than three quarters of gene promoters are marked by active histone modifications and experience transcription initiation by Pol II action. However, only half of them produce detectable transcripts. For instance, H3K4me3 is present at 80% of the annotated promoters in ES cells.[65] Many of these do not produce full-length transcripts detectable with conventional methods. In many cases these promoters are also enriched for the repressive mark H3K27me3 and hence show bivalency. Bivalent domains as described above mostly coincide with differentiation-associated genes in ES cells (Fig. 4). Interestingly, about half of the identified bivalent domains have binding sites for at least one of the three pluripotency-associated transcription factors Oct4, Nanog and Sox2.[19,73] Bivalent domains generally show PcG occupancy. However genome wide mapping in ES cells has revealed that although PRC1 and PRC2 are present simultaneously at many promoters,[74,75] there are promoters that are not co-occupied by the two complexes. In ES cells, bivalent domains can be subdivided into two groups based on co-occupancy of both PRC1 and PRC2 or by PRC2 alone.[43] Interestingly, promoters "co-occupied" by both complexes can retain PcG-mediated chromatin structure more efficiently upon differentiation. Noncanonical PRC2 complexes are also present in hES cell. Such complexes contain EZH1, a homolog of EZH2. EZH1-containing complexes appear to be selectively targeted to key developmental genes, the repression of which may be critical for preventing ES cell differentiation.[76]

The importance of PRC2 (and H3K27me3) in ES cells has been illustrated by loss-of-function experiments. Loss of PRC2 components leads to defects in ES cell differentiation.[76-78] Nevertheless, mouse ES cells lacking Eed, Suz12 or Ezh2 can be derived from the respective homozygous knockout blastocysts and propagated in vitro.[76-78] The active role of H3K27me3 in ES cells was further demonstrated by recent studies suggesting that Jarid2 (which is closely related to the Jarid1 family) regulates the enzymatic activity and the occupancy of PCR2 and hence it fine-tunes H3K27me3 levels in pluripotency.[79-82] Jarid2 is known to be highly expressed in ES and iPS cells[83] and is essential for differentiation, but appears dispensable for ESC self-renewal and maintenance.[79,80]

More recently, efforts have been made to classify bivalent domains based on histone modifications other than H3K4me3 and H3K27me3. In ES cells, H3K9me3 was detected in a subset of bivalent promoters of genes encoding developmental regulators.[84] Future studies will need to confirm the presence and role of these trivalent structures, since previous studies have failed to detect this distribution.[30,85] In a similar manner, it has been reported that the histone variant H2A.Z colocalizes with PcG proteins and more specifically with Suz12, the core component of PCR2.[86] Though, an independent study, using human cells (U2OS cells) argues that euchromatic genes that are not transcribed do not carry H2A.Z on their promoters.[87]

Finally, in addition to H3K4me3 as a histone mark found to occupy both active and inactive (bivalent) promoters, H3K56ac has been shown to have the same occupancy capabilities in hES cells.[88] Interestingly the binding of NANOG, SOX2 and OCT4 associated with K56Ac more often than H3K4me3, suggesting that K56Ac is involved in the human core transcriptional network of pluripotency.[88] Functional studies are required to confirm the claims of this study.

Histone Marks Associated with Silent Genomic Regions

Genomic regions that are associated with gene silencing, including transposon and repetitive elements, frequently possess the histone repressive marks H3K9me3 and H4K20me3.[30,89] H3K64me3 (H3K64 lies within the globular domain of H3) is a newly identified repressive mark, enriched at pericentric heterochromatin. Interestingly, H3K64me3 is higher in mES cells compared to differentiated cells,[90] consistent with the notion that epigenetic patterns at repeats show substantial change during differentiation possibly contributing to ES cell plasticity.[89,90]

Functional studies have further confirmed an important role of these histone marks in ES cells. On one hand, loss of the H3K9 methyltransferase Suv39h in mES cells leads to prominent enrichment of transcripts corresponding to all repeat classes.[89] On the other hand, the H3K9 de-methylases, Jmjd1a and Jmjd2c have been identified as Oct4 targets.[91] Depletion of Jmjd1a/2c in ES cells results in differentiation induction and a global increase in H3K9me2/3, confirming that these two demethylases and probably H3K9 methylation levels are linked to the maintenance of pluripotency.[91]

DNA Methylation

Functional Relevance of DNA Methylation in ES Cells

ES cells operate the DNA methylation machinery in a way that is very distinct from somatic cells. On the one hand, the de novo methyltransferases—Dnmt3a2 and Dnmt3b1 (which are isoforms of Dnmt3a and Dnmt3b respectively) in particular—are expressed at very high levels. On the other hand, DNA methylation is largely absent from gene-regulatory elements. Even more so, ES cells can tolerate complete loss of DNA methylation.[92,93] Dnmt1,[93] Dnmt3a/3b[92,93] or Dnmt 1, 3a and 3b[49,93] deficient mES cells are able to survive and maintain their self-renewal capacity despite significant loss of DNA methylation. They retain pluripotency, but can only differentiate if Dnmt function is restored.[92,93] In contrast, reduction of CpG methylation by inactivating Dnmts in somatic cells results in growth defects, cell death, activation of retrotransposons and genome instability,[94-97] indicating that CpG methylation plays a fundamental role in basic cellular functions of mammalian cells.

ES cells also express high levels of the gene Dnmt3l but its function in ES cells is not fully understood.[98] Dnmt3l has dual functions of binding unmethylated histone H3 tail and activating the DNA methyltransferases,[56] suggesting that high CpG density promoters in ES cells are possibly protected from de novo methylation through the H3K4 methyl marks.

In an attempt to further understand how DNA methylation regulates gene expression in ES cells, mES cells that lack all three Dnmts (TKO cells)[49] were subjected to genome-wide expression analysis.[99] Genes that were up-regulated in response to hypomenthylation, showed an overrepresentation of tissue specific genes, such as transcription factors and signaling molecules; testis and oocyte-specific genes were highly enriched in the TKO cell line as well. Interestingly, only 5% of the up-regulated genes overlap with genes possessing PcG and only 1.7% Nanog/Oct4/Sox2 bound genes overlap with the up-regulated genes in the demethylated ES cells. The latter indicates that DNA methylation and the transcriptional network governed by Oct4, Sox2 and Nanog may play distinct roles in pluripotent ES cells.

DNA Methylation Patterns in ES Cells

Recent technological advances enabled the generation of the first two complete human DNA methylomes at a base-pair resolution.[48] About 76% of CpGs assayed were found to be methylated in hES cells. Similar to previous studies in the mouse,[100] the methylation levels of CpGs in wild-type ES cells display a bimodal distribution, with most genomic regions being either ‘largely unmethylated’ or ‘largely methylated’. The methylation status of CpGs is highly correlated with the local CpG density. CpGs in regions of high CpG density (>7% over 300 bp) tend to be unmethylated, whereas CpGs in low-density regions (<5%) tend to be methylated[100] (Fig. 4). It is worth noting that high-CpG-density promoters (HCPs) are associated with two classes of genes: ubiquitous ‘housekeeping’ genes and highly regulated ‘key developmental’ genes.

On the other hand, low-CpG-density promoters (LCPs) are generally associated with tissue-specific genes.[101] In ES cells, CpGs located in LCPs are mostly methylated with the exception of LCPs enriched with H3K4me3 or H3K4me2, which have significantly reduced methylation levels. CpGs within distal regulatory regions such as enhancers, silencers and boundary elements display the same anticorrelation and only show levels of DNA methylation if lacking H3K4me1/2. These data highlight that histone methylation patterns are better predictors of DNA methylation levels than CpG density alone. These results are in agreement with previous, less comprehensive studies[99,102] and support the notion that DNA methylation and histone modifications are interconnected.[103]

DNA methylation patterns within repetitive regions in mES cells have also been revealed. CpGs located in long terminal repeats (LTRs) and long interspersed elements (LINEs) are generally hypermethylated, even in CpG rich contexts. In contrast, CpG in short interspersed elements (SINEs) show a correlation between methylation levels and CpG density that is comparable to nonrepetitive sequences.[100] Loss of DNA methylation leads to small increase of repetitive transcript levels[89] in contrast to loss of H3K9 methylation levels (see above).

Non-CpG Methylation in ES Cells

The recently completed human methylomes discussed above,[48] in agreement with previous studies in mES cells[46,49] and the early embryo,[104] have highlighted the presence of non-CpG methylation in hES cells. Methylation in non-CpG contexts showed enrichment in gene bodies and depletion in protein binding sites and enhancers. Interestingly, non-CpG methylation disappeared upon induced differentiation of ES cells and was restored in induced pluripotent stem cells.[48] Non-CpG methylation has only been found in mammalian ES cells that highly express de novo Dnmts and no function has been reported for it. Further studies are required to elucidate the biological importance of nonCpG methylation in ES cells if any.

In addition to non-CpG methylation, recent studies have confirmed the existence of 5-hydroxymethylcytosine (5hmC) in certain cell types including mES cells.[105] The levels of 5hmC are catalyzed by the protein ten-eleven translocation 1 (TET1) (Fig. 3). It has been proposed that 5hmC can be converted back to unmethylated cytosine, possibly via DNA repair mechanisms and hence 5hmC may act as an intermediate in active DNA demethylation (Fig. 3). It should be noted that current technologies for DNA methylation profiling cannot discriminate between 5mC and 5hmC and hence investigating the role of 5hmC is not trivial at the moment.

CONCLUSION

Recent advances in genome-wide technologies have allowed the unraveling of the epigenetic patterns unique to ES cells (Fig. 4) whereas functional studies have started to identify the key regulators of these patterns. However, despite their unique distribution, the role of epigenetic marks in ES cells will still need further investigation.

ES cells lacking critical epigenetic factors, including Dnmts, histone methyltansferase and demethylases, chromatin remodeling factors and PcG proteins[106] are viable and do not loose their property of self-renewal, implying that epigenetic regulation may be dispensable for maintaining ES cell identity. These observations together with the fact that bivelant domains are not present in ES cells only, have lead to the argument that epigenetic mechanisms may only contribute to the stability of pluripotency.[107] Consistent with this argument it has been proposed that the "epigenome" functions in buffering variability in transcription;[108] it is probably a control of "noise" that may accompanies the "hyper-transcription" observed in ES cells.[61] Nevertheless, ES cells deficient in epigenetic factors often exhibit defective differentiation, which may reflect roles for these components in differentiation-associated processes and subsequent commitment in lineage-specific states.[106]

These facts indicate that the functional role of the epigenome may be dynamic during development. It is probable that changes in epigenetic patterns and in expression of key epigenetic modulators during differentiation may reflect changes in the "epigenome" rationale, from a "buffer" in pluripotent cells to a "gatekeeper" of lineage-specific states.[3]

In order to further understand the role of the epigenome in pluripotency and lineage commitment, it is important to combine the studies reviewed in this chapter, which focus on the role of the epigenome in ES cells, with studies designed to understand mechanisms regulating early mammalian development. The recently developed epigenomics technologies that can be applied to a small number of cells,[109,110] will be crucial for such studies. Elucidating the mechanisms of epigenetic regulation in different cellular states during development will undoubtedly contribute to the field of stem cell biology and improve our approach to regenerative medicine.

ACKNOWLEDGEMENTS

We are grateful to Sigrid Hart (The Broad Institute, Cambridge, MA) for designing figure 2. We also thank the members of the Meissner lab and especially Natalie Jäger and Casey Gifford for comments on the manuscript. AM is supported by grants from the NIH, the Massachusetts Life Science Center and the Pew Charitable Trusts.

REFERENCES

1. Waddington C. The epigenotype. Endeavour 1942; 1:18-20.
2. Allis CD, Jenuwein T, Reinberg D. Epigenetics, x, 502 p. (Cold Spring Harbor Laboratory Press, Cold Spring Harbor, N.Y., 2007).
3. Hemberger M, Dean W, Reik W. Epigenetic dynamics of stem cells and cell lineage commitment: digging Waddington's canal. Nat Rev Mol Cell Biol 2009; 10:526-37.
4. Turner BM. Cellular memory and the histone code. Cell 2002; 111:285-91.
5. Turner BM. Defining an epigenetic code. Nat Cell Biol 2007; 9:2-6.
6. Spivakov M, Fisher AG. Epigenetic signatures of stem-cell identity. Nat Rev Genet 2007; 8:263-71.
7. Evans MJ, Kaufman MH. Establishment in culture of pluripotential cells from mouse embryos. Nature 1981; 292:154-6.
8. Martin GR. Isolation of a pluripotent cell line from early mouse embryos cultured in medium conditioned by teratocarcinoma stem cells. Proc Natl Acad Sci USA 1981; 78:7634-8.
9. Stojkovic M, Lako M, Stojkovic P et al. Derivation of human embryonic stem cells from day-8 blastocysts recovered after three-step in vitro culture. Stem Cells 2004; 22:790-7.
10. Thomson JA, Itskovitz-Eldor J, Shapiro SS et al. Embryonic stem cell lines derived from human blastocysts. Science 1998; 282:1145-7.
11. Yu J, Thomson JA. Pluripotent stem cell lines. Genes Dev 2008; 22:1987-97.
12. Murry CE, Keller G. Differentiation of embryonic stem cells to clinically relevant populations: lessons from embryonic development. Cell 2008; 132:661-80.
13. Cowan CA, Klimanskaya I, McMahon J et al. Derivation of embryonic stem-cell lines from human blastocysts. N Engl J Med 2004; 350:1353-6.
14. Sidhu KS, Ryan JP, Tuch BE. Derivation of a new human embryonic stem cell line, endeavour-1 and its clonal propagation. Stem Cells Dev 2008; 17:41-51.
15. Bao S et al. Epigenetic reversion of post-implantation epiblast to pluripotent embryonic stem cells. Nature 2009; 461:1292-5.

16. Matsui Y, Zsebo K, Hogan BL. Derivation of pluripotential embryonic stem cells from murine primordial germ cells in culture. Cell 1992; 70:841-7.
17. Takahashi K, Yamanaka S. Induction of pluripotent stem cells from mouse embryonic and adult fibroblast cultures by defined factors. Cell 2006; 126:663-76.
18. Jaenisch R, Young R. Stem cells, the molecular circuitry of pluripotency and nuclear reprogramming. Cell 2008; 132:567-82.
19. Boyer LA, Lee TI, Cole MF et al. Core transcriptional regulatory circuitry in human embryonic stem cells. Cell 2005; 122:947-56.
20. Wang J, Rao S, Chu J et al. A protein interaction network for pluripotency of embryonic stem cells. Nature 2006; 444:364-8.
21. Luger K, Mader AW, Richmond RK et al. Crystal structure of the nucleosome core particle at 2.8 A resolution. Nature 1997; 389:251-60.
22. Fraser P, Bickmore W. Nuclear organization of the genome and the potential for gene regulation. Nature 2007; 447:413-7.
23. Campos EI, Reinberg D. Histones: annotating chromatin. Annu Rev Genet 2009; 43:559-99.
24. Bernstein BE, Meissner A, Lander ES. The mammalian epigenome. Cell 2007; 128:669-81.
25. Morrison AJ, Shen X. Chromatin remodelling beyond transcription: the INO80 and SWR1 complexes. Nat Rev Mol Cell Biol 2009; 10:373-84.
26. Li B, Carey M, Workman JL. The role of chromatin during transcription. Cell 2007; 128:707-19.
27. Gutierrez J et al. Chromatin remodeling by SWI/SNF results in nucleosome mobilization to preferential positions in the rat osteocalcin gene promoter. J Biol Chem 2007; 282:9445-57.
28. Kouzarides T. Chromatin modifications and their function. Cell 2007; 128:693-705.
29. Bernstein BE, Mikkelsen TS, Xie X et al. A bivalent chromatin structure marks key developmental genes in embryonic stem cells. Cell 2006; 125:315-26.
30. Mikkelsen TS, Ku M, Jaffe DB et al. Genome-wide maps of chromatin state in pluripotent and lineage-committed cells. Nature 2007; 448:553-60.
31. Pietersen AM, van Lohuizen M. Stem cell regulation by polycomb repressors: postponing commitment. Curr Opin Cell Biol 2008; 20:201-7.
32. Henikoff S. Nucleosome destabilization in the epigenetic regulation of gene expression. Nat Rev Genet 2008; 9:15-26.
33. Sterner DE, Berger SL. Acetylation of histones and transcription-related factors. Microbiol Mol Biol Rev 2000; 64:435-59.
34. Sims RJ, 3rd, Nishioka K, Reinberg D. Histone lysine methylation: a signature for chromatin function. Trends Genet 2003; 19:629-39.
35. Nowak SJ, Corces VG. Phosphorylation of histone H3: a balancing act between chromosome condensation and transcriptional activation. Trends Genet 2004; 20:214-20.
36. Shilatifard A. Chromatin modifications by methylation and ubiquitination: implications in the regulation of gene expression. Annu Rev Biochem 2006; 75:243-69.
37. Nathan D, Ingvarsdottir K, Sterner DE et al. Histone sumoylation is a negative regulator in Saccharomyces cerevisiae and shows dynamic interplay with positive-acting histone modifications. Genes Dev 2006; 20:966-76.
38. Hassa PO, Haenni SS, Elser M et al. Nuclear ADP-ribosylation reactions in mammalian cells: where are we today and where are we going? Microbiol Mol Biol Rev 2006; 70:789-829.
39. Cuthbert GL, Daujat S, Snowden AW et al. Histone deimination antagonizes arginine methylation. Cell 2004; 118:545-53.
40. Nelson CJ, Santos-Rosa H, Kouzarides T. Proline isomerization of histone H3 regulates lysine methylation and gene expression. Cell 2006; 126:905-16.
41. Schuettengruber B, Chourrout D, Vervoort M et al. Genome regulation by polycomb and trithorax proteins. Cell 2007; 128:735-45.
42. Elderkin S, Maertens GN, Endoh M et al. A phosphorylated form of Mel-18 targets the Ring1B histone H2A ubiquitin ligase to chromatin. Mol Cell 2007; 28:107-20.
43. Ku M, Koche RP, Rheinbay E et al. Genomewide analysis of PRC1 and PRC2 occupancy identifies two classes of bivalent domains. PLoS Genet 2008; 4:e1000242.
44. Bird A. DNA methylation patterns and epigenetic memory. Genes Dev 2002; 16:6-21.
45. Grandjean V, Yaman R, Cuzin F et al. Inheritance of an epigenetic mark: the CpG DNA methyltransferase 1 is required for de novo establishment of a complex pattern of nonCpG methylation. PLoS One 2007; 2:e1136.
46. Ramsahoye BH, Biniszkiewicz D, Lyko F, Clark V et al. Non-CpG methylation is prevalent in embryonic stem cells and may be mediated by DNA methyltransferase 3a. Proc Natl Acad Sci USA 2000; 97:5237-42.
47. Finnegan EJ, Peacock WJ, Dennis ES. DNA methylation, a key regulator of plant development and other processes. Curr Opin Genet Dev 2000; 10:217-23.

48. Lister R, Pelizzola M, Dowen RH et al. Human DNA methylomes at base resolution show widespread epigenomic differences. Nature 2009; 462:315-22.
49. Meissner A, Gnirke A, Bell GW et al. Reduced representation bisulfite sequencing for comparative high-resolution DNA methylation analysis. Nucleic Acids Res 2005; 33:5868-77.
50. Morgan HD, Santos F, Green K et al. Epigenetic reprogramming in mammals. Hum Mol Genet 2005; 14 Spec No 1:R47-58.
51. Klose RJ, Bird AP. Genomic DNA methylation: the mark and its mediators. Trends Biochem Sci 2006; 31:89-97.
52. Bestor TH. The DNA methyltransferases of mammals. Hum Mol Genet 2000; 9:2395-402.
53. Goll MG, Bestor TH. Eukaryotic cytosine methyltransferases. Annu Rev Biochem 2005; 74:481-514.
54. Okano M, Bell DW, Haber DA et al. DNA methyltransferases Dnmt3a and Dnmt3b are essential for de novo methylation and mammalian development. Cell 1999; 99:247-57.
55. Aapola U, Kawasaki K, Scott HS et al. Isolation and initial characterization of a novel zinc finger gene, DNMT3L, on 21q22.3, related to the cytosine-5-methyltransferase 3 gene family. Genomics 2000; 65:293-8.
56. Ooi SK, Qiu C, Bernstein E et al. DNMT3L connects unmethylated lysine 4 of histone H3 to de novo methylation of DNA. Nature 2007; 448:714-7.
57. Meshorer E, Misteli T. Chromatin in pluripotent embryonic stem cells and differentiation. Nat Rev Mol Cell Biol 2006; 7:540-6.
58. Francastel C, Schubeler D, Martin DI et al. Nuclear compartmentalization and gene activity. Nat Rev Mol Cell Biol 2000; 1:137-43.
59. Arney KL, Fisher AG. Epigenetic aspects of differentiation. J Cell Sci 2004; 117:4355-63.
60. Meshorer E, Yellajoshula D, George E et al. Hyperdynamic plasticity of chromatin proteins in pluripotent embryonic stem cells. Dev Cell 2006; 10:105-16.
61. Efroni S, Duttagupta R, Cheng J et al. Global transcription in pluripotent embryonic stem cells. Cell Stem Cell 2008; 2:437-47.
62. Liang J, Wan M, Zhang Y, et al. Nanog and Oct4 associate with unique transcriptional repression complexes in embryonic stem cells. Nat Cell Biol 2008; 10:731-9.
63. Kaji K, Caballero IM, MacLeod R et al. The NuRD component Mbd3 is required for pluripotency of embryonic stem cells. Nat Cell Biol 2006; 8:285-92.
64. Gaspar-Maia A, Alajem A, Polesso F et al. Chd1 regulates open chromatin and pluripotency of embryonic stem cells. Nature 2009; 460:863-8.
65. Guenther MG, Levine SS, Boyer LA et al. A chromatin landmark and transcription initiation at most promoters in human cells. Cell 2007; 130:77-88.
66. Pasini D, Hansen KH, Christensen J et al. Coordinated regulation of transcriptional repression by the RBP2 H3K4 demethylase and Polycomb-Repressive Complex 2. Genes Dev 2008; 22:1345-55.
67. Chen X, Xu H, Yuan P et al. Integration of external signaling pathways with the core transcriptional network in embryonic stem cells. Cell 2008; 133:1106-17.
68. Fazzio TG, Huff JT, Panning B. Chromatin regulation Tip(60)s the balance in embryonic stem cell self-renewal. Cell Cycle 2008; 7:3302-6.
69. Torres-Padilla ME, Parfitt DE, Kouzarides T et al. Histone arginine methylation regulates pluripotency in the early mouse embryo. Nature 2007; 445:214-8.
70. Wu Q, Bruce AW, Jedrusik A et al. CARM1 is required in embryonic stem cells to maintain pluripotency and resist differentiation. Stem Cells 2009; 27:2637-45.
71. Heintzman ND, Hon GC, Hawkins RD et al. Histone modifications at human enhancers reflect global cell-type-specific gene expression. Nature 2009; 459:108-12.
72. Heintzman ND, Ren B. Finding distal regulatory elements in the human genome. Curr Opin Genet Dev 2009; 19:541-9.
73. Boiani M, Scholer HR. Regulatory networks in embryo-derived pluripotent stem cells. Nat Rev Mol Cell Biol 2005; 6:872-84.
74. Boyer LA, Plath K, Zeitlinger J et al. Polycomb complexes repress developmental regulators in murine embryonic stem cells. Nature 2006; 441:349-53.
75. Lee TI, Jenner RG, Boyer LA et al. Control of developmental regulators by Polycomb in human embryonic stem cells. Cell 2006; 125:301-13.
76. Shen X, Liu Y, Hsu YJ et al. EZH1 mediates methylation on histone H3 lysine 27 and complements EZH2 in maintaining stem cell identity and executing pluripotency. Mol Cell 2008; 32:491-502.
77. Chamberlain SJ, Yee D, Magnuson T. Polycomb repressive complex 2 is dispensable for maintenance of embryonic stem cell pluripotency. Stem Cells 2008; 26:1496-505.
78. Pasini D, Bracken AP, Hansen JB et al. The polycomb group protein Suz12 is required for embryonic stem cell differentiation. Mol Cell Biol 2007; 27:3769-79.
79. Peng JC, Valouev A, Swigut T et al. Jarid2/Jumonji coordinates control of PRC2 enzymatic activity and target gene occupancy in pluripotent cells. Cell 2009; 139:1290-302.

80. Shen X, Kim W, Fujiwara Y et al. Jumonji modulates polycomb activity and self-renewal versus differentiation of stem cells. Cell 2009; 139:1303-14.
81. Li G, Margueron R, Ku M et al. Jarid2 and PRC2, partners in regulating gene expression. Genes Dev. 2010 Feb 15;24(4):368-80.
82. Pasini D, Cloos PA, Walfridsson J et al. JARID2 regulates binding of the Polycomb repressive complex 2 to target genes in ES cells. Nature. 2010 Jan 14. [Epub ahead of print].
83. Mikkelsen TS, Hanna J, Zhang X et al. Dissecting direct reprogramming through integrative genomic analysis. Nature 2008; 454:49-55.
84. Bilodeau S, Kagey MH, Frampton GM et al. SetDB1 contributes to repression of genes encoding developmental regulators and maintenance of ES cell state. Genes Dev 2009; 23:2484-9.
85. Squazzo SL, O'Geen H, Komashko VM et al. Suz12 binds to silenced regions of the genome in a cell-type-specific manner. Genome Res 2006; 16:890-900.
86. Creyghton MP, Markoulaki S, Levine SS et al. H2AZ is enriched at polycomb complex target genes in ES cells and is necessary for lineage commitment. Cell 2008; 135:649-61.
87. Hardy S, Jacques PE, Gévry N et al. The euchromatic and heterochromatic landscapes are shaped by antagonizing effects of transcription on H2A.Z deposition. PLoS Genet 2009; 5:e1000687. .
88. Xie W, Song C, Young NL et al. Histone h3 lysine 56 acetylation is linked to the core transcriptional network in human embryonic stem cells. Mol Cell 2009; 33:417-27.
89. Martens JH, O'Sullivan RJ, Braunschweig U et al. The profile of repeat-associated histone lysine methylation states in the mouse epigenome. EMBO J 2005; 24:800-12.
90. Daujat S, Weiss T, Mohn F et al. H3K64 trimethylation marks heterochromatin and is dynamically remodeled during developmental reprogramming. Nat Struct Mol Biol 2009; 16:777-81.
91. Loh YH, Zhang W, Chen X et al. Jmjd1a and Jmjd2c histone H3 Lys 9 demethylases regulate self-renewal in embryonic stem cells. Genes Dev 2007; 21:2545-57.
92. Jackson M, Krassowska A, Gilbert N et al. Severe global DNA hypomethylation blocks differentiation and induces histone hyperacetylation in embryonic stem cells. Mol Cell Biol 2004; 24:8862-71.
93. Tsumura A, Hayakawa T, Kumaki Y et al. Maintenance of self-renewal ability of mouse embryonic stem cells in the absence of DNA methyltransferases Dnmt1, Dnmt3a and Dnmt3b. Genes Cells 2006; 11:805-14.
94. Jackson-Grusby L, Beard C, Possemato R et al. Loss of genomic methylation causes p53-dependent apoptosis and epigenetic deregulation. Nat Genet 2001; 27:31-9.
95. Gaudet F, Hodgson JG, Eden A et al. Induction of tumors in mice by genomic hypomethylation. Science 2003; 300:489-92.
96. Bourc'his D, Bestor TH. Meiotic catastrophe and retrotransposon reactivation in male germ cells lacking Dnmt3L. Nature 2004; 431:96-9.
97. Dodge JE, Okano M, Dick F et al. Inactivation of Dnmt3b in mouse embryonic fibroblasts results in DNA hypomethylation, chromosomal instability and spontaneous immortalization. J Biol Chem 2005; 280:17986-91. .
98. Nimura K, Ishida C, Koriyama H et al. Dnmt3a2 targets endogenous Dnmt3L to ES cell chromatin and induces regional DNA methylation. Genes Cells 2006; 11:1225-37.
99. Fouse SD, Shen Y, Pellegrini M et al. Promoter CpG methylation contributes to ES cell gene regulation in parallel with Oct4/Nanog, PcG complex and histone H3 K4/K27 trimethylation. Cell Stem Cell 2008; 2:160-9.
100. Meissner A, Mikkelsen TS, Gu H et al. Genome-scale DNA methylation maps of pluripotent and differentiated cells. Nature 2008; 454:766-70.
101. Weber M, Hellmann I, Stadler MB et al. Distribution, silencing potential and evolutionary impact of promoter DNA methylation in the human genome. Nat Genet 2007; 39:457-66.
102. Mohn F, Weber M, Rebhan M et al. Lineage-specific polycomb targets and de novo DNA methylation define restriction and potential of neuronal progenitors. Mol Cell 2008; 30:755-66.
103. Cedar H, Bergman Y. Linking DNA methylation and histone modification: patterns and paradigms. Nat Rev Genet 2009; 10:295-304.
104. Haines TR, Rodenhiser DI, Ainsworth PJ. Allele-specific nonCpG methylation of the Nf1 gene during early mouse development. Dev Biol 2001; 240:585-98.
105. Tahiliani M, Koh KP, Shen Y et al. Conversion of 5-methylcytosine to 5-hydroxymethylcytosine in mammalian DNA by MLL partner TET1. Science 2009; 324:930-5.
106. Niwa H. How is pluripotency determined and maintained? Development 2007; 134:635-46.
107. Silva J, Smith A. Capturing pluripotency. Cell 2008; 132:532-6.
108. Chi AS, Bernstein BE. Developmental biology. Pluripotent chromatin state. Science 2009; 323:220-1.
109. Gu H, Bock C, Mikkelsen TS et al. Genome-scale DNA methylation mapping of clinical samples at single-nucleotide resolution. Nat Methods. 2010 Feb;7(2):133-6.
110. Goren A, Ozsolak F, Shoresh N et al. Chromatin profiling by directly sequencing small quantities of immunoprecipitated DNA. Nat Methods 7:47-9.

CHAPTER 4

AUTOSOMAL LYONIZATION OF REPLICATION DOMAINS DURING EARLY MAMMALIAN DEVELOPMENT

Ichiro Hiratani and David M. Gilbert*

Abstract: It has been exactly 50 years since it was discovered that duplication of the eukaryotic genome follows a defined temporal order as cells progress through S-phase. While the mechanism of this replication-timing program still remains a mystery, various correlations of this program with both static and dynamic properties of chromatin render it an attractive forum to explore previously impenetrable higher-order organization of chromosomes. Indeed, studies of DNA replication have provided a simple and straightforward approach to address physical organization of the genome, both along the length of the chromosome as well as in the context of the 3-dimensional space in the cell nucleus. In this chapter, we summarize the 50-years history of the pursuit for understanding the replication-timing program and its developmental regulation, primarily in mammalian cells. We begin with the discovery of the replication-timing program, discuss developmental regulation of this program during X-inactivation in females as well as on autosomes and then describe the recent findings from genome-wide dissection of this program, with special reference to what takes place during mouse embryonic stem cell differentiation. We make an attempt to interpret what these findings might represent and discuss their potential relevance to embryonic development. In doing so, we revive an old concept of "autosomal Lyonization" to describe "facultative heterochromatinization" and irreversible silencing of individual replication domains on autosomes reminiscent of the stable silencing of the inactive X chromosome, which takes place at a stage equivalent to the postimplantation epiblast in mice.

*Corresponding Author: David M. Gilbert—Department of Biological Science, Florida State University, Tallahassee, Florida, USA. Email: gilbert@bio.fsu.edu

The Cell Biology of Stem Cells, edited by Eran Meshorer and Kathrin Plath.
©2010 Landes Bioscience and Springer Science+Business Media.

INTRODUCTION

With all phenomena in nature there are two major questions asked: the underlying mechanism (how?) and its biological significance (why?). All eukaryotic cells replicate their DNA in a specific temporal sequence but both the mechanism of this "replication timing program" and its biological significance remain a mystery. Understanding the mechanism might allow us to manipulate replication timing and query significance, but even with the power of molecular genetics in budding and fission yeasts, very little mechanistic insights have been gained.[1] Many studies point to regulation at the level of large chromosomal domains or subnuclear compartments, making replication timing an excellent gateway into the higher order structure and functional organization of chromosomes, albeit refractory to traditional molecular and biochemical approaches. By corollary, understanding significance might provide clues as to mechanism. Some studies have suggested evolutionary roles for replication timing in focusing mutation rates and/or suppressing recombination,[2,3] while others have suggested housekeeping roles such as the overall coordination of replicating large genomes in the presence of limited metabolic precursors.[4] In addition, a longstanding correlation between early replication timing and transcriptional competence has been substantiated by recent genome-wide studies,[1,5] but it is not clear whether transcription drives early replication or vice versa.

If replication timing were related to transcriptional competence, it should be developmentally regulated. In this chapter, we summarize the evidence for developmental regulation of replication timing in mammals historically from its original inception as a property associated with the process of X-inactivation in females[6] to very recent studies verifying an unanticipated degree of autosomal replication timing changes taking place at the level of megabase-sized chromosomal "replication domains."[7,8] In particular, we revive an old concept of "autosomal Lyonization" to describe "facultative heterochromatinization" and irreversible silencing of individual replication domains on autosomes reminiscent of the stable silencing of the inactive X chromosome (Xi),[9] which takes place at the epiblast stage in mice.[8] Moreover, comparative studies have revealed that replication timing programs and the changes that occur during development are evolutionarily conserved, to a greater extent than either the positions of replication origins or the overall GC content of chromosomal isochores.[10] These observations suggest positive selection for a replication program that is not arbitrarily dividing the genome into temporally separated segments for housekeeping purposes but is intimately related to chromosome structure and function.

REPLICATION TIMING PROGRAM: AN ELUSIVE MEASURE OF GENOME ORGANIZATION

Early Experiments

Early studies of DNA replication led to the discovery of key concepts in chromosomal organization. J. Herbert Taylor's synthesis of tritiated thymidine in 1953 allowed the first glimpses of DNA synthesis in living cells[11] at a time when the structure of DNA was just being resolved. Taylor's series of thymidine labeling experiments not only provided the first demonstration of the semi-conservative antiparallel nature of DNA replication (more than one year prior to Messelson and Stahl[12])[13], but also revealed a specific temporal

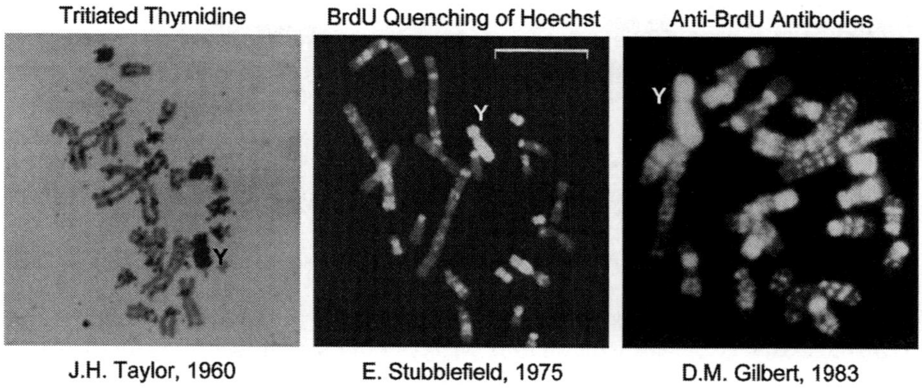

Figure 1. Brief Pulse-Labeling of DNA Synthesis Highlights Megabase-Sized "Replication Domains". Chinese hamster cells were pulse labeled for 10 minutes with either tritiated thymidine (left) or BrdU (right) during late S-phase or labeled continuously with BrdU except for 1 hour in late S-phase during which cells were labeled with thymidine (middle) and then chased with unlabeled medium into mitosis. Metaphase spreads were analyzed by autoradiography (left), BrdU quenching of Hoechst dye (middle) or indirect immunofluorescence using anti-BrdU antibodies (right). These cytogenetic methods demonstrated that megabase-sized segments of the genome are labeled in very short periods of time, producing banding patterns that were characteristic for each chromosome and varied during S-phase, with euchromatic R bands replicating early and heterochromatic G bands replicating late. Note that the heterochromatic Y chromosome is almost entirely labeled in a 10-minute period late in S-phase. With known rates of replication fork movement, this could only be achieved by the nearly synchronous firing of clusters of replication origins. Figure was adapted from J.H. Taylor[15] (left-© Taylor, 1960. Originally published in *The Journal of Biophysical and Biochemical Cytology*. 7: 455-463), E. Stubblefield[18] (middle) and D.M. Gilbert[102] (right) with permissions from *Journal of Biophysical and Biochemical Cytology*[15] *and Chromosoma.*[18]

program to the replication of DNA in the chromosomes of both plant[14] and animal[15] cells. In particular, by pulse-labeling Chinese hamster cells (whose chromosomes are easily distinguished by size) and then examining metaphase chromosomes at various times after the pulse, he found that different segments of chromosomes replicate within specific time intervals during S-phase (Fig. 1). From this study,[15] Taylor concluded that: "parts of chromosomes have a genetically controlled sequence in duplication, which may have some functional significance." Exactly 50 years later, that functional significance remains a total mystery.

By the 1970s, these coordinately labeled segments of chromosomes were found to be similar in appearance and size to chromomeric banding patterns of chromosomes seen using banding methods such as Giemsa staining. When it was discovered that incorporation of BrdU into DNA (instead of tritiated thymidine) could squelch the fluorescence of Hoechst dye,[16] a novel chromosome banding method ("replication banding") was developed that avoided the use of radioactivity and long autoradiography exposures (Fig. 1). In general, the transcriptionally active, GC-rich, R-bands were found to be early replicating, while the transcriptionally inactive, AT-rich, G-bands were late,[17] although the alignment was not absolute.[18,19] These results supported the hypothesis that heterochromatin is late replicating, which was originally proposed by Lima-de-Faria based on studies of grasshopper sex chromatin.[20] The finding that the replication time of individual chromosome segments is related to their transcriptional activity raised the possibility that coordinately replicated segments may represent not only structural but also functional units of chromosomes.

This was an attractive hypothesis, but if replication were related to transcription, one might expect to find different banding patterns in different cell types. Unfortunately, comparisons of replication banding patterns in different cell types failed to detect such differences.[21] Of course, the resolution of such studies could not rule out the existence of localized changes.[10] Moreover, studies in frog embryos[22] and studies of mammalian X-inactivation (discussed below) demonstrated that replication timing could change during development in a manner correlated with transcriptional activity.

The Lessons from X Chromosome Inactivation

In his 1960 study, Taylor noticed that, in female cells, the X chromosomes replicated asynchronously.[15] Coincidentally, in 1961, Mary Lyon proposed her famous hypothesis that the cytological manifestation of X-inactivation, the Barr body,[23] appears coincident with its genetic inactivation in early development and that both the structural and functional alterations of the homolog randomly chosen for inactivation are stably maintained in all subsequent somatic generations.[24] Taylor then went on to verify that the Barr body was late replicating in female human cells and that in cells with several X chromosomes all the Barr bodies replicated late.[6] These findings introduced a completely novel and mysterious notion: the fact that the two genetically identical X chromosomes behave differently meant that homologous chromosomes can be either heterochromatic or euchromatic in the same cell, leading to the unavoidable conclusion that replication timing is determined epigenetically—not by sequences alone.

Over the next two decades, the appearance of a late replicating Xi during early embryonic stages was used as a reliable cytological marker for X-inactivation, as it is one of the most conserved features of X-inactivation.[25] In fact, while late replication of the Xi seen in placental mammals (eutherians) is conserved in marsupials and at least partly in the egg-laying platypus (monotremes),[26,27] eutherian X-inactivation features such as the *Xist* gene, enrichment of "repressive" histone modifications and possibly promoter DNA methylation are either missing or not reported to date in marsupials and monotremes.[26,28] Upon random X-inactivation in the mouse embryo proper, a late-replicating Xi emerges at the postimplantation epiblast stage,[29] which precedes de novo DNA methylation of gene promoters on the Xi.[30] Later, the development of in vitro differentiation systems for embryonic stem cells (ESCs) allowed for the temporal order of events to be more precisely determined. ESC differentiation studies suggest *Xist* coating of the Xi to be the earliest event upon random X-inactivation, followed by an exclusion of RNA polymerase II from the Xi, then by the loss of "active" histone modifications and the acquisition of "repressive" histone marks.[31] Xi's switch to late replication either coincides with or occurs shortly after changes in histone modifications (based on metaphase spread analyses[32,33]), whereas de novo promoter DNA methylation occurs much later.[32] While it is tempting to speculate a causal role for earlier events in regulating the Xi's switch to late replication, this idea is difficult to reconcile with the fact that *Xist* and "repressive" histone modifications are missing from the Xi in marsupials[28] and yet a switch to late replication is observed.[26] Indeed no report to date has demonstrated a causal role for chromatin modifiers in regulating replication timing of the Xi. Interestingly, however, using a mouse ESC differentiation model with an inducible *Xist* transgene, Wutz et al demonstrated that the time point of commitment for X-inactivation is independent of transcriptional down-regulation but is temporally closely associated with a nearly chromosome-wide shift to late replication of the Xi.[32,34]

Compared to the late replicating state of the Xi in the embryo proper, the situation is somewhat different in the extra-embryonic lineages, in which the paternal X chromosome (Xp) is inactivated in an imprinted manner. Both in trophectoderm and primitive endoderm, the Xp temporarily becomes replicated precociously very early in S-phase, earlier than any autosomes.[35] The switch from synchronous to asynchronous replication timing of the X chromosomes takes place just after these extra-embryonic tissues are specified, at embryonic day (E) 3.5 in trophectoderm and E4.0–4.5 in primitive endoderm.[35] Then, around E6.0–6.4, the Xp in these two lineages (primitive endoderm and trophectoderm become visceral endoderm and extra-embryonic ectoderm, respectively, at this stage) switches to late replication and stably maintains its late replication state thereafter.[29] This abrupt nearly chromosome-wide switch may occur within a single cell cycle.[36] These observations indicate that the unusual precocious replication of the Xp in the extra-embryonic lineages is transient, whereas the switch to late replication of the Xi is conserved between embryonic and extra-embryonic lineages and is stably maintained during development. Interestingly, the emergence of both trophectoderm at E3.5 and primitive endoderm at E4.0 represent the first obvious manifestations of lineage commitment, segregating these extra-embryonic lineages from the embryo proper. Thus, the first two major cell fate transitions during early embryogenesis accompany alteration in replication-timing regulation of the Xi. It is not known, however, what the subsequent, abrupt switch to late replication at E6.0–6.4 in these extra-embryonic lineages represents.[29] Interactions between extra-embryonic and maternal tissues may be involved.

Replication Timing Landscape on Autosomes

Are replication timing changes unique to the X chromosome or are there equivalent events on autosomes during differentiation that escape cytological detection? Might there be a similar program to "Lyonize" (i.e.,facultatively heterochromatinize) individual replicons or clusters of replicons[9] that are too small to visualize microscopically? In the 1970s, Carl Schildkraut and Walt Fangman pioneered the use of molecular methods to investigate the temporal order of replication in mammals[37] and budding yeast,[38] respectively. Together, these studies provided strong evidence that a precisely regulated replication-timing program is a conserved property of all eukaryotic cells. In the 1980s, the replication times of a few dozen genes became cataloged in different stable cell lines,[39-42] unambiguously identifying autosomal replication timing differences. The emerging rule of thumb was that if a gene were transcriptionally active, it would be early replicating, while late replicating genes were always inactive. In fact, a study comparing active and inactive clusters of *Xenopus* 5S rDNA genes that were known to compete for the same transcription factors revealed that the active gene clusters replicate substantially earlier, suggesting a model in which early replication could provide a competitive advantage for access to limiting quantities of activating factors at the replication fork ("first come, first served").[43] This model has yet to be refuted or substantiated.

The extent of developmental changes, however, had been elusive. Due to technical difficulties of working with differentiating cell cultures, replication timing studies through the early 2000s were limited to a few dozen gene loci in established, usually transformed, cell lines. Hence, it remained possible that many of the replication differences observed between cell lines resulted from genetic or epigenetic changes accumulated during long term culture. The advent of directed ESC differentiation systems sufficiently homogeneous to perform molecular analyses permitted the first direct demonstration of differentiation-induced replication timing changes of autosomal loci in 2004.[44,45] However, replication timing of a fraction of ~100 genes analyzed was regulated and with only neural differentiation pathway analyzed, it was difficult to conclude whether developmental replication timing changes were frequent or rare. Various studies led to the conclusion that many genes replicate at the same time in all cell types, consistent with the cytogenetic studies.[46,47] Thus, these reports clearly provided evidence for differentiation-induced replication timing changes, but the small sampling demanded the use of genome-wide approaches to statistically determine the extent of changes.

The first genome-wide replication timing analysis was performed in budding yeast.[48] Unexpectedly, no correlation was found between replication timing and transcription, a finding that has since been corroborated in fission yeast.[49] Shortly thereafter, a series of microarray-based studies in *Drosophila* and mammalian cells provided evidence for a strong correlation between early replication and transcriptional activity in these higher eukaryotes,[7,50-58] suggesting that this relationship might be restricted to metazoa.[59] This correlation, along with the fact that gene expression programs change during differentiation, raised the possibility that a considerable degree of replication-timing changes might take place during development. However, the first report that examined more than one cell line found only 1% difference in replication timing across chromosome 22 in human fibroblasts vs lymphoblasts.[53] Moreover, several of these studies found that replication timing correlated strongly with static sequence features of mammalian chromosomes such as GC content and gene density.[7,52-56] Indeed, as recently as in 2008, many investigators had concluded that replication-timing changes are very rare and hence their significance to development came into question.[60-62] Still, the resolution and limited genomic coverage of existing studies and the paucity of data comparing cell lines left this fundamental question unanswered.

The advent of high-density oligonucleotide microarrays to query the genome at sufficient probe density, combined with ESC differentiation systems, offered an unprecedented opportunity to potentially induce and study changes in the replication-timing program during major cell fate transitions. In 2008, we were able to achieve such an analysis during differentiation of mouse ESCs to neural precursor cells (NPCs).[7] This study revealed that several polymorphic ESC lines showed virtually identical replication timing profiles, with clearly delineated patterns of coordinately replicated megabase-sized chromosome domains. Upon differentiation of ESCs to NPCs, changes took place across nearly 20% of the genome.[7] Combined with a follow-up study,[8] we have constructed replication profiles for 22 cell lines representing 10 different cell types that model differentiation of three germ layers during early mouse development (Fig. 2). Results revealed cell-type specific replication-timing profiles resulting from extensive developmental changes affecting nearly half the genome.[8] Replication timing changes occur coordinately across 400–800 kb segments of chromosomes within larger regions of constitutive replication, explaining why they escape cytological detection. Moreover, although the correlation between early replication and GC content or gene density is always positive and the most GC- or AT-rich genomic segments remain early or late replicating respectively in all cell types, it is the segments with intermediate GC content and gene density that change replication timing and these changes can substantially alter the overall degree to which replication timing correlates with static sequence features.[8]

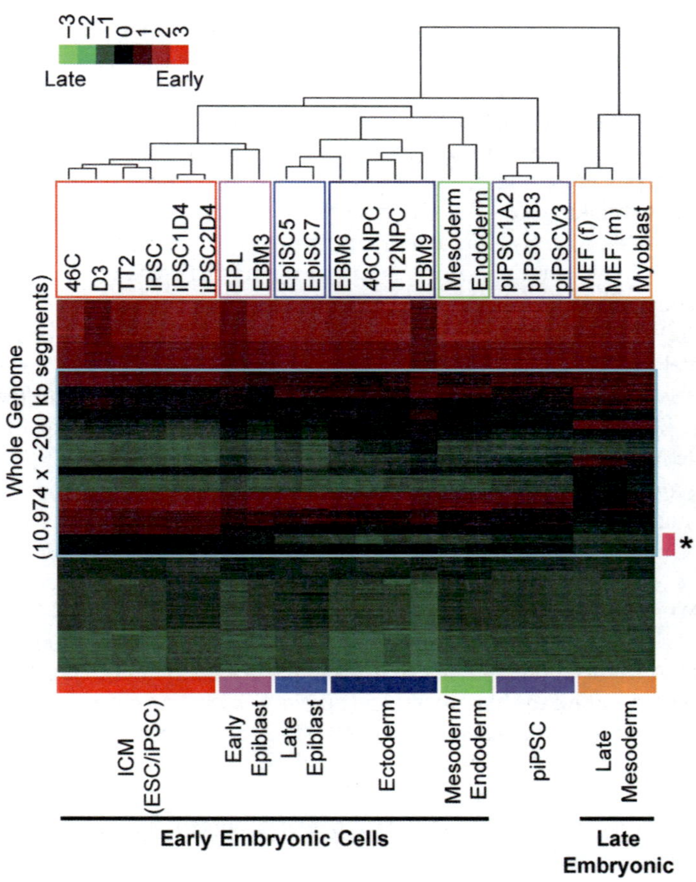

Figure 2. Relationship between cell culture models of mouse embryogenesis based on replication timing profiles. Hierarchical clustering of 22 mouse cell lines based on replication timing profiles obtained by microarrays.[8] The dendrogram reveals an epigenetic separation of cell types representing the late epiblast (EpiSCs) from the early epiblast (EPL and EBM3) as well as the ICM [ESCs (46C, D3, TT2) and fully reprogrammed iPSCs]. EpiSCs were more related to committed germ layer cell types of the early embryo [ectoderm (EBM6), neurectoderm (46CNPC, TT2NPC and EBM9), nascent mesoderm and endoderm]. Three partially reprogrammed iPSC (piPSC) lines were distinct from late embryonic cell types (MEFs and myoblasts), but were also distinct from ICM, epiblast or early germ layer cell types, forming an independent branch. The asterisk on the right indicates genomic segments that complete lineage-independent EtoL changes by the postimplantation epiblast stage (which roughly corresponds to 155 Mb total). Late replication of these segments was stably maintained in all downstream lineages and not reversed in piPSCs, which also exhibited difficulty in transcriptional reprogramming of genes within these segments. Figure was adapted from Hiratani et al with permission from *Genome Research*.[8]
Methods: Cells were pulse-labeled with BrdU, separated into early and late S-phase fractions by flow cytometry and BrdU-substituted DNA from each fraction was immunoprecipitated with an anti-BrdU antibody. The early and late replicating DNA samples were differentially labeled and cohybridized to whole-genome oligonucleotide microarrays. The ratio of the abundance of each probe in the early and late fraction ["replication timing ratio" = $\log_2$(Early/Late)] was then used to generate a replication timing profile for the entire genome at a density of one probe every 5.8 kb. Then, the whole genome was divided into 10,974 ~200-kb segments and their average replication timing ratios were compared between cell lines by hierarchical clustering. The heatmap shows the replication-timing ratios [= Log_2(Early/Late)] of 10,974 ~200-kb segments, with red and green representing early and late replication, respectively. Segments framed in blue shows those with significant differential between any cell types, which represent 45% of the genome.

AN EVOLUTIONARILY CONSERVED EPIGENETIC FINGERPRINT

It is now clear that replication-timing differences are extensive during early mammalian development. Moreover, comparison of two *Drosophila* cultured cell lines derived from embryonic or imaginal disc tissue also revealed approximately 20% differences in replication timing,[57] suggesting that extensive developmental changes are common in higher eukaryotes. But are these changes meaningful to the development of the animal or are they merely stochastic events with very little consequence to the fitness of the organism? Our current understanding of the mechanisms regulating replication does not permit a direct manipulative approach to this question. An indirect alternative is to evaluate whether the replication timing programs of individual cell types have been positively selected during evolution. To this end, we extended our analyses to differentiating human ESCs. Consistent with the mouse data, we found that multiple human ESC lines displayed nearly identical replication timing profiles that changed across approximately 20% of the genome during differentiation to NPCs (T. Ryba, I.H. and D.M.G, *unpublished*). As in mice, changes in replication timing generally occurred coordinately across 400–800 kb chromosome domains, suggesting a conserved unit size of replication timing changes that most likely involves a coordinated regulation of at least 2-3 replicons. However, human ESCs differed substantially in their replication timing profiles from mouse ESCs within regions of conserved synteny (T. Ryba, I.H. and D.M.G, *unpublished*). In fact, they aligned much more closely with stem cells derived from the postimplantation mouse epiblast, the EpiSCs (epiblast-derived stem cells),[63,64] providing a genome-wide support for the hypothesis that human ESCs represent an epiblast-like state that is downstream from the inner cell mass(ICM)-like state that mouse ESCs represent (T. Ryba, I.H. and D.M.G, *unpublished*). In addition, human cells showed a significantly lower correlation of early replication to GC content and GC content was significantly less well conserved than replication timing between human and mouse (T. Ryba, I.H. and D.M.G, *unpublished*).

Furthermore, recent unpublished comparisons of distantly related fission (N. Rhind, *personal communication*) and budding (K. Lindstrom and B. Brewer, *personal communication)* yeast species find a lack of conserved replication origin positions but a remarkable conservation of the replication-timing program. This is consistent with many observations suggesting that replication timing is independent of where replication initiates. For instance, the human beta-globin locus frequently replicates from one of two closely spaced origins while the mouse locus uses many widely dispersed origins and yet replication timing is conserved.[65] Moreover, the replication time of chromosomal domains is re-established in each cell cycle at a time point in G1 (TDP; timing decision point) prior to and independent of origin site specification, which takes place later in G1 at the ODP (origin decision point).[66-69] Altogether, these findings indicate that the replication-timing program is under considerably stronger positive selection during evolution than either overall GC content or replication origin positions.

The significance of this program and whether or not it is reflecting a mechanistic linkage to another chromosome property that is the direct object of the selective pressure remains to be determined. For example, early replication correlates positively with overall transcriptional activity domain-wide and with histone modifications associated with transcriptionally active chromatin (H3K4me3, H3K36me3) and inactive chromatin (H3K9me2) (albeit not with known inactive modifications H3K9me3 or H3K27me3).[7,70] However, ablation of several chromatin-modifying enzymes (Mll, Mbd3, Eed, Suv39h1/h2,

G9a, Dnmt1/3a/3b, Dicer) has surprisingly modest effects on replication timing.[1,70-72] Meanwhile, a recent study in fission yeast showed that Swi6, an HP1 (heterochromatin protein (1) ortholog enriched in heterochromatic domains, regulates replication timing through the loading of the replication initiation factor, Sld3.[73] Although no mammalian Sld3 orthologs have been reported and HP1 does not appear to regulate the late replication timing of pericentric heterochromatin in mice,[71] this study raises the intriguing possibility that domain-wide chromatin factors could regulate replication timing through initiation factor accessibility. For example, indirect observations suggest that a competition between histone H1 vs HMG-I/Y proteins across large chromosomal segments could regulate replication timing during differentiation.[74]

REPLICATION TIMING AS A QUANTITATIVE INDEX OF 3-DIMENSIONAL GENOME ORGANIZATION

During X-inactivation, the switch to late replication leads to an almost synchronous late replication of the entire X chromosome, which becomes highly condensed and localized to the periphery of the nucleus (Fig. 3A), forming what is known as a Barr body in a process that used to be called "Lyonization."[23] This relationship between replication timing and subnuclear position is not confined to the X chromosome, but is also reflected in the positions of individual autosomal replication domains. Domains that replicate at different times during S-phase are localized to different compartments within the nucleus that can be visualized by pulse labeling with nucleotide analogs and staining with antibodies against them (Fig. 3B), giving rise to the appearance of punctate labeled sites known as "replication foci."[1] These replication foci are clearly not artifacts of fixation as they can also be observed in living cells labeled with fluorescent nucleotides.[75] In virtually every animal cell examined, the first half of S-phase consists of hundreds of dispersed internally localized sites of replication that dramatically transition into more clustered peripherally localized foci near the middle of S-phase.[75,76] When these labeled foci are chased through subsequent cell cycles, the labeled segments do not mix, separate or change in shape, size or intensity, suggesting that the DNA that replicates together remains together as a stable structural and functional unit of interphase chromosome structure.[77] Quantitative microscopic methods in human cells estimate close to 1 Mb of DNA is replicated within each focus.[78] It is tempting to speculate that the 400–800 kb units of replication change observed during both mouse and human cell differentiation are the molecular equivalents of replication foci, although this remains a very difficult hypothesis to test.

If a mechanism resembling "Lyonization" were occurring on autosomes, then we would expect autosomal replication-timing changes to be accompanied by changes in subnuclear position. Indeed, this was found to be the case for all seven loci tested during neural differentiation.[7,8,79] Interestingly, whereas replication-timing changes occurred over the course of several cell cycles before stabilizing, subnuclear repositioning was relatively abrupt, occurring primarily during the time when replication-timing changes traversed the mid-late stages of S-phase (Fig. 3C).[8] For example, genes found to undergo significant shifts to later replication but confined to the first half of S-phase remained in the interior of the nucleus, whereas even smaller shifts to late replication that traverse the mid-late stages of S-phase were accompanied by movement toward the periphery.[8] Moreover, genes replicated throughout the first half of S-phase have an equal probability of being expressed, whereas the strongest correlation between transcription and replication

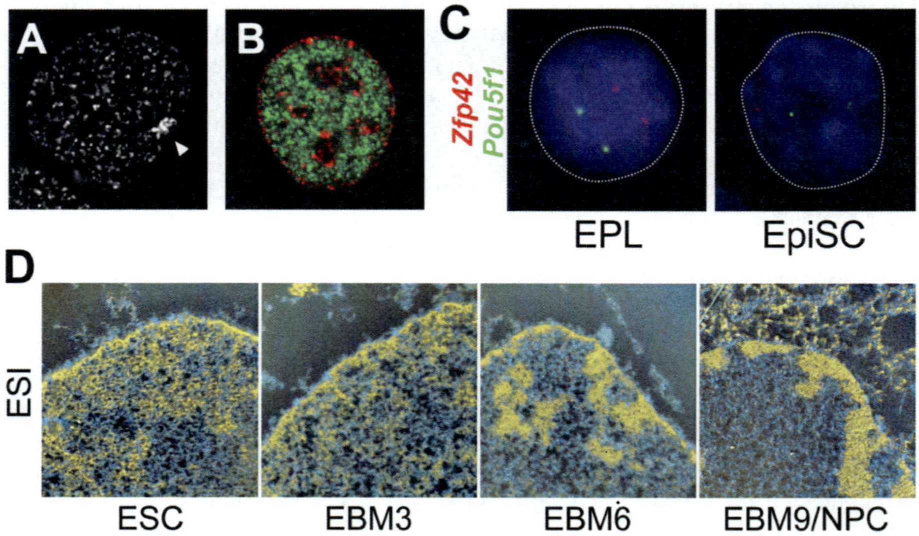

Figure 3. Subnuclear genome organization revealed by studies of DNA replication. A) Late-replicating, condensed inactive X chromosome (Xi, arrowhead) at the nuclear periphery in MEF cells revealed by a 10-minute BrdU pulse labeling during mid-late stages of S-phase. Figure was adapted from Wu et al with permission from Journal of Cell Biology.[71] B) Sites of DNA replication in early S (green, a 10-minute CldU pulse in early S) vs late S-phase (red, a 10-minute IdU pulse in late S of the same cell cycle) revealed by a "pulse-chase-pulse" experiment in mouse C127 fibroblast cells. Note that DNA replication during late S-phase takes place preferentially at the nuclear periphery and nucleolar periphery, whereas during early S-phase, it takes place in the interior of the nucleus excluding these two subnuclear compartments. C) Representative 2D DNA-FISH of Zfp42 (also known as Rex2) and Pou5f1 (also known as Oct4) loci in EPL cells and EpiSCs, which model early and late epiblast stages in mice, respectively.[8] Zfp42 locus is a representative lineage-independent EtoL locus that completes its EtoL change during the transition from early to late epiblast equivalent stage (see Figure 2, asterisk), during which its timing change traverses the mid-late stages of S-phase. By contrast, Pou5f1 locus is constitutively early replicating. Zfp42 locus (red signals) is repositioned from the interior toward the nuclear periphery in EpiSCs but not EPL cells, whereas Pou5f1 locus (green signals) maintains its internal positioning in both cell types. Figure was adapted from Hiratani et al with permission from Genome Research.[8] D) Electron spectroscopic imaging (ESI) analysis of nuclei during ESC differentiation.[8] Images from left to right are: ESC, EBM3 (day 3 differentiated cells), EBM6 (day 6 differentiated) and EBM9 (day 9 NPCs). Relative levels of phosphorus and nitrogen levels were used to delineate chromatin (yellow) vs protein and ribonucleoprotein (blue).[81] ESC nucleus is a relatively uniform meshwork of 10 nm chromatin fibers with a low degree of chromatin compaction along the nuclear envelope or throughout the nucleoplasm. EBM3 showed a landscape very similar to ESCs. However, note that in EBM6, a dramatic accumulation of compact chromatin was evident near the nuclear periphery, the boundaries of which became further sharpened in EBM9. The EBM3–EBM6 transition roughly corresponds to the early to late epiblast transition, based on gene expression, replication timing profiles and subnuclear position analysis of several gene loci.[8] Figure was adapted from Hiratani et al with permission from *Genome Research*.[8]

time is found for genes replicating in the mid-late stages of S-phase.[1] Together, these results predict that genes replicating in the second half of S-phase will be located near the periphery of the nucleus. Given that studies of gene position in the nucleus are currently very laborious, these results imply that replication-timing profiling can provide a genome-wide prediction of genes that change position during differentiation. In fact, we have recently discovered that spatial proximity of chromatin as measured by Hi-C analysis (A novel form of chromosome conformation capture analysis, or 3C, applied genome-wide)[80] shows a strikingly high correlation to replication timing profiles in a cell-type specific manner (T. Ryba, I.H. and D.M.G., *unpublished*).

REPLICATION TIMING REVEALS AN EPIGENETIC TRANSITION: AUTOSOMAL LYONIZATION AT THE EPIBLAST STAGE

Perhaps most importantly for our appreciation of the significance of replication timing changes, these studies have unveiled a close association of replication timing changes with cell fate transitions during early mouse development (Fig. 2).[7,8] Replication timing changes were coordinated with changes in transcription, with expression of weak CpG-poor promoters showing the strongest relationship. The earliest events during mouse development include a distinct set of early-to-late (EtoL) replication timing changes completed during the postimplantation epiblast stage (Fig. 2, asterisk), coincident with repositioning of EtoL loci toward the nuclear periphery (Fig. 3C) and Xi's shift to late replication.[8] Moreover, by electron microscopy using an analytical technique called electron spectroscopic imaging (ESI),[81] a dramatic chromatin conformation change in the nucleus was revealed, with the nuclei showing an emergence of compact chromatin mass near the nuclear periphery during the transition from early to late epiblast equivalent stage (Fig. 3D),[8] coincident with radial subnuclear repositioning of EtoL loci (Fig. 3C). This reorganization was evident in EpiSCs, demonstrating that it is prior to down-regulation of Oct4/Nanog/Sox2 and germ layer commitment.[8] This suggests that the epigenetic landscape that these core pluripotency circuitry factors[82] must act upon is considerably different in late epiblast (EpiSCs) vs ICM/early epiblast [ESCs or early primitive ectoderm-like (EPL) cells[83]]. In contrast to EtoL changes, late-to-early (LtoE) changes occurred later during germ layer commitment in a lineage-dependent manner to generate cell-type specificity (Fig. 2). Taken together, there are extensive changes before and after the epiblast stage, corresponding to lineage-independent and lineage-dependent changes during development, respectively. Moreover, these results suggest that the replication timing program is specific to a given differentiation state, reflects global organization of chromatin within the nucleus and changes in this program represent or reflect epigenetic commitment of cells during key cell fate transitions. In such a way, replication profiling has revealed previously unappreciated epigenetic distinctions between closely related cell culture models that represent early vs late epiblast cells.

An epigenetic distinction between ICM/early epiblast vs late epiblast cell culture models is consistent with the fact that they exhibit major phenotypic differences[84] despite showing only small differences in gene expression.[85] First, unlike pre-implantation ICM cells, the postimplantation epiblast cells fail to colonize the blastocyst, despite the expression of many 'pluripotency' marker genes. Second, the success rate in isolating ESCs from epiblast in the permissive 129 mouse strain also seems to decline precipitously between E5.0 and E6.0.[84] Third, in vitro models of early and late epiblast cells demonstrate that the latter cell types have progressed beyond an as-yet unidentified epigenetic barrier that is difficult to overcome upon nuclear reprogramming. That is, an early epiblast model, EPL cells, have lost the ability to contribute to chimeric mice formation, but can readily revert back to the ESC state by culturing in ESC medium containing LIF, upon which they can contribute to chimeric mice.[83] In contrast, a late postimplantation epiblast model, EpiSCs, have lost the ability to easily revert back to the ESC state.[63,64] In fact, generating

ESC-like iPSCs (induced pluripotent stem cells) from EpiSCs is as inefficient (0.1-1%) as from other somatic cell types.[86] By these criteria, epiblast development appears to accompany a major cell fate transition that is not accompanied by major transcriptional differences but is reflected by significant changes in the replication timing profile and spatial genome organization (Figs. 2 and 3).

As discussed earlier, the first two major cell fate decisions during early mammalian embryogenesis accompany alterations in replication timing of the Xi, upon the emergence of trophectoderm at E3.5 and primitive endoderm at E4.0, both representing the divergence of extra-embryonic lineages from the embryo proper. In this regard, the postimplantation epiblast, in which the Xi's signature shift to late replication is first observed in the embryo proper, may represent the next major cell fate transition after the divergence of these two extra-embryonic lineages. We speculate that it is an important determination step for the late epiblast cells to first shut down their reversibility to the ICM state. Obviously, it is not only the female embryos but also male embryos that go through these series of cell fate decisions during early embryogenesis. Thus, it is reasonable to speculate that Xi's unique replication behavior is a part of a larger scale "Lyonization" event not limited to X-inactivation but that involves autosomes and may be related to the emergence of compact chromatin near the nuclear periphery (reminiscent of Barr body formation near the nuclear periphery[23]). In this sense, it is of interest that the lineage-independent autosomal EtoL changes at the epiblast stage (Fig. 2, asterisk) take place primarily in GC-poor/LINE-1-rich chromosomal segments, which is a hallmark sequence feature of the X chromosome.[8] Thus, Xi may be simply manifesting a putative "default" mode of behavior associated with cell fate changes that is somehow shared with chromosomal segments possessing GC-poor/LINE-1-rich sequence properties. If this were the case, it follows that the remaining active X is the one that bears an imprint to escape such regulation. Indeed, this is what Lyon and Rastan had proposed in 1984.[87] They argued that, in the case of imprinted X-inactivation in the extra-embryonic lineages, the experimental data fits better with a scenario in which the maternal X that stays active is the one that bears an imprint that preserves its activity, rather than the opposite scenario in which the paternal X that undergoes inactivation is the one that bears an inactivation imprint.

REPLICATION TIMING AND CELLULAR REPROGRAMMING: FURTHER SUPPORT FOR AUTOSOMAL LYONIZATION

iPSCs derived from adult somatic cells hold great promise for regenerative medicine in the 21st century, but they also provide an opportunity for understanding the nuclear reprogramming process. iPSCs share a replication-timing profile indistinguishable from that of ESCs in mice, consistent with the conclusion that replication-timing profiles reflect cell identity.[7,8] Thus, dissecting how replication-timing program is altered as somatic cells are reprogrammed back to pluripotent ESCs may reveal novel insight into the reprogramming process. In particular, partially reprogrammed iPSC lines (piPSCs), which are clonal cell lines that emerge from reprogramming experiments based on selection by morphology or reporter gene expression, provide a unique opportunity to view an intermediate state of the replication timing reprogramming process and to assess the reprogramming efficiency of different chromosomal regions. piPSCs fail to express many pluripotency genes and cannot contribute to chimeric mice formation, suggesting that they are blocked at an intermediate stage of the reprogramming process.[88,89]

Replication timing profiling of three independent piPSC lines derived from mouse embryonic fibroblasts (MEFs) revealed that their profiles were distinct from ESCs/iPSCs, early epiblast-like EPL cells, EpiSCs or nascent cells of the three germ layers (Fig. 2).[8] The three piPSCs were very similar to each other, suggesting that they were trapped at a common epigenetic state despite having independent retroviral integration site.[88] Interestingly, the majority of chromosomal segments that had experienced lineage-independent EtoL replication changes at the epiblast stage maintained their late replicating state in piPSCs (Fig. 2, asterisk),[8] which in females included the late-replicating Xi, underscoring the stability of these EtoL switches that were completed in the epiblast. In contrast, replication-timing changes that occurred later during development in a lineage-dependent (i.e., MEF-specific) manner were more readily reprogrammed. Furthermore, in these same piPSCs, expression of genes located within the lineage-independent EtoL switching segments showed the least similarity to ESCs when transcription profiles were compared.[8] Likewise, the Xi fails to become transcriptionally reactivated in female piPSCs.[88,89] Overall, these results suggest that many autosomal replication timing switches and in particular the EtoL replication timing switches at the epiblast stage coincident with X-inactivation, are stable epigenetic changes that are as difficult to reprogram as X-inactivation, supporting the notion of "autosomal Lyonization."

MAINTENANCE AND ALTERATION OF REPLICATION TIMING PROGRAM AND ITS POTENTIAL ROLES

As mentioned earlier, replication timing is re-established during early G1-phase at the TDP.[67] Intriguingly, this is coincident with the repositioning of chromosomal domains in the nucleus after mitosis.[67,90] The precise timing of TDP during G1 varies between cell types but is typically 1-3 hours into G1 in mammals[67] and at some point between mitosis and START in budding yeast.[91] In a parallel line of studies, chromatin mobility has been shown to be relatively high during the first 1-2 hours of G1-phase, after which it is locally constrained through the remainder of interphase.[92,93] Moreover, inducible targeting of loci to the nuclear lamina requires passage through mitosis and takes place during late telophase to early G1-phase.[94,95] Together, early G1 period seem to offer a temporal window for 3D organization of chromosomes to be re-established during each cell cycle, or alternatively, for a novel 3D chromosome organization pattern to be established, which in turn might dictate the replication timing program executed in the upcoming S-phase. Hence, this cell cycle regulation may offer a point of intervention for developmentally regulated changes in replication timing. It is also possible that the regulation of G1-phase length itself may influence the extent to which nuclei are reorganized before replication initiates and in turn affect the replication-timing program. Indeed, G1 length is highly variable between cell types and lengthens upon differentiation of mouse ESCs when a large degree of replication-timing changes is observed.[8]

Regarding the roles of the replication timing program, it should be emphasized that chromatin is assembled at the replication fork, providing a convenient window of opportunity to regulate this assembly process. Indeed, when reporter plasmids are injected into early or late S-phase mammalian nuclei, they assembled into hyper- or hypo-acetylated chromatin, respectively, providing evidence for different chromatin structure assembly at different times during S-phase.[96] Taking advantage of the fact that bovine papilloma virus plasmids replicate at different times in consecutive cell cycles,[97] the same authors recently showed that tightly packaged, late replicating chromatin becomes loosely packaged when the mini-circle is replicated early in the subsequent cell cycle.[98] Hence, reports using an artificial experimental system support a positive feedback loop whereby replication timing dictates chromatin states that in turn regulate replication timing in the subsequent cell cycle.[46] While in vivo evidence for recruitment of different sets of chromatin modifiers to replication forks at different times during S-phase remains scarce, this represents an attractive scenario for stable epigenetic inheritance of a given chromatin state.

CONCLUSION

The precise role of a replication-timing program and why this program is developmentally regulated remains to be elucidated. However, temporal regulation of genome duplication and the existence of multi-replicon domains are conserved from humans to budding and fission yeasts.[49,99,100] DNA replication is centrally linked to many basic cellular processes that are regulated during the cell cycle and development and defects in replication timing have been observed in various disease models.[101] Recent studies allow us to conclude that widespread developmental replication timing switches occur in flies,[57] mice,[7] and humans (T. Ryba, I.H. and D.M.G., *unpublished*). Moreover, there is significant conservation of the replication-timing program when regions of conserved synteny from similar cell types are compared between human and mouse (T. Ryba, I.H. and D.M.G., *unpublished*). As discussed, the first two major cell fate decisions during early mouse development accompany changes in the replication timing program of the Xi.[35] We have proposed that the postimplantation epiblast may experience the next major cell fate transition through a process of "Lyonization," involving a change in the replication timing program of Xi as well as autosomes that is stably maintained thereafter.[8] Whatever their role, whether causal or reflective, replication-timing programs are cell-type specific and appear to be an integral part of cell identity. In such ways, studies of DNA replication, more than 50 years after the discovery of the double helical structure and successful visualization of replicating DNA in the nucleus, continue to provide new insights into the organization of chromosomes and its changes during differentiation.

ACKNOWLEDGEMENTS

We would like to thank N. Rhind, K. Lindstrom and P. Tesar for helpful discussions and J. Rathjen for discussions and critical reading of the manuscript. We apologize for the limited discussion of nonmammalian model organisms and asynchronous replication in this chapter due to topic and space constraints. Research in the Gilbert lab is supported by NIH grants GM83337 and GM085354.

REFERENCES

1. Hiratani I, Takebayashi S, Lu J et al. Replication timing and transcriptional control: beyond cause and effect—part II. Curr Opin Genet Dev 2009; 19:142-9.
2. Stamatoyannopoulos JA, Adzhubei I, Thurman RE et al. Human mutation rate associated with DNA replication timing. Nat Genet 2009; 41:393-5.
3. Chang BH, Smith L, Huang J et al. Chromosomes with delayed replication timing lead to checkpoint activation, delayed recruitment of Aurora B and chromosome instability. Oncogene 2007; 26:1852-61.
4. Tabancay AP Jr, Forsburg SL. Eukaryotic DNA replication in a chromatin context. Curr Top Dev Biol 2006; 76:129-84.
5. MacAlpine DM, Bell SP. A genomic view of eukaryotic DNA replication. Chromosome Res 2005; 13:309-26.
6. Morishima A, Grumbach MM, Taylor JH. Asynchronous duplication of human chromosomes and the origin of sex chromatin. Proc Natl Acad Sci USA 1962; 48:756-63.
7. Hiratani I, Ryba T, Itoh M et al. Global reorganization of replication domains during embryonic stem cell differentiation. PLoS Biol 2008; 6:e245.
8. Hiratani I, Ryba T, Itoh M et al. Genome-Wide Dynamics of Replication Timing Revealed by In Vitro Models of Mouse Embryogenesis. Genome Res 2010; 20:155-69.
9. Holmquist GP. Role of replication time in the control of tissue-specific gene expression. Am J Hum Genet 1987; 40:151-73.
10. Pope BD, Hiratani I, Gilbert DM. Domain-Wide Regulation of DNA Replication Timing During Mammalian Development. Chromosome Research 2010; 18:127-36.
11. Taylor JH. Intracellular localization of labeled nucleic acid determined by autoradiographs. Science 1953; 118:555-7.
12. Meselson M, Stahl FW. The Replication of DNA in Escherichia Coli. Proc Natl Acad Sci USA 1958; 44:671-82.
13. Taylor JH, Woods PS, Hughes WL. The Organization and Duplication of Chromosomes as Reveals by Autoradiographic Studies Using Tritium-Labeled Thymidine. Proc Natl Acad Sci USA 1957; 43:122-8.
14. Taylor JH. The mode of chromosome duplication in Crepis capillaris. Exp Cell Res 1958; 15:350-7.
15. Taylor JH. Asynchronous duplication of chromosomes in cultured cells of Chinese hamster. J Biophys Biochem Cytol 1960; 7:455-64.
16. Latt SA. Microfluorometric detection of deoxyribonucleic acid replication in human metaphase chromosomes. Proc Natl Acad Sci USA 1973; 70:3395-9.
17. Latt SA. Fluorescent probes of chromosome structure and replication. Can J Genet Cytol 1977; 19:603-23.
18. Stubblefield E. Analysis of the replication pattern of Chinese hamster chromosomes using 5-bromodeoxyuridine suppression of 33258 Hoechst fluorescence. Chromosoma 1975; 53:209-21.
19. Latt SA. Fluorescence analysis of late DNA replication in human metaphase chromosomes. Somatic Cell Genet 1975; 1:293-321.
20. Lima De Faria A. Incorporation of tritiated thymidine into meiotic chromosomes. Science 1959; 130:503-4.
21. Epplen JT, Siebers JW, Vogel W. DNA replication patterns of human chromosomes from fibroblasts and amniotic fluid cells revealed by a Giemsa staining technique. Cytogenet Cell Genet 1975; 15:177-85.
22. Stambrook PJ, Flickinger RA. Changes in chromosomal DNA replication patterns in developing frog embryos. J Exp Zool 1970; 174:101-13.
23. Barr ML, Bertram EG. A morphological distinction between neurones of the male and female and the behaviour of the nucleolar satellite during accelerated nucleoprotein synthesis. Nature 1949; 163:676.
24. Lyon MF. Gene action in the X-chromosome of the mouse (Mus musculus L.). Nature 1961; 190:372-3.
25. Heard E, Disteche CM. Dosage compensation in mammals: fine-tuning the expression of the X chromosome. Genes Dev 2006; 20:1848-67.
26. Deakin JE, Chaumeil J, Hore TA et al. Unravelling the evolutionary origins of X chromosome inactivation in mammals: insights from marsupials and monotremes. Chromosome Res 2009; 17:671-85.
27. Ho KK, Deakin JE, Wright ML et al. Replication asynchrony and differential condensation of X chromosomes in female platypus (Ornithorhynchus anatinus). Reprod Fertil Dev 2009; 21:952-63.
28. Koina E, Chaumeil J, Greaves IK et al. Specific patterns of histone marks accompany X chromosome inactivation in a marsupial. Chromosome Res 2009; 17:115-26.
29. Takagi N, Sugawara O, Sasaki M. Regional and temporal changes in the pattern of X-chromosome replication during the early post-implantation development of the female mouse. Chromosoma 1982; 85:275-86.
30. Lock LF, Takagi N, Martin GR. Methylation of the Hprt gene on the inactive X occurs after chromosome inactivation. Cell 1987; 48:39-46.
31. Chow J, Heard E. X inactivation and the complexities of silencing a sex chromosome. Curr Opin Cell Biol 2009; 21:359-66.

32. Keohane AM, O'Neill LP, Belyaev ND, Lavender JS, Turner BM. X-Inactivation and histone H4 acetylation in embryonic stem cells. Dev Biol 1996; 180:618-30.
33. Chaumeil J, Okamoto I, Guggiari M et al. Integrated kinetics of X chromosome inactivation in differentiating embryonic stem cells. Cytogenet Genome Res 2002; 99:75-84.
34. Wutz A, Jaenisch R. A shift from reversible to irreversible X inactivation is triggered during ES cell differentiation. Mol Cell 2000; 5:695-705.
35. Sugawara O, Takagi N, Sasaki M. Correlation between X-chromosome inactivation and cell differentiation in female preimplantation mouse embryos. Cytogenet Cell Genet 1985; 39:210-9.
36. Snow MHL. Gastrulation in the mouse: growth and regionalization of the epiblast. J Embryol Exp Morphol 1977; 42:293-303.
37. Balazs I, Brown EH, Schildkraut CL. The temporal order of replication of some DNA cistrons. Cold Spring Harb Symp Quant Biol 1974; 38:239-45.
38. Burke W, Fangman WL. Temporal order in yeast chromosome replication. Cell 1975; 5:263-9.
39. Hatton KS, Dhar V, Brown EH et al. Replication program of active and inactive multigene families in mammalian cells. Mol Cell Biol 1988; 8:2149-58.
40. Goldman MA, Holmquist GP, Gray MC et al. Replication timing of genes and middle repetitive sequences. Science 1984; 224:686-92.
41. Epner E, Reik A, Cimbora D et al. The beta-globin LCR is not necessary for an open chromatin structure or developmentally regulated transcription of the native mouse beta-globin locus. Mol Cell 1998; 2:447-55.
42. Taljanidisz J, Popowski J, Sarkar N. Temporal order of gene replication in Chinese hamster ovary cells. Mol Cell Biol 1989; 9:2881-9.
43. Gilbert DM. Temporal order of replication of Xenopus laevis 5S ribosomal RNA genes in somatic cells. Proc Natl Acad Sci USA 1986; 83:2924-8.
44. Hiratani I, Leskovar A, Gilbert DM. Differentiation-induced replication-timing changes are restricted to AT-rich/long interspersed nuclear element (LINE)-rich isochores. Proc Natl Acad Sci USA 2004; 101:16861-6.
45. Perry P, Sauer S, Billon N et al. A dynamic switch in the replication timing of key regulator genes in embryonic stem cells upon neural induction. Cell Cycle 2004; 3:1645-50.
46. Gilbert DM. Replication timing and transcriptional control: beyond cause and effect. Curr Opin Cell Biol 2002; 14:377-83.
47. Azuara V, Brown KE, Williams RR et al. Heritable gene silencing in lymphocytes delays chromatid resolution without affecting the timing of DNA replication. Nat Cell Biol 2003; 5:668-74.
48. Raghuraman MK, Winzeler EA, Collingwood D et al. Replication dynamics of the yeast genome. Science 2001; 294:115-21.
49. Hayashi M, Katou Y, Itoh T et al. Genome-wide localization of preRC sites and identification of replication origins in fission yeast. EMBO J 2007; 26:1327-39.
50. Schubeler D, Scalzo D, Kooperberg C et al. Genome-wide DNA replication profile for Drosophila melanogaster: a link between transcription and replication timing. Nat Genet 2002; 32:438-42.
51. MacAlpine DM, Rodriguez HK, Bell SP. Coordination of replication and transcription along a Drosophila chromosome. Genes Dev 2004; 18:3094-105.
52. Woodfine K, Fiegler H, Beare DM et al. Replication timing of the human genome. Hum Mol Genet 2004; 13:191-202.
53. White EJ, Emanuelsson O, Scalzo D et al. DNA replication-timing analysis of human chromosome 22 at high resolution and different developmental states. Proc Natl Acad Sci USA 2004; 101:17771-6.
54. Jeon Y, Bekiranov S, Karnani N et al. Temporal profile of replication of human chromosomes. Proc Natl Acad Sci USA 2005; 102:6419-24.
55. Karnani N, Taylor C, Malhotra A et al. Pan-S replication patterns and chromosomal domains defined by genome-tiling arrays of ENCODE genomic areas. Genome Res 2007; 17:865-76.
56. Farkash-Amar S, Lipson D, Polten A et al. Global organization of replication time zones of the mouse genome. Genome Res 2008; 18:1562-70.
57. Schwaiger M, Stadler MB, Bell O et al. Chromatin state marks cell-type- and gender-specific replication of the Drosophila genome. Genes Dev 2009; 23:589-601.
58. Desprat R, Thierry-Mieg D, Lailler N et al. Predictable Dynamic Program of Timing of DNA Replication in Human Cells. Genome Res 2009; 19:2288-99.
59. Gilbert DM. Replication timing and metazoan evolution. Nat Genet 2002; 32:336-7.
60. Schmegner C, Hameister H, Vogel W et al. Isochores and replication time zones: a perfect match. Cytogenet Genome Res 2007; 116:167-72.
61. Grasser F, Neusser M, Fiegler H et al. Replication-timing-correlated spatial chromatin arrangements in cancer and in primate interphase nuclei. J Cell Sci 2008; 121:1876-86.
62. Costantini M, Bernardi G. Replication timing, chromosomal bands and isochores. Proc Natl Acad Sci USA 2008; 105:3433-7.

63. Brons IG, Smithers LE, Trotter MW et al. Derivation of pluripotent epiblast stem cells from mammalian embryos. Nature 2007; 448:191-5.
64. Tesar PJ, Chenoweth JG, Brook FA et al. New cell lines from mouse epiblast share defining features with human embryonic stem cells. Nature 2007; 448:196-9.
65. Aladjem MI. Replication in context: dynamic regulation of DNA replication patterns in metazoans. Nat Rev Genet 2007; 8:588-600.
66. Wu J-R, Gilbert DM. A distinct G1 step required to specify the chinese hamster DHFR replication origin. Science 1996; 271:1270-2.
67. Dimitrova DS, Gilbert DM. The spatial position and replication timing of chromosomal domains are both established in early G1-phase. Mol Cell 1999; 4:983-93.
68. Li F, Chen J, Solessio E et al. Spatial distribution and specification of mammalian replication origins during G1 phase. J Cell Biol 2003; 161:257-66.
69. Dimitrova DS, Prokhorova TA, Blow JJ et al. Mammalian nuclei become licensed for DNA replication during late telophase. J Cell Sci 2002; 115:51-9.
70. Yokochi T, Poduch K, Ryba T et al. G9a Selectively Represses a Class of Late-Replicating Genes at the Nuclear Periphery. Proc Natl Acad Sci USA 2009; 106:19363-8.
71. Wu R, Singh PB, Gilbert DM. Uncoupling global and fine-tuning replication timing determinants for mouse pericentric heterochromatin. J Cell Biol 2006; 174:185-94.
72. Jorgensen HF, Azuara V, Amoils S et al. The impact of chromatin modifiers on the timing of locus replication in mouse embryonic stem cells. Genome Biol 2007; 8:R169.
73. Hayashi MT, Takahashi TS, Nakagawa T et al. The heterochromatin protein Swi6/HP1 activates replication origins at the pericentromeric region and silent mating-type locus. Nat Cell Biol 2009; 11:357-62.
74. Flickinger R. Replication timing and cell differentiation. Differentiation 2001; 69:18-26.
75. Panning MM, Gilbert DM. Spatio-temporal organization of DNA replication in murine embryonic stem, primary and immortalized cells. J Cell Biochem 2005; 95:74-82.
76. Wu R, Terry AV, Singh PB et al. Differential subnuclear localization and replication timing of histone H3 lysine 9 methylation states. Mol Biol Cell 2005; 16:2872-81.
77. Gilbert DM, Gasser SM. Nuclear Structure and DNA Replication. In: DePamphilis ML, ed. DNA Replication and Human Disease. Cold Spring Harbor, New York: Cold Spring Harbor Laboratory Press, 2006.
78. Berezney R, Dubey DD, Huberman JA. Heterogeneity of eukaryotic replicons, replicon clusters and replication foci. Chromosoma 2000; 108:471-84.
79. Williams RR, Azuara V, Perry P et al. Neural induction promotes large-scale chromatin reorganisation of the Mash1 locus. J Cell Sci 2006; 119:132-40.
80. Lieberman-Aiden E, van Berkum NL, Williams L et al. Comprehensive Mapping of Long-Range Interactions Reveals Folding Principles of the Human Genome. Science 2009; 326:289-93.
81. Bazett-Jones DP, Li R, Fussner E et al. Elucidating chromatin and nuclear domain architecture with electron spectroscopic imaging. Chromosome Res 2008; 16:397-412.
82. Jaenisch R, Young R. Stem cells, the molecular circuitry of pluripotency and nuclear reprogramming. Cell 2008; 132:567-82.
83. Rathjen J, Lake JA, Bettess MD et al. Formation of a primitive ectoderm like cell population, EPL cells, from ES cells in response to biologically derived factors. J Cell Sci 1999; 112 (Pt 5):601-12.
84. Gardner RL, Brook FA. Reflections on the biology of embryonic stem (ES) cells. Int J Dev Biol 1997; 41:235-43.
85. Pfister S, Steiner KA, Tam PP. Gene expression pattern and progression of embryogenesis in the immediate post-implantation period of mouse development. Gene Expr Patterns 2007; 7:558-73.
86. Guo G, Yang J, Nichols J et al. Klf4 reverts developmentally programmed restriction of ground state pluripotency. Development 2009; 136:1063-9.
87. Lyon MF, Rastan S. Parental source of chromosome imprinting and its relevance for X chromosome inactivation. Differentiation 1984; 26:63-7.
88. Maherali N, Sridharan R, Xie W et al. Directly reprogrammed fibroblasts show global epigenetic remodeling and widespread tissue contribution. Cell Stem Cell 2007; 1:55-70.
89. Sridharan R, Tchieu J, Mason MJ et al. Role of the murine reprogramming factors in the induction of pluripotency. Cell 2009; 136:364-77.
90. Li F, Chen J, Izumi M et al. The replication timing program of the Chinese hamster beta-globin locus is established coincident with its repositioning near peripheral heterochromatin in early G1 phase. J Cell Biol 2001; 154:283-92.
91. Raghuraman M, Brewer B, Fangman W. Cell cycle-dependent establishment of a late replication program. Science 1997; 276:806-9.
92. Walter J, Schermelleh L, Cremer M et al. Chromosome order in HeLa cells changes during mitosis and early G1, but is stably maintained during subsequent interphase stages. J Cell Biol 2003; 160:685-97.

93. Thomson I, Gilchrist S, Bickmore WA et al. The radial positioning of chromatin is not inherited through mitosis but is established de novo in early G1. Curr Biol 2004; 14:166-72.
94. Kumaran RI, Spector DL. A genetic locus targeted to the nuclear periphery in living cells maintains its transcriptional competence. J Cell Biol 2008; 180:51-65.
95. Reddy KL, Zullo JM, Bertolino E et al. Transcriptional repression mediated by repositioning of genes to the nuclear lamina. Nature 2008; 452:243-7.
96. Zhang J, Xu F, Hashimshony T et al. Establishment of transcriptional competence in early and late S phase. Nature 2002; 420:198-202.
97. Gilbert DM, Cohen SN. Bovine papilloma virus plasmids replicate randomly in mouse fibroblasts throughout S-phase of the cell cycle. Cell 1987; 50:59-68.
98. Lande-Diner L, Zhang J, Cedar H. Shifts in replication timing actively affect histone acetylation during nucleosome reassembly. Mol Cell 2009; 34:767-74.
99. Versini G, Comet I, Wu M et al. The yeast Sgs1 helicase is differentially required for genomic and ribosomal DNA replication. EMBO J 2003; 22:1939-49.
100. McCune HJ, Danielson LS, Alvino GM et al. The Temporal Program of Chromosome Replication: Genomewide Replication in clb5{Delta} Saccharomyces cerevisiae. Genetics 2008; 180:1833-47.
101. Hiratani I, Gilbert DM. Replication timing as an epigenetic mark. Epigenetics 2009; 4:93-7.
102. Gilbert DM. Temporal order of DNA replication in eukaryotic cells: it's relationship to gene expression. Genetics. Stanford: Stanford University, 1989.

CHAPTER 5

PRESERVATION OF GENOMIC INTEGRITY IN MOUSE EMBRYONIC STEM CELLS

Peter J. Stambrook* and Elisia D. Tichy

Abstract Embryonic stem (ES) cells and germ cells have the potential to give rise to an entire organism. A common requirement is that both must have very robust mechanisms to preserve the integrity of their genomes. This is particularly true since somatic cells have very high mutation frequencies approaching 10-4 in vivo that would lead to unacceptable levels of fetal lethality and congenital defects. Notably, between 70% and 80% of mutational events monitored at a heterozygous endogenous selectable marker were loss of heterozygosity due to mitotic recombination, a mechanism that affects multiple heterozygous loci between the reporter gene and the site of crossing over. This chapter examines three mechanisms by which mouse embryonic stem cells preserve their genomic integrity. The first entails suppression of mutation and recombination between chromosome homologues by two orders of magnitude when compared with isogenic mouse embryo fibroblasts which had a mutation frequency similar to that seen in adult somatic cells. The second renders mouse ES cells hypersensitive to environmental challenge and eliminates damaged cells from the self-renewing population. Mouse ES cells lack a G1 checkpoint so that cells damaged by exogenous insult such as ionizing radiation do not arrest at the G1/S phase checkpoint but progress into the S phase where the damaged DNA is replicated, the damage exacerbated and the cells driven to apoptosis. The third mechanism examines how mouse ES cells repair double strand DNA breaks. Somatic cells predominantly utilize error prone nonhomologous end joining which, from a teleological perspective, would be disadvantageous for ES cells since it would promote accumulation of mutations. When ES cells were tested for the preferred pathway of double strand DNA break repair, they predominantly utilized the high fidelity homology-mediated repair pathway, thereby minimizing the incurrence of mutations during the repair process. When mouse ES cells are induced to differentiate, the predominant repair pathway switches from homology-mediated repair to nonhomologous end joining that is characteristic of somatic cells.

*Corresponding Author: Peter J. Stambrook—Department of Molecular Genetics, Biochemistry and Microbiology, University of Cincinnati College of Medicine, 231 Albert Sabin Way, Cincinnati, Ohio 45267-0524, USA. Email: stambrpj@ucmail.uc.edu

The Cell Biology of Stem Cells, edited by Eran Meshorer and Kathrin Plath.
©2010 Landes Bioscience and Springer Science+Business Media.

INTRODUCTION AND HISTORICAL PERSPECTIVE

The capacity of a single cell to give rise to multiple different cell types of an organism has been area of investigational interest for many years. A question that lingers is that of nuclear equivalence, (i.e.,whether all cells of a multicellular organism have genomes that are quantitatively and qualitatively equivalent). Nuclear equivalence was tested as early as 1902 when Spemann separated the two blastomeres following the first cleavage division of a newt embryo and showed that both blastomeres could develop into a complete embryo, supportive of functional equivalence (cited in ref. 1). This work was presaged eleven years earlier by Hans Driesch who showed that separated blastomeres of sea urchin embryos were capable of developing into normal, albeit smaller, sea urchins (cited in ref. 1).

Half a century later, a period replete with many important and insightful discoveries, two seminal experiments were described. The classic and elegant experiments by Hämmerling, using the green alga *Acetabularium* as a model organism, showed that the nucleus contained all of the information necessary to dictate cell morphology.[2] *Acetabularium* is a single cell organism that has a base that contains the nucleus, a stalk and a cap. Using two species of *Acetabularium*, *A. mediterrania* and *A. crenulata*, the former with a smooth cap and the latter with a wrinkled cap, Hämmerling showed that when the stalk and cap of *A. mediterrania* was grafted to the base of *A. crenulata,* the grafted cap was transformed from smooth to wrinkled. The reciprocal experiment also held true, indicating that the genetic information determining cap morphology was dictated by the nucleus within the grafted base. At about the same time, Briggs and King expanded on the findings of Driesch and Spemann by successfully transferring the nucleus of an undifferentiated frog blastula cell to an enucleated fertilized egg. In about one third of their attempts, the transplanted nucleus was capable of directing development to a normal embryo. When using nuclei from later staged embryos (neurula or tailbud), however, they found that none of the recipient eggs developed normally and that the majority failed to complete gastrulation,[4] suggesting a restriction in the potential to differentiate as cells mature.

In 1962, the question of nuclear equivalence and the potential pluripotency of nuclei from differentiated cells in a multicellular organism was reignited by the findings of John Gurdon. He reported that the nucleus from a *Xenopus* tadpole intestinal cell, when introduced into an enucleated *Xenopus* egg, was able to support the development of a fully-formed feeding tadpole, demonstrating that the intestinal cell nucleus retained the genetic information necessary to produce all of the cells of a complete multicellular organism.[5] It should be noted, however, that of the large number of nuclear transplant experiments performed, only slightly more than one percent of recipient eggs successfully produced a mature tadpole, leading to discussions as to whether the nuclei with demonstrated pluripotency are derived from stem cells or come from truly differentiated cells.

A major advance in cloning technology occurred in 1981 when Martin Evans in Cambridge and Gail Martin in San Francisco concomitantly and independently succeeded in culturing embryonic stem (ES) cells from mouse blastocysts.[6,7] The pluripotency of ES cells coupled with the ability to target specific gene sequences to their cognate sites by targeted homologous recombination led to cloning of mice with very specifically and selectively inactivated or modified genes. Oliver Smithies and his colleagues first used targeted homologous recombination to correct a mutant *Hprt* gene in mouse ES cells[8] and independently Capecchi and colleagues inactivated an *Hprt* gene in a similar manner.[9] Three years later reports of genetically modified mice produced by introduction of targeted ES cells into the blastocoel of recipient blastocysts appeared.[10-12] The fact that mouse ES cells were able to support the development of an intact organism and enter the germ line without any sign of other developmental or physiological problems suggested that the ES cells had retained full genomic competency.

A test for a similar degree of nuclear equivalence in mammalian somatic cells came eight years later with the development of somatic cell nuclear transfer (SCNT) which essentially mimicked the experiments described by Gurdon.[5] Ian Wilmut and colleagues[13] transplanted the nucleus derived from a sheep mammary gland cell into an enucleated egg that was activated and allowed to develop to the blastocyst stage in vitro, at which time it was implanted into the uterus of a surrogate mother. A viable lamb was born and designated Dolly. The success rate from this procedure, however, was very low with only three of 277 transplanted blastocysts producing live births and only one (Dolly) surviving to adulthood. Since the initial report, however, SCNT has been applied successfully in a range of mammals extending from mice to horses to ferrets (cited in ref. 14). SCNT has also been used to produce ES cells that then were used to correct a genetic defect in a mouse model,[15] suggesting a potential means for avoiding the requirement and attendant ethical dilemma, for using ES cells derived from fertilized eggs. Despite the successes of SCNT, problems with the technology remain. The frequency with which transplanted eggs reach blastula stage is low and in models in which animals are allowed to develop to maturity, there are frequent developmental abnormalities. In the case of Dolly, she was euthanized at half the age of the normal lifespan of a sheep. She suffered from arthritis not normally seen in animals her age and had progressive lung disease that on autopsy proved to be an adenocarcinoma. Dolly also had short telomeres characteristic of somatic cells of older animals, suggesting that either her telomeres had shortened prematurely, promoting rapid aging, or that the adult nucleus from which she had been derived already had eroded telomeres.

More recently, two laboratories independently showed that adult somatic cells, first from mice[16] and then from humans[17,18] could be induced to dedifferentiate and produce pluripotent stem cells (IPS cells) that mimic ES cells in morphology and competence to rediffentiate into many cell types following the concomitant introduction of cDNAs encoding four genes. The genes that initially seemed sufficient for IPS induction were Oct4, Sox2, Klf4 and c-Myc, but additional genes such as Nanog and other members of the Klf family appear to also be important in the induction of IPS cells. The important issue in the context of this discussion is that the data reaffirm that somatic cells retain all of the nuclear information to potentially direct differentiation into multiple different cell types. Challenges still exist to regulate epigenetic changes that must occur during the dedifferentiation process and concerns about the extent of nuclear equivalence in somatic cells remain and are similar to those raised for nuclei used for SCNT.

Somatic cells and mouse ES cells are intrinsically different in the extent to which they retain genomic integrity. Implicit in this contention is that somatic cells, which form the basis for the SCNT and IPS technologies, lack the same degree of nuclear equivalence as true ES cells derived from blastulae which may, in part, account for the low efficiency of these procedures. This proposition is consistent with the argument that somatic cells are more tolerant of deleterious mutations than germ cells and, by extrapolation, ES, cells and that somatic mutations accumulate as a function of age ultimately leading to somatic disease and death.[19-22] From an evolutionary and selfish perspective, this aging process may be viewed as described by Weil and Radman, "We have apparently evolved to spend most of our 'life capital' for the selfish objective of evolution: maintaining the species through immortality of the germline and disposing of the postreproductive soma. Although evolution may not 'care' about our post reproductive soma, we certainly do."[23] It follows that ES cells, which share the characteristic of pluripotency with germ cells, have robust mechanisms for preserving the integrity of their genomes that are different from or augment those present in somatic cells. This discussion focuses principally on mouse ES cells and describes some of the strategies observed in ES cells that promote genomic integrity.

MUTATION FREQUENCIES IN SOMATIC CELLS

Mammalian somatic cells display very high mutation frequencies in vivo, approaching 10^{-4}.[21,22,24,25] Furthermore, between 70% and 80% of mutation events at heterozygous loci involve loss of heterozygosity (LOH) which occurs as a consequence of recombination between homologous chromosomes (mitotic recombination) and which affects all heterozygous loci proximal to the cross-over site.[24-27] Mutation frequencies involving point mutations small insertions or deletions have been established using transgenic mice usually harboring prokaryotic reporters such as *lacI*,[26] *lacZ*,[28] *gpt*[29] and the *c11* gene of λ phage,[30] among others. In rodent models, the endogenous genes predominantly used as reporters are the selectable genes encoding the purine salvage enzymes hypoxanthine guanine phosphoribosyltransferase (Hprt)[27] and adenine phosphoribosyltransferase (Aprt).[25,27] Of these models, the *Aprt* model is the most effective in identifying events involving mitotic recombination which accounts for the majority of mutations at heterozygous loci in mouse and man. The strategy underlying this mouse model has been to cross mice of one strain (e.g., 129/sv) homozygous for a targeted null *Aprt* allele with mice of a second strain (e.g., C57Bl/6) with wildtype *Aprt*. The F1 mice are obligate heterozygotes at *Aprt* as well as at multiple flanking polymorphic markers characteristic of each strain. At the age of 8 to 10 weeks, the mice are sacrificed and skin fibroblasts, usually from the ear and splenic T-cells are placed into culture and immediately in medium containing either 5-fluoroadenine or 2, 6 diaminopurine (DAP), both of which are cytotoxic to cells with functional Aprt activity and which allow only cells lacking Aprt to grow in culture. Colonies deficient in Aprt activity arise in the selection medium and are picked and assessed first for LOH at *Aprt* by PCR analysis. If it is determined that there is LOH at *Aprt*, the flanking polymorphic markers are further analyzed and if necessary *Aprt* is PCR amplified and sequenced. In this manner, virtually all mutational pathways leading to Aprt deficiency can be identified. Between 70% and 80% of events involve LOH at *Aprt* and at linked flanking polymorphic markers to a point proximal to *Aprt* where retention of heterozygosity is maintained, indicative of mitotic recombination. Other events such as interstitial deletion, gene conversion, point mutation or epigenetic inactivation of the functional allele constitute between 20% and 30% of the remaining mutational events.

PROTECTION OF THE MOUSE ES CELL GENOME

While mutation in the germline is important for genetic diversity and the evolution of species, a mutation frequency or LOH in the order of 10^{-4}, such as that found in somatic cells (Fig. 1), would result in large numbers of defective embryos and births of individuals with congenital malformations.

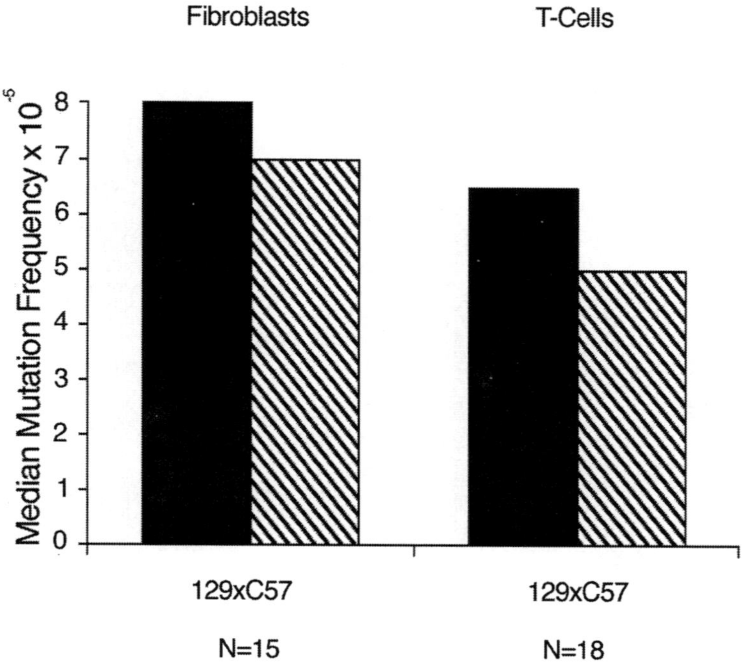

Figure 1. Mutation frequencies and frequencies of mitotic recombination in skin fibroblasts and splenic T-cells in vivo. The solid bars represent total mutation frequency and diagonally striped bars represent the proportion of events due to mitotic recombination and consequent LOH.

The question arises whether pluripotent stem cells are protected from acquiring and accumulating mutations. At least three mechanisms by which this may occur have been described in mouse ES cells. These include suppression of mutation and recombination between chromosome homologues, elevated apoptosis and preferential utilization of high fidelity homology-mediated repair of DNA double strand breaks (DSBs) rather than error prone nonhomologous end joining (NHEJ) that is predominant in somatic cells. While these mechanisms have been described for mouse ES cells, they are not necessarily the same as those found in true germ cells or even in human ES cells. They do, however, provide insight into the importance for pluripotent cells to maintain genomic stability.

The Frequency of Mutation Is Suppressed in Mouse ES Cells

As indicted above, somatic cells have high mutation frequencies that would be detrimental to the reproductive and developmental functions of germ cells and pluripotent stem cells. Using the *Aprt* mouse model, ES cells were prepared from F1 blastocysts derived from a cross between 129 strain mice homozygous for the targeted null *Aprt* allele and C3H mice with wildtype *Aprt*. Isogenic mouse embryo fibroblasts (MEFs) were prepared from 14 day old embryos for comparative purposes. The capacity for ES cells and MEFs to form Aprt or Hprt deficient colonies in DAP or in 6-thioguanine (6-GT) respectively, showed that mutation frequency at Aprt in ES cells (~10^{-6}) was about 100-fold lower than in MEFs (~10^{-4}), which approximated the in vivo mutant frequency observed in adult

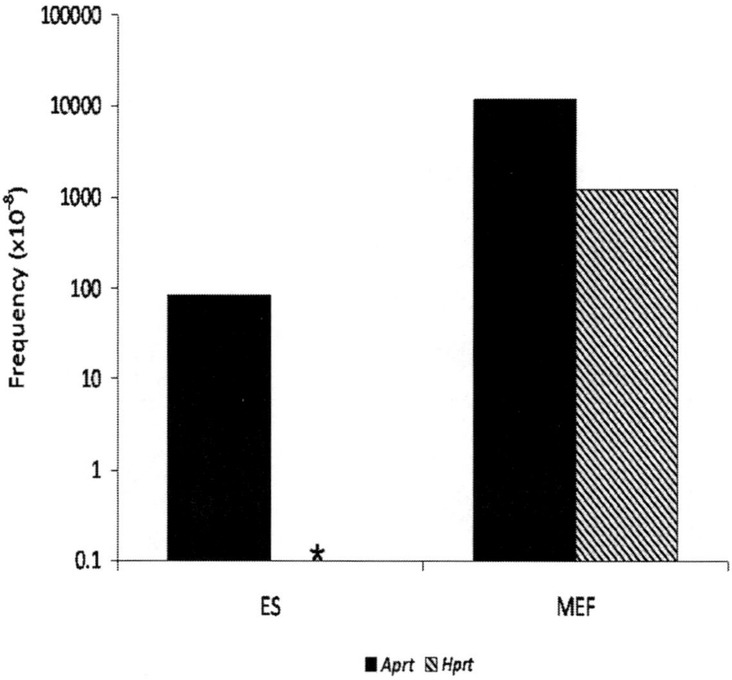

Figure 2. Mutation frequency at *Aprt* (solid bars) and at *Hprt* (diagonally striped bars) in ES cells and MEFs. The asterisk indicates that no Hprt deficient colonies were detected in 10^8 ES cells plated. The ordinate is on a logarithmic scale.

somatic cells (Fig. 2). The mutation frequencies at Hprt showed a similar trend, albeit with lower frequencies in all cases. This difference is attributed in part to the fact that *Aprt* is autosomal and is able to undergo mitotic recombination whereas *Hprt* is X-linked and is monosomic in males and effectively monosomic in females due to X-inactivation. About 60% of the mutation events detected involved LOH, but unlike MEFs and adult somatic cells, LOH was not predominantly due to mitotic recombination, but rather to nondisjunction resulting in uniparental disomy.[31] The *Aprt* gene is on mouse chromosome 8 and it may be that nondisjunction involving this chromosome is unusually high, skewing the data and interpretation.[32] Discordant data regarding mitotic recombination frequencies in mouse ES cells have been reported in a model in which genes encoding two different fluorescent proteins were targeted to the ROSA26 locus in each of the two chromosome 6 homologues.[33] Conversion from dichromatic to monochromatic fluorescence, indicative of loss of one of the alleles, occurred with a frequency of about 10^{-4}. About half of the events appear to be due to mitotic recombination since there was retention of heterozygosity at a locus proximal to ROSA26. Most of the other clones with monochromatic fluorescence had lost heterozygosity at ROSA26 and also at the centromeric locus, indicative of nondisjunction and chromosome 6 uniparental disomy. The recombination frequency is about two orders of magnitude higher than that observed with the *Aprt* model and may be reconciled by arguing that different chromosomes support very different recombination frequencies in ES cells or that insertion of fluorescent protein genes into the ROSA26

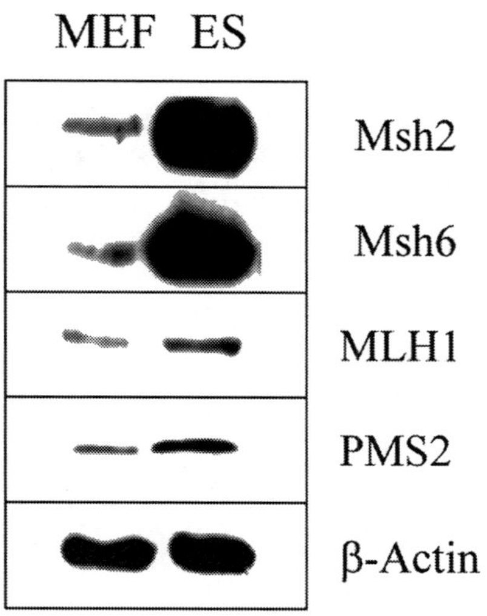

Figure 3. Abundance of mismatch repair proteins in ES cells and MEFs. Cell lysates from ES cells and MEFs were subjected to Western blots following gel electrophoresis using antibodies to Msh2, Msh6, Mlh1 and Pms2. β-actin was used as a control.

locus elevates that propensity to undergo mitotic recombination. It is noteworthy that two other studies report values that are considerably lower, but still higher than those using *Aprt* as a reporter.[34,35] A summary of several studies describing frequencies of mutation and LOH in ES cells has been described.[36]

Impairment of mismatch repair in prokaryotes enhances recombination and the potential for lateral genomic transfer between species.[37,38] The absence of effective mismatch repair produces a hyper-recombination phenotype in bacteria[37,38] and possibly also in yeast.[39] From an evolutionary perspective, it was not unreasonable to propose that mismatch repair proteins might control mitotic recombination, in part, in ES cells. Western blots of ES cell extracts show that mismatch repair proteins MSH2 and MSH6 are highly elevated in mouse ES cells compared with isogenic MEFs and that MLH1 and PMS2 are also elevated, but to a far lesser extent (Fig. 3). The role of mismatch repair in suppressing mitotic recombination has been examined using ES cells heterozygous at *Aprt* and transfected with an siRNA targeted to MSH2. Transfected cells were assessed for mutation frequency and frequency of LOH due to mitotic recombination, as was an ES cell line with an MSH2 missense mutant (G674A) which is devoid of mismatch repair activity.[40] Figure 4 shows that when mismatch repair in mouse ES cells is impaired or absent, mutation frequency and mitotic recombination are elevated. One clone in which MSH2 has been reduced by about 60% showed about a 20-fold increase in mutation frequency and an almost proportional increase in LOH due to mitotic recombination. The G674A MSH2 mutant that is mismatch repair deficient showed an even greater elevation in mutation frequency and mitotic recombination, almost reaching the level observed in MEFs. In aggregate, the data indicate that one mechanism by which mouse ES cells preserve the integrity of their genomes is by suppressing mutation in general and mitotic recombination and consequent LOH in particular, the latter being mediated in part by proteins involved in mismatch repair.

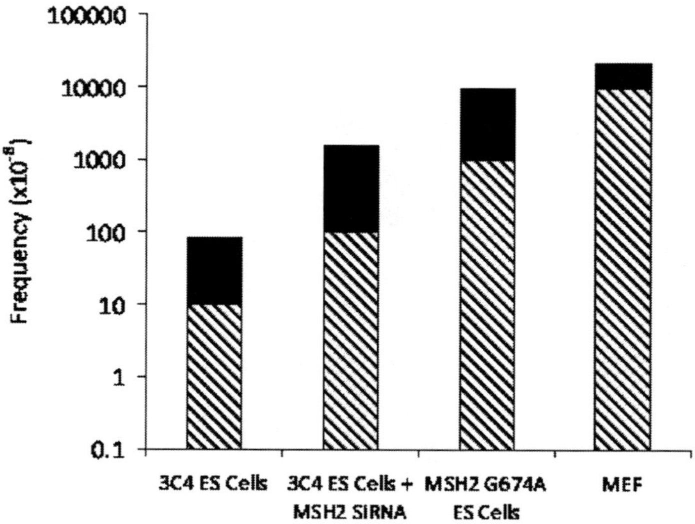

Figure 4. Effect of reducing Msh2 on mutation frequency and mitotic recombination in ES cells. ES cells heterozygous at *Aprt* were transfected with siRNA to Msh2 and were then cultured in medium containing DAP. Colonies that arose were quantified and mutation frequencies normalized to colony formation in the absence of selection. The DNAs were then analyzed by PCR for mechanism by which the cells became Aprt deficient.

ES Cell Populations Retain Pristine Genomes by Eliminating Cells with Damaged DNA

Mouse ES cells are hypersensitive to exogenous challenge and DNA damage, leading to apoptotic cell death.[41-43] This may have advantages by ridding damaged cells from the self-renewing stem cell population, thereby maintaining the genomes of the remaining cells pristine. Mouse ES cells lack a G1 checkpoint[44,45] and the two major pathways that are function in somatic cells to activate the G1 checkpoint following introduction of DNA double strand breaks are compromised in ES cells.[44] The pathways are presented as a simplified schematic in Figure 5. In brief, DNA lesions are probably sensed by the MRN complex (MRE11, RAD50, NBS1), resulting in phosphorylation and activation of ATM. When double strand DNA breaks are incurred, ATM initiates two pathways that result in the arrest of cells at a G1/S checkpoint. In one case, ATM phosphorylates p53 at serine 15, which contributes to p53 activation. P53 then induces transcription of the Cdk inhibitor p21, resulting in G1/S arrest. In the second case, ATM phosphorylates checkpoint kinase Chk2 on threonine 68 and promotes its activation. Activated Chk2, in turn, phosphorylates Cdc25A on serine 123 which contributes to its proteasome-mediated degradation. The Cdc25A bifunctional phosphatase dephosphorylates Cdk2 phosphothreonine 14 and phosphotyrosine 15 in the Cdk2/CyclinE and CyclinA complexes, promoting passage of cells from G1 into the S-phase. With diminished Cdc25A due to proteolytic degradation, or in its absence, Cdk2 is not dephosphorylated and cells are arrested in G1. Both of these pathways are compromised in mouse ES cells.[44] In the first pathway, p53 is present but is predominantly extranuclear in mouse ES cells, but not in somatic cells after irradiation. This is particularly well exemplified in Figure 6 where a colony of ES cells and a single

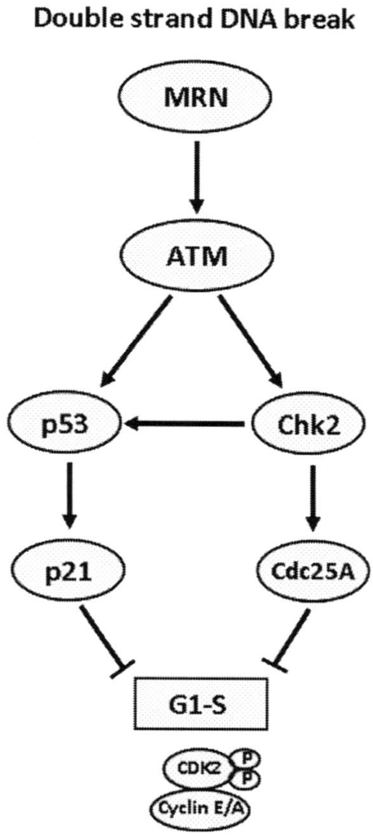

Figure 5. Schematic showing the two major signaling pathways activated by double strand DNA breaks leading to G1 arrest.

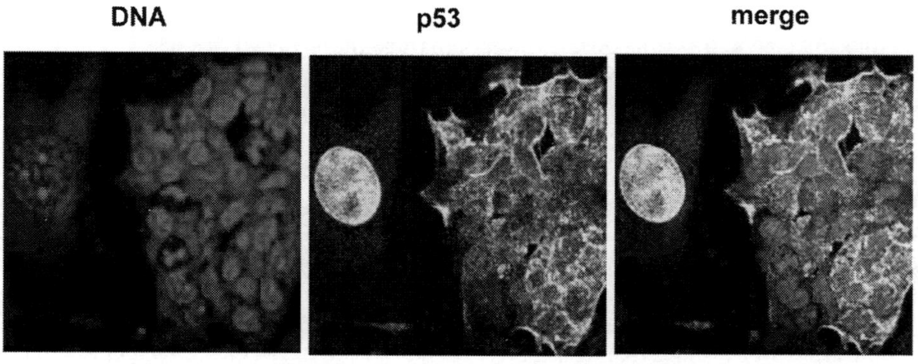

Figure 6. Immunofluorescence of ES cells and a MEF on the same coverslip for p53 after 10 Gy X-irradiation. First panel shows DRAQ5 staining of nuclei. Note nucleus of MEF in the same field of focus as a colony of ES cells. Middle panel shows p53 staining after irradiation. The last panel is a merge of the first two images. (Adapted from Hong et al, PNAS 2004; 101:14443-14448.)

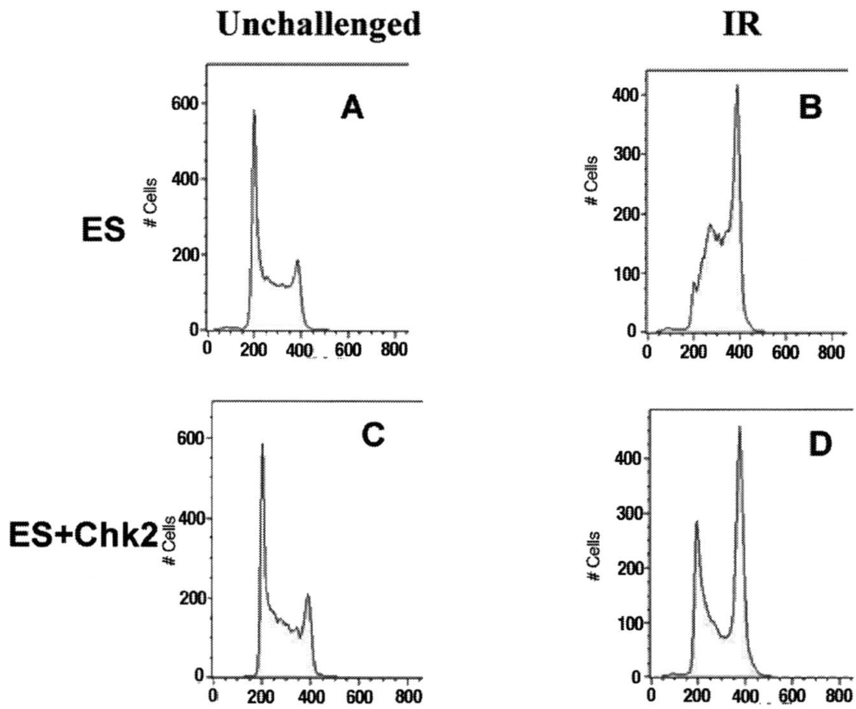

Figure 7. Wildtype ES cells (panels A and B) and ES cells transfected with a plasmid encoding Chk2 were treated with ionizing radiation or left untreated, fixed, stained with propidium iodide and fractionated by flow cytometry to assess cell cycle distribution. In the absence of radiation (panels A and C) the cellar distribution of both cell populations was similar. Following radiation, the wildtype cell population (panel B) showed a lack of cells in G1 with an accumulation of cells in S-phase and a clearly demarked G2/M cell cycle arrest. Following radiation, ES cells expressing Chk2 ectopically retained the G2/M checkpoint and additionally showed that cells also were arrested in G1/S (panel D). Adapted from Hong et al, PNAS 2004; 101:14443-14448.

MEF on the same coverslip show very different subcellular localization of p53 after being subjected to the same dose of radiation. Most of the detectable p53 is localized in the nucleus of the MEF whereas p53 is predominantly cytoplasmic in the cells of the ES cell colony. Consistent with this observation, p21 is not detectable in mouse ES cells, even after challenge by ionizing radiation.[45]

The second pathway appears to be regulated in part by the availability of Chk2. In mouse ES cells, Chk2 colocalizes with γ-tubulin at centrosomes and is therefore likely unavailable to phosphorylate its substrates such as Cdc25A. In ES cells, Cdc25A is not degraded after exposure to ionizing radiation as it is in somatic cells and Cdk2 remains predominantly in its hypophosphorylated state, allowing unimpaired transit into S-phase. When ES cells are transfected with a plasmid encoding Chk2 so that Chk2 is expressed ectopically, Cdc25A is proteolyzed after irradiation and Cdk2 is phosphorylated, predictive of a G1 arrest.[45] Consistent with this observation, ES cells expressing Chk2 ectopically and challenged with ionizing radiation display a G1 arrest (Fig. 7).

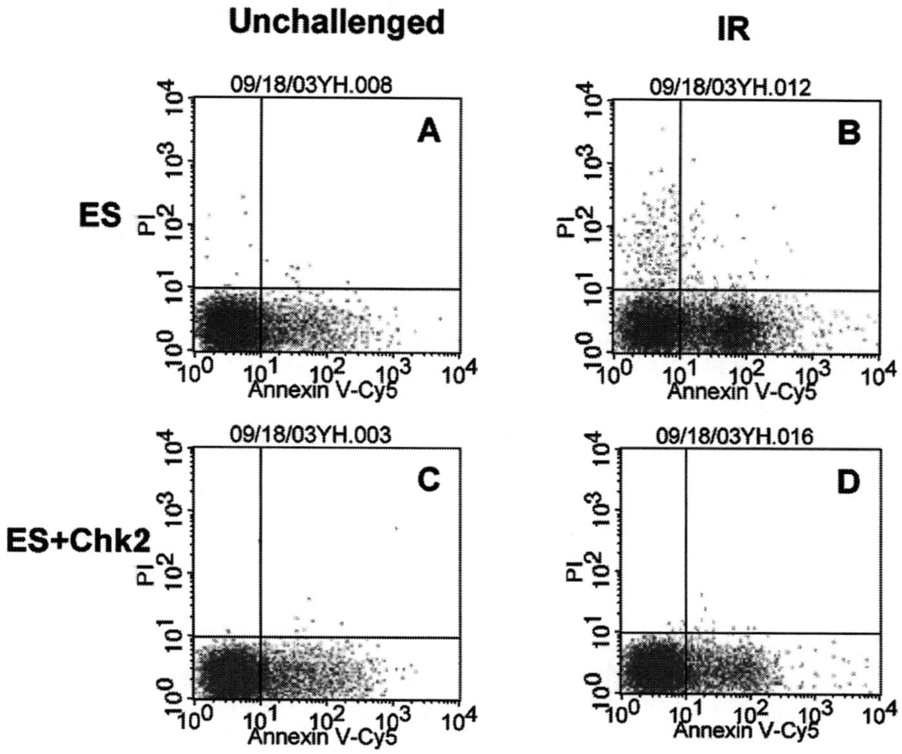

Figure 8. Wildtype ES cells (panels A and B) and ES cells transfected with a plamid encoding Chk2 were treated with ionizing radiation (panels B and D) or left untreated (panels A and C), stained with Annexin V and propidium iodide and subjected to bivariate flow analysis. Cells above the horizontal line (PI positive) are dead and cells to the right of the vertical line are Annexin V positive (apoptotic). Adapted from Hong et al, PNAS 2004; 101:14443-14448.

The operating hypothesis is that the absence of a G1 checkpoint allows damaged mouse ES cells to enter S-phase, where damaged DNA is replicated and the damage exacerbated with consequent cell death. Restoration of a G1 arrest in ES cells expressing Chk2 ectopically following exposure to ionizing radiation predicts that these cells should be protected from cell death. The prediction is borne out as described in Figure 8 where wildtype ES cells and ES cells expressing Chk2 ectopically were subjected to ionizing radiation and stained with propidium iodide as a vital stain and with Annexin V as an early marker of apoptosis. Populations of unchallenged mouse ES cells and ES cells expressing ectopic Chk2 are about 11% and 16% Annexin V positive, respectively, with very few dead (PI positive) cells. Following irradiation, about 40% of wildtype ES cells are Annexin V positive and a significant number are dead. In contrast, cells expressing ectopic Chk2 show no change in the number of apoptotic or dead cells (Fig. 8). The data again support the contention that the absence of a G1 checkpoint provides a mechanism to eliminate cells with damaged DNA from a population and thereby preserve the genomic integrity of the population as a whole.

Mouse ES Cells Preferentially Utilize High-Fidelity Homology-Mediated Repair Rather Than Nonhomologous End-Joining to Repair Double Strand DNA Breaks

When cells are exposed to certain therapeutics such as etoposide, a topoisomerase II inhibitor and radiomimetic, they sustain DNA damage including double strand DNA breaks.[46,47] The two major mechanisms by which this form of damage is repaired are the error-prone nonhomologous end joining (NHEJ) pathway and the high fidelity mechanism of homology-mediated repair. Although both mechanisms are operative in somatic cells, NHEJ is predominant and is active in all phases of the cell cycle.[48] When considering the importance of genomic integrity for germ and ES cells, any error-prone mechanism of DNA repair would intuitively appear to be disadvantageous. To test the possibility that NHEJ is not the predominant pathway for repairing double strand DNA breaks in ES cells, the abundance of participating proteins in each of the pathways has been compared by Western Blots between ES cells and isogenic MEFs. Notably, all of the proteins involved with homology-mediated repair are significantly elevated in extracts from ES cells compared with MEFs. In contrast, the levels of proteins involved with NHEJ are variable between ES cells and MEFs (Fig. 9). Of interest is that DNA Ligase IV, a rate-limiting enzyme in NHEJ is substantially lower in ES cells than in MEFs.

The abundance of proteins, while informative, is not necessarily indicative of function. The relative activities of both pathways have been examined in somatic cells and ES cells using a set of three reporter plasmids that collectively distinguish between homology-mediated repair and NHEJ. The plasmid for homology-mediated repair was pDR-GFP[49] which has two tandem nonfunctional GFP genes separated by a puromycin resistance marker which allows for selection of stable transfectants. The 5′ gene is inactivated by insertion of an I-Sce1 restriction site that produces a frame-shift and a premature termination signal. The

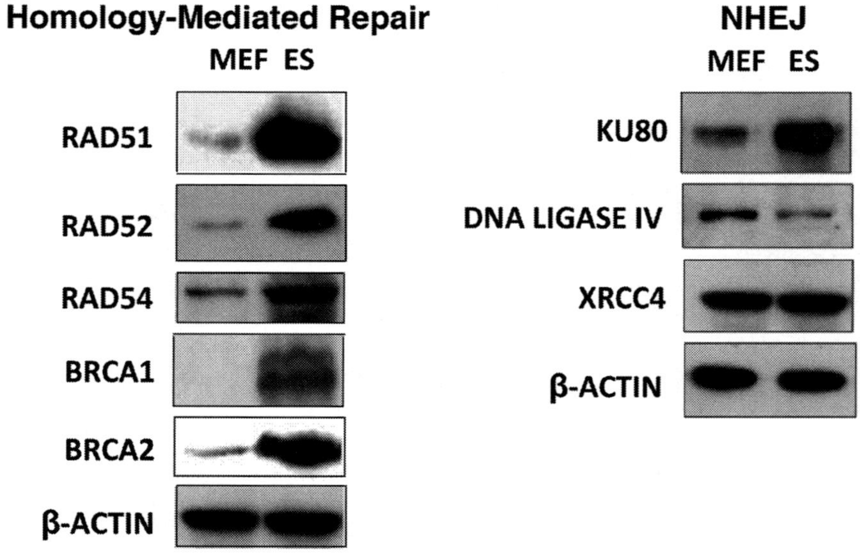

Figure 9. Whole cell extracts prepared from ES cells and from MEFs were probed by Western blots for relative abundance of proteins involved in homology-mediated repair (left panel) and for NHEJ (right panel).

downstream GFP is truncated at both ends to render it inactive. A double strand break can be initiated at the I-Sce1 restriction site by the I-Sce1 enzyme. Repair of the GFP gene to functional status, mediated by recombination between the two nonfunctional fragments produces fluorescent cells that can be quantified. To assess NHEJ, two additional reporter plasmids have been used. The plasmid pEGFP-PEM1-AD2[50] also uses restoration of GFP fluorescence as a measure of activity, but the strategy allows fluorescence only after repair of a double strand break by NHEJ. In this reporter construct, two halves of the *GFP* gene are separated by an intron retaining its splice signals derived from the *PEM* homeobox gene. Embedded within the *PEM* intron is an adenoviral AD2 gene that retains its own very strong splice acceptor and splice donor sites that perturb productive splicing and prevent restoration of GFP fluorescence. The AD2 gene is also flanked by paired Hind III sites and I-Sce1 sites in inverted orientation. When used in a transient transfection assay, the AD2 sequence is removed by digestion with I-Sce1 enzyme prior to transfection of cells. This also removes the its accompanying AD2 splice sites and leaves noncomplementary ends that can only be repaired by processing and NHEJ. When repaired, the PEM sequence is spliced out, the two GFP halves are brought into close proximity and the cells with repaired GFP are recognized and quantified by their fluorescence. The third reporter plasmid, designated pINV-CD4,[51] is based on the expression of the surface marker CD4 following repair of a double strand break lesion by NHEJ. The parent plasmid contains a CMV promoter driving expression of the histocompatibility surface antigen H-2Kd followed by a CD8 cDNA in inverted orientation and a promoterless CD4 cDNA. The H-2Kd and inverted CD8 cDNAs are flanked by inverted I-Sce1 sequences that will delete the H-2Kd and CD8 sequences and yield noncomplementary ends when digested with I-Sce1 enzyme. When the ends are rejoined by NHEJ the CD4 surface marker is brought into close apposition to the CMV promoter and expressed. Cells in which the reporter plasmid has been repaired by NHEJ can be sorted and quantified by flow cytometry based on CD4 expression. The first and third reporter plasmids have been used in both stable and transient transfections.

As previously reported, somatic cells use predominantly NHEJ to repair double strand DNA breaks. Using the plasmids described above, this also appears to be true of MEFs, but not mouse ES cells, unless they are induced to differentiate. These data are summarized in Figure 10. Since MEFs and ES cells that had been induced to differentiate by culture

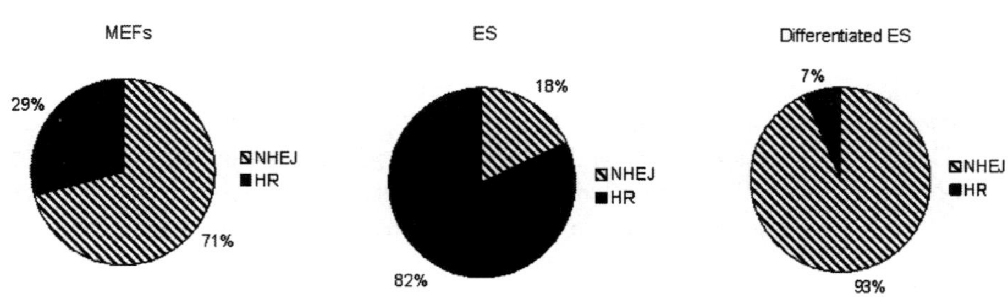

Figure 10. Relative activities of double strand break repair by NHEJ (striped sectors) and by homology-mediated repair (solid sectors) in MEFs, mouse ES cells and ES cells that had been induced to differentiate following culture in all-trans retinoic acid (ATRA). Cells of each population were electroporated with plasmid pDR-GFP after linearization with I-Sce1 to assess relative levels of recombination-mediated repair and with plasmid pEGFP-PEM1-AD2, also after I-Sce1 digestion to assess relative levels of NHEJ. In each case the assays were performed by transient transfection and quantification of cells that had acquired GFP fluorescence.

with all-trans retinoic acid (ATRA) cannot easily be selected for stable reporter plasmid integrants, they were assayed for repair pathway utilization by transient transfection. Each cell population was electroporated with pDR-GFP after linearization by I-Sce1 digestion to assess homology-mediated repair. Repair by NHEJ was assessed by electroporation with pEGFP-PEM1-AD2 after I-Sce1 digestion. In both cases, efficiency of electroporation was determined by cotransfection with plasmid pDs-Red which emits red fluorescence. Efficacy of repair by each mechanism was based on the proportion of red fluorescent cells also had restored GFP-derived green fluorescence. The data confirmed that somatic cells utilize NHEJ predominantly for repair of double strand DNA breaks and demonstrate that ES cells preferentially utilize recombination-mediated repair for the same purpose. When ES cells are induced to differentiate, they switch their preferred pathway from recombination-mediated repair to NHEJ. Accompanying this switch is a reduction in the abundance of proteins, such as Rad51, that are involved in the recombination-mediated process. There is also a significant increase in the level of DNA Ligase IV, which is particularly striking since all efforts to elevate the abundance of this protein in ES cells by overexpression or inhibition of its degradation were unsuccessful and led to rapid cell death. Thus, increased levels of DNA ligase IV that are not tolerated by ES cells appear to naturally accompany ES cells as they become differentiated and may help regulate NHEJ activity in somatic cells.

CONCLUSION

Historically, nuclei from somatic cells have been reported to have the capacity to support the development of a complete organism following transplantation into an enucleated egg, lending credence to the proposition that somatic cell nuclei retain sufficient integrity or "equivalence" to do so. The emergence IPS cell technology confirms the potential of somatic cell nuclei to undergo a dedifferentiation program and become pluripotent. The frequency of success with these techniques is usually very low, leading to concerns that the genomes of somatic nuclei, particularly from older individuals, are not equivalent to those of germ cells or ES cells and that the successes may come from a subset of cells that are not fully differentiated or that still maintain some stem-like properties. The low success frequencies may also be due, in part, to accumulated mutations in somatic cell nuclei that render many of the nuclei incapable of supporting complete conversion to pluripotency or to eroded telomeres that cannot always be restored. While these issues still remain to be resolved, they do call attention to the requirement for germ and ES cells to preserve the integrity of their genomes.

The present discussion has highlighted three mechanisms by which mouse ES cells or ES cell populations maintain genomic integrity. This does not imply that these mechanisms are unique to mouse ES cells or that these are the only mechanisms that can contribute to preservation of genomic stability. For example, human ES cells have more efficient mechanisms than somatic cells to repair types of DNA damage such as intrastrand crosslinks and 8-hydroxydeoxyguanosine lesions due to oxidative damage as well as double strand breaks.[52] Mouse ES cells have more effective stress response mechanisms to render them more resistant to oxidative stress than fibroblasts.[43] Human ES cells similarly have elevated expression levels of antioxidant genes *SOD2* and *GPX2* and lower levels of reactive oxygen species (ROS) than differentiated cells, which may contribute to reduced levels of DNA damage due to endogenous sources.[53] While less amenable to experimental manipulation and analysis than cells in culture, there is confirmatory evidence for protective strategies in vivo. Male germ cells in mice, for example, characteristically undergo extensive apoptosis[54,55] that occurs in massive waves. These occur during the time of primordial germ cell migration to the gonads and during the first round of spermatogenesis about 10 days postnatally,[56] perhaps as a means to rid the germ cell population of damaged cells. It is also noteworthy that dividing spermatogonia are hypersensitive to radiation whereas the supporting Sertoli cells are far less so.[57] Thus, Dawkins' concept[58] reiterated by Weill and Radman "...that we indeed live as disposable somas, slaves of our germline genome..."[23] is consistent with ever-increasing evidence that pluripotent cells, particularly cells of the germline, have developed multiple strategies for maintaining properties of self-renewal, pluripotency and essentially immortality.

ACKNOWLEDGEMENTS

The authors thank Drs. James Stringer, Alvaro Puga, El Mustapha Bahassi and Carolyn Price for their careful reading of the manuscript and their many helpful suggestions. This work was supported by grants from the NIEHS and the Center for Environmental Genetics at the University of Cincinnati.

REFERENCES

1. Ebert JD. Cell interactions: the roots of a century of research. Biol Bull 1985; 168:80-87.
2. Hämmerling J. "Nucleo-cytoplasmic relationships in the development of Acetabularia". J Intern Rev Cytol 1953; 2:475-498.
3. Briggs R, King TJ. Transplantation of living nuclei from blastula cells into enucleated frogs' eggs. Proc Natl Acad Sci USA 1952; 38:455-463.
4. Briggs R, King TJ. Nuclear transplantation studies on the early gastrula (Rana pipiens). Develop Biol 1960; 2:252-270.
5. Gurdon JB. The developmental capacity of nuclei taken from intestinal epithelium cells of feeding tadpoles. J Embryol Exp Morphol 1962; 10:622-640.
6. Evans M, Kaufman M. Establishment in culture of pluripotential cells from mouse embryos. Nature 1981; 292:154-156.
7. Martin G. Isolation of a pluripotent cell line from early mouse embryos cultured in medium conditioned by teratocarcinoma stem cells. Proc Natl Acad Sci USA 1981; 78:7634-7638.
8. Doetschman T, Gregg RG, Maeda N et al. Targeted correction of a mutant HPRT gene in mouse embryonic stem cells. Nature 1987; 330:576-578.
9. Thomas KR, Capecchi MR. Site-directed mutagenesis by gene targeting in mouse embryo-derived stem cells. Cell 1987; 51:503-512.
10. Koller BH, Hagemann LJ, Doetschman T et al. Germ-line transmission of a planned alteration made in a hypoxanthine phosphoribosyltransferase gene by homologous recombination in embryonic stem cells. Proc Natl Acad Sci USA 1989; 86:8927-8931.
11. Zijlstra M, Li E, Sajjadi F et al. Germ-line transmission of a disrupted beta 2-microglobulin gene produced by homologous recombination in embryonic stem cells. Nature 1989; 342:435-438.
12. Thomas KR, Capecchi MR. Targeted disruption of the murine int-1 protooncogene resulting in severe abnormalities in midbrain and cerebellar development. Nature 1990; 346:847-850.
13. Wilmut I, Schnieke AE, McWhir J et al. "Viable offspring derived from fetal and adult mammalian cells". Nature 1997; 385:810-813.
14. Campbell KHS, Fisher P, Chen WC et al. Somatic cell nuclear transfer: Past, present and future perspectives. Theriogenology 2007; 68(Suppl 1):S214-S231.
15. Rideout III WM, Hochedlinger K, Kyba M et al. Correction of a genetic defect by nuclear transplantation and combined cell and gene therapy. Cell 2002; 109:17-27.
16. Takahashi K, Yamanaka S. Induction of pluripotent stem cells from mouse embryonic and adult fibroblast cultures by defined factors. Cell 2006; 126:663-676.

17. Takahashi K, Tanabe K, Ohnuki M et al. Induction of pluripotent stem cells from adult human fibroblasts by defined factors. Cell 2007; 131(5):861-872.
18. Yu J, Vodyanik MA, Smuga-Otto K et al. Induced pluripotent stem cell lines derived from human somatic cells. Science 2007; 318:1917-1920.
19. Szilard L. On the nature of the aging process. Proc Natl Acad Sci USA 1959; 45:30-45.
20. Kirkwood TB, Proctor CJ. Somatic mutations and ageing in silico. Mech Ageing Dev 2003; 124:85-92.
21. Suh Y, Vijg J. Maintaining Genetic integrity in aging: a zero sum game. Antioxid Redox Signal 2006; 8:559-571.
22. Stambrook PJ. Do embryonic stem cells protect their genomes? Mech Ageing Dev 2007; 329:313-326.
23. Weil JC, Radman M. How good is our genome? Philos Trans R Soc London B 2004; 359:95-98.
24. Shao C, Deng L, Henegariu O et al. Mitotic recombination produces the majority of recessive fibroblast variants in heterozygous mice. Proc Natl Acad Sci USA 1999; 96:9230-9235.
24. Stambrook PJ, Shao C, Stockelman M et al. High frequency in vivo loss of heterozygosity is primarily a consequence of mitotic recombination. Cancer Res 1997; 157:1188-1193.
25. Stambrook PJ, Shao C, Stockelman M et al. Tischfield JA.APRT: a versatile in vivo resident reporter of local mutation and loss of heterozygosity. Environ Mol Mutagen 1996; 28:471-482.
26. Kohler SW, Provost GS, Fieck A et al. Analysis of spontaneous and induced mutations in transgenic mice using a lambda ZAP/lacI shuttle vector. Environ Mol Mutagen 19991; 18:316-321.
27. Van Sloun PP, Wijnhoven SW, Kool HJ et al. Determination of spontaneous loss of heterozygosity mutations in Aprt heterozygous mice. Nucleic Acids Res 1998; 26:4888-4894.
28. Gossen J, de Leeuw WJ, Verwest A et al. High somatic mutation frequencies in a LacZ transgene integrated on the mouse X-chromosome, Mutat Res 1991; 250:423-429.
29. Nohmi T, Katoh M, Suzuki H et al. A new transgenic mouse mutagenesis test system using spi- and 6-thioguanine selections, Environ Mol Mutagen 1996; 28:465-470.
30. Jakubczak JL, Merlino G, French JE et al. Analysis of genetic instability during mammary tumor progression using a novel selection-based assay for in vivo mutations in a bacteriophage lambda transgene target. Proc Natl Acad Sci USA 1996; 93:9073-9078.
31. Cervantes RB, Stringer JR, Shao C et al. Embryonic stem cells and somatic cells differ in mutation frequency and type. Proc Natl Acad Sci USA 2002; 99:3586-3590.
32. Liu X, Wu H, Loring J et al. Trisomy eight in ES cells is a common potential problem in gene targeting and interferes with germ line transmission. Dev Dyn 1997; 209:85-91.
33. Larson JS, Yin M, Fischer JM et al. Expression and loss of alleles in cultured mouse embryonic fibroblasts and stem cells carrying allelic fluorescent protein genes. BMC Mol Biol 2006; 7:36.
34. Yusa K, Horie K, Condo G et al. Genome-wide phenotype analysis in ES cells by regulated disruption of Bloom's syndrome gene. Nature 2004; 429:896-899.
35. Luo G, Santoro IM, McDaniel LD et al. Cancer predisposition caused by elevated mitotic recombination in Bloom mice. Nat Genet 2000; 26:424-429.
36. Fischer JM, Stringer JR. Visualizing loss of heterozygosity in living mouse cells and tissues. Mutat Res 2008; 645:1-8.
37. Vulic M, Lenski RE, Radman M. Mutation, recombination and incipient speciation of bacteria in the laboratory. Proc Natl Acad Sci USA 1999; 96:7348-7351.
38. Matic I, Taddei F, Radman M. No genetic barriers between Salmonella enterica serovar typhimurium and Escherichia coli in SOS-induced mismatch repair-deficient cells. J Bacteriol 2000; 182:5922-5924.
39. Manivasakam P, Rosenberg SM, Hastings PJ. Poorly repaired mismatches in heteroduplex DNA are hyperrecombinagenic in Saccharomyces cerevisiae. Genetics 1996; 142:407-416.
40. Lin DP, Wang Y, Scherer SJ et al. An Msh2 point mutation uncouples DNA mismatch repair and apoptosis. Cancer Res 2004; 64:517-522.
41. Van Sloun PPH, Jansen JG, Weeda G et al. The role of nucleotide excision repair in protecting embryonic stem cells from genotoxic effects of UV-induced DNA damage. Nucleic Acids Res 1999; 27:3276-3282.
42. Roos WP, Christmann M, Fraser ST et al. Mouse embryonic stem cells are hypersensitive to apoptosis triggered by the DNA damage O6-methylguanine due to high E2F1 regulated mismatch repair. Oncogene 2007; 26:186-197.
43. Saretzki G, Armstrong L, Leake A et al. Stress defense in murine embryonic stem cells is superior to that of various differentiated murine cells. Stem Cells 2004; 22:962-971.
44. Aladjem MI, Spike BT, Rodewald LW et al. ES cells do not activate p53-dependent stress responses and undergo p53-independent apoptosis in response to DNA damage. Curr Biol 1998; 8:145-155.
45. Hong Y, Stambrook PJ. Restoration of an absent G1 arrest and protection from apoptosis in embryonic stem cells after ionizing radiation. Proc Natl Acad Sci USA 2004; 101:14443-14448.
46. Ferguson LR, Baguley BC. Topoisomerase II enzymes and mutagenicity. Environ Molec Mutagenesis 1994; 24:245-261.

47. Mclendon AK, Osheroff N. Topoisomerase II, genotoxicity and cancer. Mutat Res 2007; 623:83-97.
48. Mao Z, Bozzella M, Seluanov A et al. Comparison of nonhomologous end joining and homologous recombination in human cells. DNA Repair 2008; 7:1765-1771.
49. Pierce AJ, Johnson RD, Thompson LH et al. XRCC3 promotes homology-directed repair of DNA damage in mammalian cells. Genes Dev 1999; 13:2633-2638.
50. Seluanov A, Mittelman D, Pereira-Smith OM et al. DNA end joining becomes less efficient and more error-prone during cellular senescence. Proc Natl Acad Sci USA 2004; 101:7624-7629.
51. Guirouilh-Barbat J, Huck S, Bertrand P et al. Impact of the KU80 Pathway on NHEJ-Induced Genome Rearrangements in Mammalian Cells. Molec Cell 2004; 14:611-623.
52. Maynard S, Swistowska AM, Lee JW et al. Human embryonic stem cells have enhanced repair of multiple forms of DNA damage. Stem Cells 2008; 26:2266-2274.
53. Saretzki G, Walter T, Atkinson S et al. Downregulation of multiple stress defense mechanisms during differentiation of human embryonic stem cells. Stem Cells 2008; 26:455-464.
54. Brinkworth MH, Weinbauer GF, Schlatt S et al. Identification of male germ cells undergoing apoptosis in adult rats. Journal of Reproduction and Fertility 1995; 105:25-33.
55. Richburg JH. The relevance of spontaneous- and chemically induced alterations in testicular germ cell apoptosis to toxicology. Toxicol Let 2000; 112-113:79-86.
56. Rodriguez I, Ody C, Araki K et al. An early and massive wave of germinal cell apoptosis is required for the development of functional spermatogenesis. EMBO J 1997; 16:2262-2270.
57. Hasegawa M, Zhang Y, Niibe H et al. Resistance of differentiating spermatogonia to radiation-induced apoptosis and loss in p53-deficient mice. Radiat Res 1998; 149:263-270.
58. Dawkins R. The Selfish Gene Oxford University Press 1990.

CHAPTER 6

TRANSCRIPTIONAL REGULATION IN EMBRYONIC STEM CELLS

Jian-Chien Dominic Heng and Huck-Hui Ng*

Abstract: Transcriptional regulation is a pivotal process that confers cellular identity and modulates the biological activities within a cell. In embryonic stem cells (ESCs), the intricate interplay between transcription factors and their targets on the genomic template serves as building blocks for the transcriptional network that governs self-renewal and pluripotency. At the heart of this complex network is the transcription factor trio, Oct4, Sox2 and Nanog, which constitute the ESC transcriptional core. Regulatory mechanisms such as autoregulatory and feedforward loops support the ESC transcriptional framework and serve as homeostatic control for ESC maintenance. Large-scale studies such as loss of function RNAi screens and transcriptome analysis have led to the identification of more players that support pluripotency. In addition, genome-wide localization studies of transcription factors have further unraveled the interconnectivity within the ESC transcriptional circuitry. Transcription factors also work in concert with epigenetic factors and together, this crosstalk between transcriptional and epigenetic regulation maintains the homeostasis of ESC. This chapter provides an overview of the significance of transcriptional regulation in ESC and traces the recent advances made in dissecting the ESC transcriptional regulatory network.

INTRODUCTION

Development in multicellular organisms is marked by the intricate and complex process of cell differentiation in which cells gradually lose their developmental plasticity to take on specialized functions. Despite possessing identical sets of genetic material, the cells of each individual organism exhibit clear distinctions in their morphologies and cellular functions. Such differences in cell specialization are, for the most part, brought

*Corresponding Author: Huck-Hui Ng—Genome Institute of Singapore, 60 Biopolis Street , #02-01, Genome Building, Singapore 138672.
Email: nghh@gis.a-star.edu.sg

The Cell Biology of Stem Cells, edited by Eran Meshorer and Kathrin Plath.
©2010 Landes Bioscience and Springer Science+Business Media.

forth by the distinct and dynamic regulation of information embedded within genomic DNA. This controlled expression of genes is regulated by a class of proteins known as transcription factors. In humans, transcription factors belong to the largest protein family[1] and they are DNA binding proteins that bind to regulatory elements such as enhancers, silencers or promoters of genes. These proteins act in context-dependent manners (i.e., either activating or repressing genes) to modulate gene expression. Upon the establishment of transcription factor-DNA binding at gene regulatory regions, RNA polymerases (RNA Pol) and other factors are recruited to gene promoter regions to initiate transcription, leading to the production of messenger RNA (mRNA).[2]

Besides engaging in direct interactions with DNA, transcription factors also recruit other transcription modulators to the genome. Modulators such as epigenetic modifiers alter the structures and properties of chromatin by posttranslationally modifying histone proteins or remodeling the nucleosome structure. These interactions result in variations in the accessibility of gene promoters and thus add new dimensions of complexity to transcriptional regulation. For instance, though transcription factors have numerous binding sites in the genome, they can only bind to target selective genomic sites and they are unable to dock at genomic sites where the chromatin is inaccessible.

The unique compendium of mRNA molecules that is synthesized in a cell is called the transcriptome. The complexity of transcriptomes varies not only between different cells within an organism, but also between different organisms. Interestingly, complexity in transcriptional regulation appears to be correlated to complexity of organisms.[3] Transcriptional regulation dictated by transcription factors and coregulators, coupled to the fine-tuned epigenetic changes undertaken by epigenetic modifiers is one of the mechanisms to establish unique specification and identity of a cell within an organism.

EMBRYONIC STEM CELLS AS A MODEL TO STUDY TRANSCRIPTIONAL REGULATION

Stem cells have the capacity to self-renew and differentiate into various cell types. As such, they are ideal models for the elucidation of transcriptional regulatory mechanisms which orchestrate developmental potency and self-renewal. Different types of stem cells vary in their developmental potentials. For example, hematopoietic stem cells are multipotent as they can differentiate into different blood cells while germs cells are only unipotent. Amongst the diverse range of stem cells that can be cultured, embryonic stem cells (ESCs) stand out as one of those that are pluripotent. Pluripotency is defined as the ability of a cell to differentiate into cell types stemming from the three major embryonic lineages, the endoderm, the ectoderm and the mesoderm. ESCs, which are obtained from pre-implantation blastocysts, possess tightly controlled transcriptional regulatory networks that maintain the cells in their self-renewing and pluripotent state and poise them for differentiation. Since ESCs can be cultured indefinitely in vitro and propagated in large amount with relatively good homogeneity, they are amenable to molecular and biochemical studies that address the mechanisms behind self-renewal. In addition, the ability of ESCs to differentiate into cells of all three germ layers also makes them good models to study the molecular events defining cellular differentiation.

TRANSCRIPTION FACTORS GOVERNING ESC PLURIPOTENCY

Since the isolation of ESCs, progress has been made in identifying the transcription factors that are important in maintaining these cells. Oct4, expressed from the *Pou5f1* gene, is a key transcription factor in ESCs that is downregulated upon differentiation. Oct4 belongs to the POU (Pit/Oct/Unc) family of homeodomain proteins and falls within the class of Octamer transcription factors which bind to a 8 basepair DNA sequence. Other than cells of the inner cell mass (ICM), Oct4 is also expressed in epiblasts, as well as primordial germ cells (PGCs).[4] Following gastrulation, Oct4 expression is restricted to the germ cells. *Pou5f1*-null embryos develop abnormal ICM cells which are not pluripotent but instead have a greater propensity to express trophoblast markers and subsequently die at the peri-implantation stage of development.[5] Similarly, when Oct4 expression is repressed or ablated in ESCs, cells lose their self-renewing state and tend to spontaneously differentiate to the trophectodermal lineage.[5,6] In agreement with this, Oct4 represses lineage-specific genes such as *Cdx2* (Fig. 1), a gene that is required for the development of the trophectodermal lineage.[7] It is crucial that ESCs maintain Oct4 at an appropriate level in order to remain pluripotent. When expression of Oct4 either increases or decreases 50% from the normal expression level, ESCs are induced to differentiate.[6] While ESCs with Oct4 downregulation differentiate to the trophectodermal lineage, ESCs with an overexpression of Oct4 tend to differentiate to multiple cell types, including cells from the primitive endodermal lineage.[6]

Another transcription factor important in ESCs is Sox2, a SRY (Sex determining Region-Y)-related transcription factor that possesses a DNA domain known as the high mobility group (HMG) box. Besides its expression in ESCs, Sox2 is also expressed in other cells such as neural progenitor cells (NPCs). *Sox2*-null mutants fail to form proper epiblasts[8] whereas depletion of Sox2 in ESCs results in trophectodermal differentiation,[9] a phenotype akin to Oct4 depletion in ESCs.[5,6] Interestingly, the introduction of Oct4 is able to rescue the self-renewal of *Sox2*-null ESCs.[9] Thus, Sox2 is suggested to preserve ESC stability by maintaining Oct4 expression at appropriate levels.[9] On the other hand, a slight increase in Sox2 can induce differentiation of mouse ESCs towards the neural lineage.[10] Other than being a transcriptional target of Sox2, Oct4 forms a heterodimer with Sox2 (Fig. 1) in which both transcription factors synergistically and cooperatively bind to Oct4-Sox2 binding elements to regulate several ESC-specific genes[11-17] including themselves.[18,19] Existing data of the crystal structure of Sox2 and Oct1 binding to DNA[20,21] have enabled a better understanding of how Sox2 and Oct4 bind to DNA. For example, homology modeling based on the crystal structure of the ternary complex formed by the HMG domain of Sox2 and the POU domain of Oct1 assembled onto the fibroblast growth factor 4 (Fgf4) enhancer was used to construct a structural model for the Oct4-Sox2-DNA complex.[20] Although Sox2 has a pivotal role in regulating gene expression in ESCs, it is striking to find that the Oct-Sox enhancers were still activated in *Sox2*-null cells. This suggests that other Sox proteins may be involved in activating Oct-Sox regulatory elements.[9] Consistent with this finding, other Sox proteins such as Sox4, Sox11 and Sox15 were found to be able to bind Oct-Sox elements.[9] Similar to Sox2, Sox15 forms a heterodimer with Oct4 and binds several Oct-Sox elements, albeit with a weaker binding affinity.[22] However, Sox15 has differing roles from Sox2 as shown by ESCs devoid of Sox15 which are normal[22] while ESC lacking Sox2 tend to differentiate into trophoctodermal cells.[9]

Nanog is another key transcription factor in ESCs. Nanog is specifically expressed in ICM cells and PGCs. Nanog, a homeodomain transcription factor, is known to function

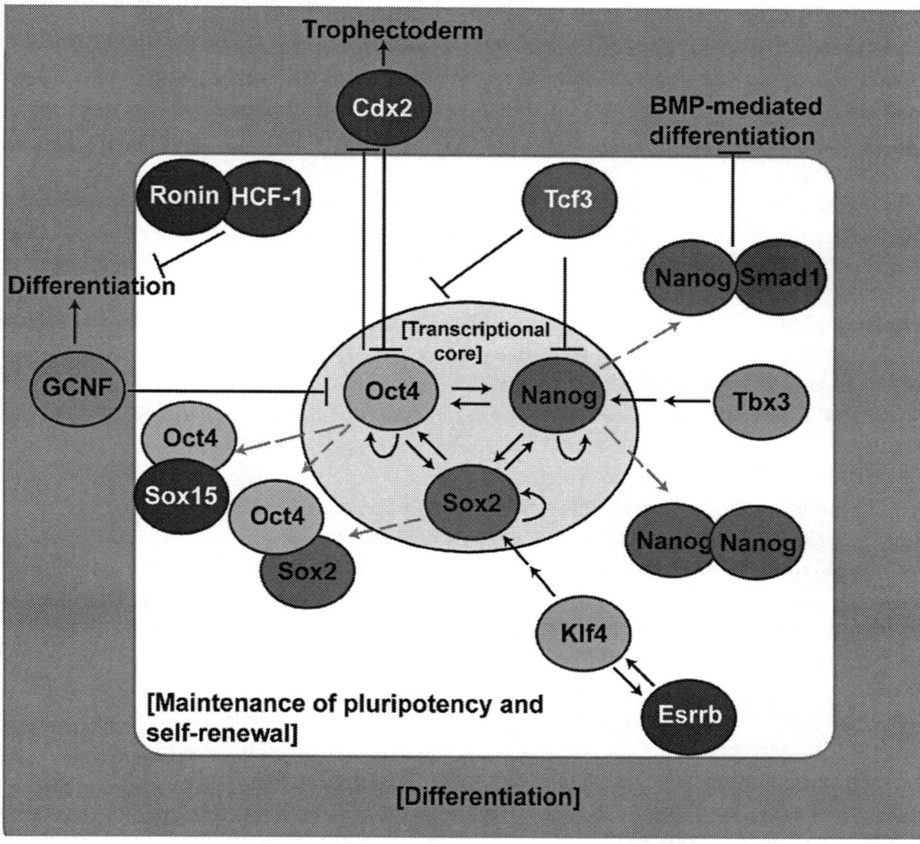

Figure 1. Model depicting ESC transcriptional regulatory network. The transcriptional core of ESC transcriptional regulatory network comprises Oct4, Sox2 and Nanog (in oval region). These factors participate in regulatory mechanisms such as autoregulatory (curved arrows) and feedforward loops. Reciprocal regulation may also occur as exemplified by the factors within the transcriptional core and between Esrrb and Klf4 (bi-directional arrows). Transcription factors can partner with other factors or with themselves (dashed arrows) to regulate other genes. For instance, Oct4 can form a heterodimer with either Sox2 or Sox15 while Nanog forms a homodimer with itself. Core regulators also participate in repressing differentiation. For example, Nanog can bind to Smad1 to inhibit BMP-mediated differentiation, while Oct4 can repress Cdx2 expression. Furthermore, Ronin interacts with HCF-1 (in addition to other factors) to inhibit differentiation. On the other hand, differentiation-related factors such as GCNF and Cdx2 can inhibit Oct4 expression. Factors/genes located within the white region are associated with the maintenance of pluripotency and self-renewal whereas factors/genes located within the green region are associated with differentiation. A color version of this figure is available at www.landesbioscience.com/curie.

as a dimer (Fig. 1) and this dimerization is important for its function in preserving ESC self-renewal and pluripotency.[23,24] Furthermore, dimerization of Nanog is vital for its interaction with other pluripotency-related proteins.[24] Nanog was discovered from a screen for novel pluripotency factors that could sustain ESCs in the absence of leukemia inhibitory factor (LIF).[25] From an *in silico* analysis of genes specifically expressed in ESCs and preimplantation embryos, another independent group also found Nanog to be important in maintaining pluripotency in ESCs independent of the LIF-Stat3 pathway.[26] While embryos devoid of *Nanog* were unable to form epiblasts, *Nanog*-null ICM differentiates to parietal endodermal-like cells.[26] Similar to Oct4, Nanog is downregulated upon ESC differentiation and ESCs deficient in Nanog differentiate[27] into cells of the extraembryonic endoderm lineage.[26] Hence, both transcription factors, Nanog and Oct4 are critical in ensuring pluripotency in ESCs.[27,28] Strikingly, contrary to previous findings, *Nanog*-null ESCs were shown to maintain self renewal with sustained Oct4 and Sox2 expressions levels.[29] Heterogeneity in Nanog expression has been reported in mouse ESCs.[29-32]Chambers et al showed that a subpopulation of ESCs does not express Nanog and these Nanog-deficient cells were able to self-renew with Oct4 and Sox2 expression still maintained.[29] However, the self-renewing capacity of these Nanog deficient cells is greatly reduced and they have a higher tendency to undergo primitive endodermal differentiation.[29] Intriguingly, the ESCs with down-regulated Nanog were capable of re-expressing Nanog. Nanog nonetheless plays an important role in ESC with other pluripotency-related factors and has been suggested to be important for mediating a pluripotent ground state in ESCs.[33] That is, Nanog is essential for establishing pluripotency but is dispensable for the maintenance of self-renewal and pluripotency in ESCs.

In addition to Oct4, Sox2 and Nanog, many other transcription factors contribute in supporting the pluripotent and self-renewing framework in ESCs. For instance, krüppel-like factors (Klf), which are zinc finger transcription factors, have a defining role in maintaining self-renewal in ESCs.[34] Overexpression of Klf2 and Klf4 can promote the self-renewal of ESCs.[35] Klf5 has also been implicated in the maintenance of mouse ESCs.[36] Similar to Sox2, transcriptional redundancy is also observed in the Klf family of proteins. For example, Klf5, in addition to Klf2 and Klf4 not only share overlapping binding sites but they also possess redundant roles in ESCs as shown by the triple knockdown of these three Klf transcripts.[34]

Ronin, a protein that has a zinc finger binding motif, was recently discovered to mediate pluripotency in mouse ESCs.[37] This factor is also important in embryogenesis as evidenced by the peri-implantation lethality of *Ronin*-null embryos. *Ronin* knockout ESCs were found to be nonviable and the overexpression of Ronin can inhibit ESC differentiation. Interestingly, Dejosez et al found that ectopic expression of Ronin was able to rescue the phenotype of *Pou5f1* knockdown in ESCs.[37] In addition, the authors postulated that Ronin maintains pluripotency by repressing genes responsible for differentiation. Ronin potentially interacts with a key transcriptional regulator known as host cell factor-1 (HCF-1) (Fig. 1) and other proteins to form a multimeric complex that represses gene expression.

Large-scale loss of function RNAi screening experiments have identified more transcription factors that support ESC pluripotency or self-renewal. From a RNAi screen, Ivanova et al identified two other transcription factors, Tbx3 and Esrrb to be important in preserving ESC self-renewal.[27] Interestingly, Nanog overexpression was able to rescue the differentiation phenotype induced by the knockdown of these self-renewal regulators, indicating that Nanog can bypass the lack of these factors.[27] In another independent study, depletion of several potential target genes bound by Oct4 and Nanog also revealed Esrrb to be important in maintaining ESC pluripotency.[28]

Besides transcription factors, large-scale RNAi screens have also identified factors other than transcription factors to be important in the maintenance of ESCs. The screen conducted by Ivanova and colleagues found a cofactor of Akt signaling, Tcl1, to be also important in ensuring ESC self-renewal.[27] Loh et al identified Rif1, a telomeric protein, in addition to the aforementioned Esrrb, to play a role in supporting the pluripotency framework in ESCs.[28] Employing a genome-wide RNAi screening strategy, a group has reported two transcriptional coregulators, Trim28 and Cnot3 to be essential for self-renewal.[38] More recently, a genome-wide RNAi studies have identified components of Paf1C, a complex associated with RNA Pol II, as modulators of Oct4,[39] while another RNAi sreen has identified Chd1, a chromatin remodelling factor to be important in maintaining open chromatin in ESCs.[40]

With more players identified for the maintenance of pluripotency, the stage is set for understanding the relationship of these transcription factors and how they interact with the genomic templates to specify ESC-specific gene expression.

TRANSCRIPTIONAL REGULATORY NETWORK

Transcription factors do not necessarily act on target genes independently but work in concert with other transcription factors to mediate gene regulation. The transcriptional partnership between Oct4 and Sox2 is one such example. The intricate interplay and partnership of transcription factors and their binding to diverse gene targets establish a complex transcriptional network that support and maintain cellular identity.

Mapping protein interactions of key transcription factors can provide invaluable insights into how the transcription regulatory networks regulate ESC identity. For example, utilizing affinity purification approach coupled to mass spectrometry, proteins that physically interact with Nanog were identified.[41] Protein partners of Nanog that were identified include Sall4, Rif1, Oct4, Dax1, Hdac2, Nac1 and Zfp281. Another group has also independently demonstrated Sall4 as an interacting partner of Nanog.[42] Besides interacting with factors to mediate pluripotency and self-renewal, Nanog can also interact with proteins to inhibit differentiation. For example, Nanog interacts with Smad1 to prevent BMP-mediated ESC differentiation[43] (Fig. 1) and it also binds and represses nuclear factor kappaB (NF-κB) proteins which are associated with the promotion of differentiation, concomitantly working in tandem with Stat3.[44]

Dissecting the transcriptional regulatory network can be very challenging and indeed requires the use of various genomics methods to elucidate the ESC transcriptome and genome-wide binding profiles of transcription factors, in which these and other techniques will be introduced in the following sections.

TECHNOLOGIES FOR DISSECTING THE TRANSCRIPTIONAL REGULATORY NETWORK

Various technological platforms have allowed us to study the transcriptome of ESCs in depth. One common and widely used platform to analyze the transcriptome of cells is that of the DNA microarray. DNA microarray involves the high throughput analysis of gene expression based on hybridization of expressed transcripts to probes. This method allows us to not only identify genes that are highly expressed in ESCs but also study gene expression changes upon ESC differentiation or specific gene depletion in ESCs. Microarray analysis can be used in conjunction with other strategies to better identify other transcriptional nodes in ESCs. For example, in the study by Ivanova et al, an integrative approach was adopted by coupling microarray analysis of rapidly downregulated genes in differentiating ESCs to loss of function RNAi screening. Such integrative methods are indeed powerful approaches to dissect transcriptional regulatory networks.

Though microarray analysis is a robust approach that can be used to study gene expression, it is unable to differentiate between every spliced isoform of a gene and more importantly does not allow us to discover novel genes. Other techniques have been employed to comprehensively unravel the ESC transcriptome and such studies reveal that numerous rare and novel gene transcripts exists in ESCs.[45,46] Other disadvantages of the microarray method include the challenging task involved in comparing data between different microarray platforms and the inevitable false negatives and positives generated by the hybridization technique employed in this methodology. In this regard, utilization of other genome-wide and higher throughput approaches with deeper coverage should be considered. For example, the recently developed RNA-sequencing (RNA-seq) method employs deep-sequencing technologies to better characterize transcriptomes.[47,48] RNA-seq involves the high-throughput sequencing of cDNA using the next-generation sequencing platforms which allow for deeper coverage, differentiation between isoforms of a gene and more precise quantitation of transcript expression. Interestingly, rapid advances in RNA-seq have allowed the sequencing to be performed directly on mRNA instead of cDNA.[49] These advantages greatly overcome the limitations of microarray and the usage of improved transcriptome analysis platforms such as RNA-seq will indeed allow us to better characterize the transcriptomes of ESCs.

Protein-DNA interaction occurs when a transcription factor binds to a particular genomic sequence. Chromatin Immunoprecipitation (ChIP) assay is a powerful methodology that can be employed to study this in vivo interaction of a transcription factor with the genomic DNA that it binds to. Hence, ChIP is widely used to study the docking of transcription factors to regulatory regions of genes such as promoters and enhancers. Briefly, ChIP involves the cross-linking of proteins to DNA using formaldehyde. After which, chromatin is fragmented and antibodies pull down specific proteins. The cross-linkages are then reversed and the immunoprecipitation-enriched chromatin is isolated (Fig. 2). Given the rapid advances in technology, it is feasible to carry out genome-wide transcription factor binding analyses based on various high throughput ChIP approaches (Fig. 2), which will be further elaborated below.

THE CORE TRANSCRIPTIONAL REGULATORY NETWORK: Oct4, Sox2 AND Nanog

The transcriptional network in ESCs entails a complex and intricate interplay of transcription factors and their cognate targets. At the heart of this transcriptional network, the transcription factor trio of Oct4, Sox2 and Nanog forms the transcriptional core that preserves ESC self-renewal and pluripotency (Fig. 1).

Loh et al employed the ChIP-PET technology to map the global binding profiles of Oct4 and Nanog in mouse ESCs to further dissect the architecture of ESC transcriptional regulatory networks.[28] ChIP-PET method entails the sequencing of vectors in which sequences of both ends of ChIP-enriched DNA called paired-end diTags (PETs) have been cloned. Genome-scale binding analyses revealed that Oct4, Sox2 and Nanog share a high degree of overlap in their binding to target genes.[28,50] Boyer et al employed ChIP-chip assay to study the binding of OCT4, SOX2 and NANOG at promoters in human ESCs.[50] In the ChIP-chip approach, ChIP DNA is amplified and hybridized to microarray

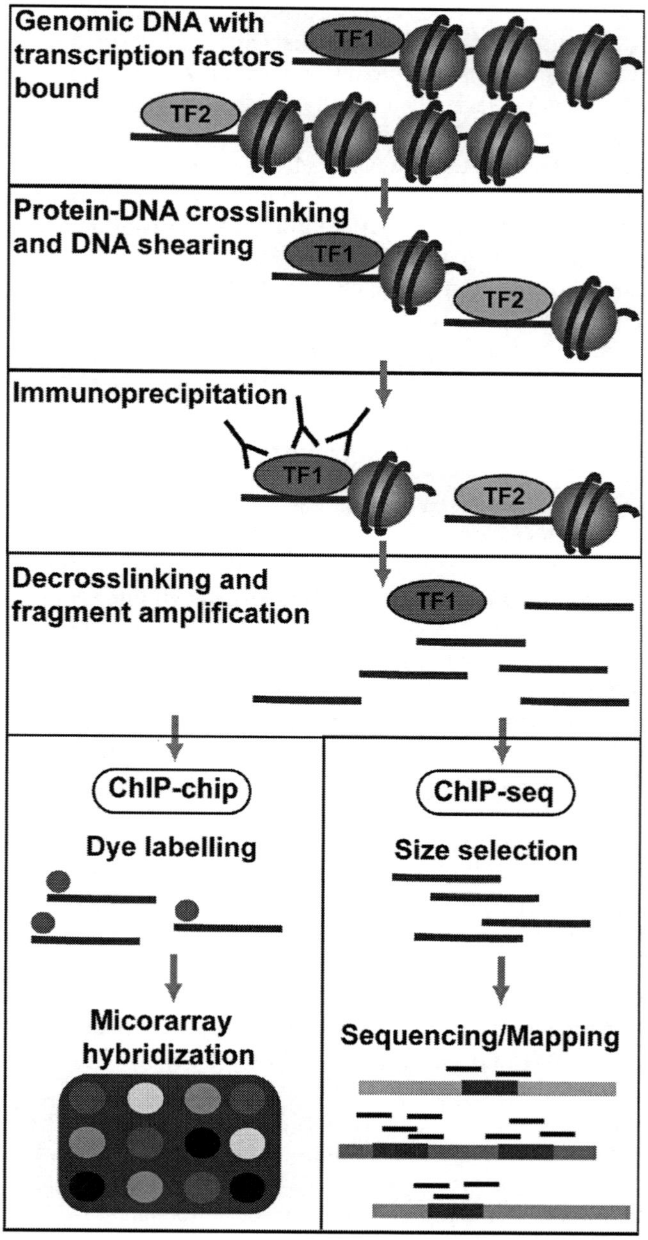

Figure 2. Genome-wide ChIP-based approaches to map transcription factor binding sites. Proteins are cross-linked to DNA by formaldehyde, followed by shearing of the genomic DNA into fragments either by sonication or nuclease treatment. Protein of interest is then immunoprecipitated with a specific antibody. Next, cross-links between the protein and DNA are reversed before amplification is performed. Following ChIP, global profiling techniques such as ChIP-chip and ChIP-seq can be employed to study the global binding profiles of a transcription factor. In ChIP-chip, ChIP is coupled to microarray technology whereby amplified fragments are dye-labeled and hybridized to a microarray. As for ChIP-seq, size fractionation of the amplified DNA is first performed followed by sequencing and mapping.

(Fig. 2). Consistent findings from both studies are that these factors co-occupy many target genes. In addition, both groups found that Oct4, Nanog and Sox2 can autoregulate themselves by binding to their own promoters thus forming autoregulatory, feedforward and feedback loops of regulation.[18,19,28,50] Autoregulation, which can be either positive or negative in nature, is a simple but effective way for a factor to control its own expression. Such regulatory systems have a stabilizing effect in ESCs yet keeping them poised for differentiation. Regulatory loops stemming from the transcription factor core of Oct4, Sox2 and Nanog will subsequently affect other downstream genes, which may also possess similar homeostatic regulatory control.

In summary, the transcription factor core of Oct4, Sox2 and Nanog regulates and supports the pluripotent framework of ESC by employing regulatory mechanisms such as feedforward and autoregulatory loops (Fig. 1).

EXPANDED TRANSCRIPTIONAL REGULATORY NETWORK

Kim et al investigated the promoter binding sites of nine transcription factors (Oct4, Sox2, Klf4, c-Myc, Nanog, Dax1, Rex1, Zfp281 and Nac1) to further elucidate the transcriptional network of ESCs.[51] A modified method of ChIP-chip (Fig. 2) known as in vivo biotinylation ChIP (bioChIP)-chip was employed. Interestingly, the data show co-occupancy of many genes by different transcription factors. Those genes bound by more than four are generally transcriptionally active in mouse ESCs, whereas those bound by a few transcription factors are transcriptionally repressed or inactive. Furthermore, this study also found that Oct4, Sox2, Klf4, Nanog, Dax1 and Zfp281 form a cluster that share many common gene targets while c-Myc and Rex1 form another distinct cluster.[51]

Another method of genome-wide ChIP known as ChIP-sequencing (ChIP-seq), which integrates ChIP with massively parallel DNA sequencing was later developed (Fig. 2).[52-54] Unlike ChIP-chip which is restricted by a predetermined and limited number of probes on the array, ChIP-seq enables the investigators to perform genome-wide and unbiased interrogation of transcription factor binding sites. Moreover, ChIP-seq is advantageous over ChIP-PET as it does not require special modifications to ChIP-enriched DNA and tedious cloning before sequencing is performed. Due to the depth of sampling, ChIP-seq analysis generates high resolution binding site datasets. Chen et al employed this ChIP-seq approach to map the binding sites of 13 transcription factors (Oct4, Sox2, Klf4, c-Myc, Nanog, STAT3, Smad1, Zfx, n-Myc, Klf4, Esrrb, Tcfcp2l1, E2f1 and CTCF) and 2 transcriptional coregulators, p300 and Suz12 in mouse ESCs. The study also reported the dense binding of multiple transcription factors to numerous genomic regions.[52] These genomic "hotspots" are also known as multiple transcription-factor-binding loci (MTL). Interestingly, it was found that two major clusters of transcription factors bind to MTLs. The first cluster was composed of the transcription factor trio Nanog-Oct4-Sox2 in addition to Stat3 and Smad1 while the second cluster is represented by c-Myc, n-Myc, Zfx and E2F1 (Fig. 3). The separate clustering of c-Myc from that of Nanog, Oct4 and Sox2 is in agreement with the findings of Kim et al and both studies suggest that transcriptional activation by multiple factors has an additive effect required for the transcriptional expression of genes.[51,52] Transcriptional synergy may also be achieved upon this combinatorial binding of transcription factors.

One of the major advantages of ChIP-based high throughput sequencing assays is that they allow investigators to study and map transcription factor-DNA interactomes

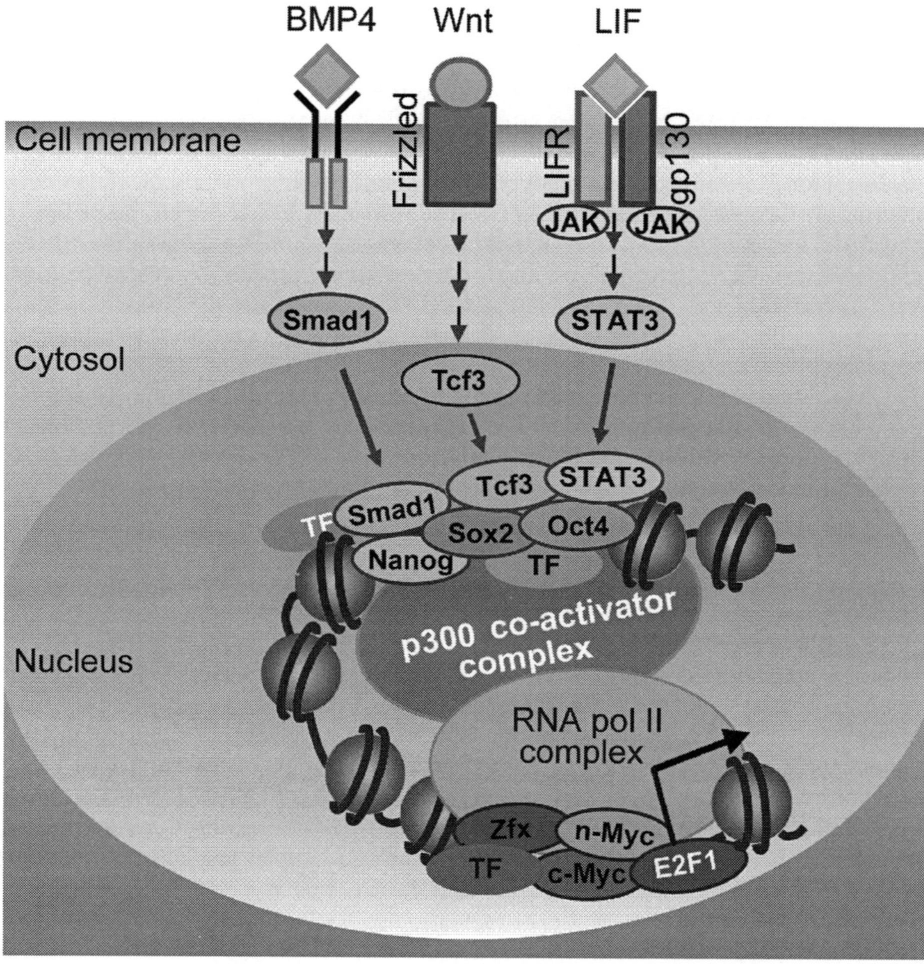

Figure 3. Enhanceosome complexes in regulating gene expression in ESCs. A cluster of transcription factors mainly comprising Oct4, Sox2, Nanog, Smad1 and Stat3 tend to colocalize at the enhancer region of genes. Another cluster consisting mainly of c-Myc, n-Myc, Zfx and E2f1 predominantly colocalize at gene promoters. TFs represent other transcription factors which may potentially interact with either cluster. The co-activator, p300 may be recruited to the Oct4-Sox2-Nanog cluster and is suggested to facilitate the interaction of the enhancer complex with the promoter complex by interacting with RNA Pol II. This interaction induces DNA looping and subsequently triggers transcriptional activation. Extrinsic signals critical in sustaining ESCs such as LIF and BMP are transmitted from cell surface receptors to transcriptional regulators such as Stat3 and Smad1, respectively which in turn propagate these signals to the nuclear machinery. Other signals such as Wnt are propagated to downstream factors such as Tcf3.

at a genome-scale level. However, it should be emphasized that genome-wide ChIP approaches such as ChIP-PET, ChIP-chip and ChIP-seq only identify target genes, which a transcription factor of interest binds to and do not demonstrate if these target genes are indeed functionally regulated by their respective bound factors. Hence, it is ideal to combine such ChIP assays with global gene expression profiling to identify bound and regulated genes.[55-57] For example, microarray analysis performed on ESCs which had altered Oct4 expression was coupled to existing genome-wide ChIP data so as to identify the downstream regulated targets of Oct4.[57] One gene that was highlighted from this study was *Tcl1*, which also appeared in Ivanova's RNAi screen.[27] Tcl1 was found to be a direct target of Oct4 and is associated with the regulation of ESC proliferation.[57] Another group analyzed available ChIP-chip data with time-course microarray datasets of differentiating ESCs as well as expression profiling of *Pou5f1* and *Sox2* knockdown ESCs to predict novel transcriptional networks that dictate undifferentiated ESC fate.[56] Also using this integrative approach, Zhou et al identified a set of factors, Esrrb, Sall4, Lrh-1, Tcf7 and Stat3 to be potential coregulators that cooperate with the core transcription factor of Oct4, Sox2 and Nanog in the gene regulatory network of ESCs.[55]

Taken together, genome-wide mapping of transcription factor binding in addition to global expression profiling has indeed provided us with a wealth of information and this has allowed us to develop a better understanding of the complex transcriptional wiring in ESCs.

ENHANCEOSOMES: TRANSCRIPTION FACTOR COMPLEX

Regulatory regions of a gene may not necessarily be bound by a single transcription factor but could be bound simultaneously by multiple transcription factors. This protein complex consisting of multiple transcription factors that bind to a short enhancer region of a gene is called an enhanceosome.[58] An enhanceosome is exemplified by the enhancer complex comprising NF-κB, interferon-regulatory factor 1 (IRF1) and activating transcription factor 2 (ATF-2)/c-JUN which binds concertedly to the human interferon-beta (IFNβ) gene that is upregulated upon viral infection.[58] Another factor, known as HMG-I, stabilizes this transcription enhancer complex and induces functional synergism with these transcription factors.[59] The presence of enhancesomes has also been postulated in the transcriptional regulation of ESCs. For instance, the reported dense binding of multiple transcription factors at MTLs by Chen et al suggests the existence of enhanceosomes in ESCs (Fig. 3). It was also found that the transcriptional co-activator, p300 is predominantly localized at enhancer regions associated with the Nanog-Oct4-Sox2 cluster.[52] Intriguingly, Nanog, Oct4 and Sox2 could be involved in recruiting p300 as knockdown of either three transcription factors reduced p300 binding.[52] Given that the Nanog-Oct4-Sox2 cluster predominantly binds to the enhancer region of genes and the c-Myc, n-Myc cluster predominantly binds to gene promoters, it is tempting to speculate that both major clusters of transcription factors may interact with the assistance of p300, which has also been previously reported to interact with c-Myc at promoter regions[60] (Fig. 3). This interaction may in turn promote DNA looping which bridges the enhancer region to the RNA Pol II complex-bound promoter region (Fig. 3). Nonetheless, establishment of the enhanceosome is an interesting transcriptional strategy that the cells may adopt to kick-start transcription and indeed a greater understanding of the enhancesome model requires further elucidation.

INTEGRATION OF SIGNALING PATHWAYS TO TRANSCRIPTIONAL NETWORK

Transcriptional regulatory networks do not stand alone but are intrinsically connected to cell signaling pathways. Such extrinsic signals that maintain ESCs are received from cell surface receptors and propagated by downstream effectors to transcription factors (Fig. 3). Chen et al not only established a transcriptional interconnectivity between the transcription factors that they profiled in ESCs, but also integrated the LIF and BMP signaling pathways to the Oct4-Sox2-Nanog cluster which shares many common targets with Stat3 and Smad1, the respective downstream effectors of both key signaling pathways (Fig. 3). Both the LIF-Stat and BMP pathways are essential to maintain self-renewal in ESCs. Paradoxically, by inhibiting the pro-differentiation FGF-MEK pathway, ESCs can be maintained independent of Stat3 activation.[61] Recently it was shown that the JAK-Stat pathway acting downstream of LIF activates Klf4, which in turn activates Sox2 (Fig. 1).[30] In addition, phosphatidylinositol-3-OH kinase–Akt and mitogen-activated protein kinase pathways acting through LIF activates *Tbx3* which in turn activates Nanog (Fig. 1).[30] Sox2 and Nanog then activate *Oct4*, which critically ensures ESC pluripotency. Interestingly, ectopic expression of both Klf4 and Tbx3 were sufficient to ensure ESC pluripotency when LIF is not present.[30] Hence, Klf4 and Tbx3 are responsible for transmitting downstream signals from LIF to the core regulatory network of mouse ESCs to maintain pluripotency. In addition, it was also recently shown that the effect of Klf4 activation by LIF-Stat pathway is additively coupled to Klf2 activation by Oct4 to sustain ESCs in their self-renewing state.[30]

Unlike in mouse ESCs, the LIF-JAK-Stat pathway is dispensable for human ESC self-renewal. Instead, activin/Nodal/TGFβ and FGF are required in human ESCs. It is noteworthy that components of the LIF signaling pathway are not expressed in human ESCs. On the other hand, most components of the FGF pathway are not expressed in mouse ESCs. The transcriptome of human and mouse ESCs may be intrinsically different due to species-specificity of gene expression. Another possible explanation for the inconsistency in gene expression could be that human and mouse ESCs represent different pluripotent states of early embryonic development. For example, epiblast stem cells (EpiSCs), which are pluripotent cells derived from the late epiblast layer of the postimplantation mouse embryos[62,63] show more similarities to human ESCs rather than mouse ESCs. Similar to human ESCs, EpiSCs are sustained by activin/Nodal and FGF signaling and share similar gene expression patterns with human ESCs such as the low expression of Dax1,[63] a transcription factor that is instead associated with mouse ESC maintenance.[41] Furthermore, human ESCs share a higher degree of overlap of Oct4 targets with EpiSCs as compared to mouse ESCs.[63] Together, these differences in the transcriptomes warrant further studies to characterize these cells at the molecular levels.

Despite such differences, both mouse and human ESCs require Wnt signaling for the maintenance of pluripotency.[64] The transcription factor, T-cell factor-3 (Tcf3), is a downstream effector of Wnt (Fig. 3). Though Tcf3 co-occupies several genomic sites with Oct4 and Nanog,[65] it also represses several important pluripotency and self-renewal genes such as Oct4 and Nanog in ESCs (Fig. 1).[65-67] In addition, Tcf3 depletion causes ESCs to be less prone to differentiation.[65-67] Therefore, Tcf3 acts as a fine tuner of both self-renewal and differentiation of ESCs.

Taken together, the close connection between transcription factors and signaling pathways adds greater dimension and complexity to the transcriptional regulatory network in ESCs.

INTERFACE BETWEEN TRANSCRIPTIONAL AND EPIGENETIC REGULATION

Both genetic and epigenetic regulation are equally crucial in the maintenance of ESCs. In line with this notion, Efroni et al reported that both transcription factors and chromatin remodeling factors show an elevated level of expression in ESCs, which undergo global gene silencing upon differentiation.[68] This phenomenon is reflective of global hypertranscription whereby transcription is even observed at several noncoding regions and silent genes.[68] In concert with hypertranscription, ESCs also display hyperdynamic binding of chromatin proteins such as linker histones and heterochromatin protein 1 (HP1).[69] Chromatin modifiers are responsible for introducing changes to chromatin thus making them more plastic and accessible for transcription factors to bind to their genomic targets. Similar to hypertranscription, hyperdynamic binding of chromatin-associated proteins reduces when ESCs differentiate.[69] Together, the phenomena of both hypertranscription and hyperdynamic binding of chromatin proteins provide support that genetic regulation is tightly intertwined to epigenetic regulation whereby the complex interplay of transcription factors and epigenetic modifiers maintain ESCs in their self-renewing and pluripotent state.

In the protein interaction network mapped out by Wang et al, it was found that Oct4 and Nanog also physically interacted with epigenetic regulators such as SWI/SNF chromatin remodeling complex, polycomb group (PcG) proteins and the histone deactylase complex NuRD.[41] PcG proteins make up the Polycomb repressive complexes (PRC) such as PRC2, which is important for early development and this epigenetic modifier induces histone 3 lysine 27 (H3K27) methylation, an epigenetic mark associated with gene silencing.[70] In human ESCs, OCT4, SOX2 and NANOG cobind to several PRC2 target genes. Oct4 also forms a complex with a component of PRC1 known as Ring1b in which the binding of Ring1b to target genes is dependent on Oct4.[71] While PRC2 is responsible for conferring repressive epigenetic marks, PRC1 is suggested to bind to these repressive marks and induce conformational changes in chromatin. Both Nanog and Oct4 have also been associated with other transcriptional repressive complexes such as Sin3A and Pml.[72]

Recently it was shown that Oct4 recruits Eset to repress several genes such as *Cdx2* and *Tcfap2a,* which are specifically expressed in the trophectodermal lineage.[73-75] Eset is a histone modifier that catalyzes H3K9 methylation at target genes such as *Cdx2* and *Tcfap2a*, hence resulting in repression of these genes in ESCs.[73-75] Consistent with this finding, morula cells that were deficient in *Eset* displayed higher expression levels of *Cdx2* and *Tcfap2a.*[74] After microinjection into the embryos, ESCs deficient in *Eset* tend to incorporate into the trophectoderm.[74] In another study, Loh et al showed that Oct4 positively regulates the expression of Jmjd1a and Jmjd2c, which are H3K9 demethylases that in turn regulate genes such as *Nanog* and *Tcl1*, respectively.[76] Taken together, cross talk between genetic and epigenetic regulations forms the basis of the complex and intricate regulatory network within ESCs.

CONCLUSION

Transcriptional regulation is an essential biological activity within a cell. Since the isolation of ESCs, a great wealth of information has been generated on the transcriptional regulation that oversees ESC self-renewal and pluripotency. Rapid advances in technologies have also enabled us to undertake genome-wide studies and more comprehensively dissect transcriptional regulatory networks. However, as most studies have been performed on mouse ESCs, it will be definitely more interesting to further unravel the transcriptional network in human ESCs. A better understanding of transcriptional regulation in human ESCs should allow us to translate this knowledge towards using stem cells for regenerative medicine in the future.

ACKNOWLEDGEMENTS

We thank Yun-Shen Chan, Jia-Hui Ng and Lin Yang for critical comments to the manuscript.

REFERENCES

1. Lander ES, Linton LM, Birren B et al. Initial sequencing and analysis of the human genome. Nature 2001; 409:860-921.
2. Johnson KM, Mitsouras K, Carey M. Eukaryotic transcription: the core of eukaryotic gene activation. Curr Biol 2001; 11:R510-3.
3. Levine M, Tjian R. Transcription regulation and animal diversity. Nature 2003; 424:147-51.
4. Pesce M, Scholer HR. Oct-4: gatekeeper in the beginnings of mammalian development. Stem Cells 2001; 19:271-8.
5. Nichols J, Zevnik B, Anastassiadis K et al. Formation of pluripotent stem cells in the mammalian embryo depends on the POU transcription factor Oct4. Cell 1998; 95:379-91.
6. Niwa H, Miyazaki J, Smith AG. Quantitative expression of Oct-3/4 defines differentiation, dedifferentiation or self-renewal of ES cells. Nat Genet 2000; 24:372-6.
7. Niwa H, Toyooka Y, Shimosato D et al. Interaction between Oct3/4 and Cdx2 determines trophectoderm differentiation. Cell 2005; 123:917-29.
8. Avilion AA, Nicolis SK, Pevny LH et al. Multipotent cell lineages in early mouse development depend on SOX2 function. Genes Dev 2003; 17:126-40.
9. Masui S, Nakatake Y, Toyooka Y et al. Pluripotency governed by Sox2 via regulation of Oct3/4 expression in mouse embryonic stem cells. Nat Cell Biol 2007; 9:625-35.
10. Kopp JL, Ormsbee BD, Desler M et al. Small increases in the level of Sox2 trigger the differentiation of mouse embryonic stem cells. Stem Cells 2008; 26:903-911.
11. Rodda DJ, Chew JL, Lim LH et al. Transcriptional regulation of nanog by OCT4 and SOX2. J Biol Chem 2005; 280:24731-7.
12. Kuroda T, Tada M, Kubota H et al. Octamer and Sox elements are required for transcriptional cis regulation of Nanog gene expression. Mol Cell Biol 2005; 25:2475-85.
13. Ambrosetti DC, Scholer HR, Dailey L et al. Modulation of the activity of multiple transcriptional activation domains by the DNA binding domains mediates the synergistic action of Sox2 and Oct-3 on the fibroblast growth factor-4 enhancer. J Biol Chem 2000; 275:23387-97.
14. Ambrosetti DC, Basilico C, Dailey L. Synergistic activation of the fibroblast growth factor 4 enhancer by Sox2 and Oct-3 depends on protein-protein interactions facilitated by a specific spatial arrangement of factor binding sites. Mol Cell Biol 1997; 17:6321-9.
15. Botquin V, Hess H, Fuhrmann G et al. New POU dimer configuration mediates antagonistic control of an osteopontin preimplantation enhancer by Oct-4 and Sox-2. Genes Dev 1998; 12:2073-90.
16. Nishimoto M, Fukushima A, Okuda A et al. The gene for the embryonic stem cell coactivator UTF1 carries a regulatory element which selectively interacts with a complex composed of Oct-3/4 and Sox-2. Mol Cell Biol 1999; 19:5453-65.

17. Yuan H, Corbi N, Basilico C et al. Developmental-specific activity of the FGF-4 enhancer requires the synergistic action of Sox2 and Oct-3. Genes Dev 1995; 9:2635-45.
18. Chew JL, Loh YH, Zhang W et al. Reciprocal transcriptional regulation of Pou5f1 and Sox2 via the Oct4/Sox2 complex in embryonic stem cells. Mol Cell Biol 2005; 25:6031-46.
19. Okumura-Nakanishi S, Saito M, Niwa H et al. Oct-3/4 and Sox2 Regulate Oct-3/4 Gene in Embryonic Stem Cells. J Biol Chem 2005; 280:5307-5317.
20. Remenyi A, Lins K, Nissen LJ et al. Crystal structure of a POU/HMG/DNA ternary complex suggests differential assembly of Oct4 and Sox2 on two enhancers. Genes Dev 2003; 17:2048-59.
21. Williams DC, Jr., Cai M, Clore GM. Molecular basis for synergistic transcriptional activation by Oct1 and Sox2 revealed from the solution structure of the 42-kDa Oct1.Sox2.Hoxb1-DNA ternary transcription factor complex. J Biol Chem 2004; 279:1449-57.
22. Maruyama M, Ichisaka T, Nakagawa M et al. Differential roles for Sox15 and Sox2 in transcriptional control in mouse embryonic stem cells. J Biol Chem 2005; 280:24371-24379.
23. Mullin NP, Yates A, Rowe AJ et al. The pluripotency rheostat Nanog functions as a dimer. Biochem J 2008; 227-231.
24. Wang W, Lin C, Lu D et al. Chromosomal transposition of PiggyBac in mouse embryonic stem cells. Proc Natl Acad Sci USA 2008; 105:9290-5.
25. Chambers I, Colby D, Robertson M et al. Functional expression cloning of Nanog, a pluripotency sustaining factor in embryonic stem cells. Cell 2003; 113:643-55.
26. Mitsui K, Tokuzawa Y, Itoh H et al. The homeoprotein Nanog is required for maintenance of pluripotency in mouse epiblast and ES cells. Cell 2003; 113:631-42.
27. Ivanova N, Dobrin R, Lu R et al. Dissecting self-renewal in stem cells with RNA interference. Nature 2006; 442:533-8.
28. Loh YH, Wu Q, Chew JL et al. The Oct4 and Nanog transcription network regulates pluripotency in mouse embryonic stem cells. Nat Genet 2006; 38:431-40.
29. Chambers I, Silva J, Colby D et al. Nanog safeguards pluripotency and mediates germline development. Nature 2007; 450:1230-4.
30. Niwa H, Ogawa K, Shimosato D et al. A parallel circuit of LIF signalling pathways maintains pluripotency of mouse ES cells. Nature 2009; 460:118-22.
31. Kalmar T, Lim C, Hayward P et al. Regulated fluctuations in nanog expression mediate cell fate decisions in embryonic stem cells. PLoS Biol 2009; 7:e1000149.
32. Carter MG, Stagg CA, Falco G et al. An in situ hybridization-based screen for heterogeneously expressed genes in mouse ES cells. Gene Expr Patterns 2008; 8:181-98.
33. Silva J, Nichols J, Theunissen TW et al. Nanog is the gateway to the pluripotent ground state. Cell 2009; 138:722-37.
34. Jiang J, Chan YS, Loh YH et al. A core Klf circuitry regulates self-renewal of embryonic stem cells. Nat Cell Biol 2008; 10:353-60.
35. Hall J, Guo G, Wray J et al. Oct4 and LIF/Stat3 additively induce Kruppel factors to sustain embryonic stem cell self-renewal. Cell Stem Cell 2009; 5:597-609.
36. Ema M, Mori D, Niwa H et al. Kruppel-like factor 5 is essential for blastocyst development and the normal self-renewal of mouse ESCs. Cell Stem Cell 2008; 3:555-67.
37. Dejosez M, Krumenacker JS, Zitur LJ et al. Ronin is essential for embryogenesis and the pluripotency of mouse embryonic stem cells. Cell 2008; 133:1162-74.
38. Hu G, Kim J, Xu Q et al. A genome-wide RNAi screen identifies a new transcriptional module required for self-renewal. Genes Dev 2009; 23:837-48.
39. Ding L, Paszkowski-Rogacz M, Nitzsche A et al. A genome-scale RNAi screen for Oct4 modulators defines a role of the Paf1 complex for embryonic stem cell identity. Cell Stem Cell 2009; 4:403-15.
40. Gaspar-Maia A, Alajem A, Polesso F et al. Chd1 regulates open chromatin and pluripotency of embryonic stem cells. Nature 2009; 460:863-8.
41. Wang J, Rao S, Chu J et al. A protein interaction network for pluripotency of embryonic stem cells. Nature 2006; 444:364-8.
42. Wu Q, Chen X, Zhang J et al. Sall4 interacts with Nanog and co-occupies Nanog genomic sites in embryonic stem cells. J Biol Chem 2006; 281:24090-4.
43. Suzuki A, Raya A, Kawakami Y et al. Nanog binds to Smad1 and blocks bone morphogenetic protein-induced differentiation of embryonic stem cells. Proc Natl Acad Sci USA 2006; 103:10294-9.
44. Torres J, Watt FM. Nanog maintains pluripotency of mouse embryonic stem cells by inhibiting NFkappaB and cooperating with Stat3. Nat Cell Biol 2008; 10:194-201.
45. Araki R, Fukumura R, Sasaki N et al. More than 40,000 transcripts, including novel and noncoding transcripts, in mouse embryonic stem cells. Stem Cells 2006; 24:2522-8.
46. Sharov AA, Piao Y, Matoba R et al. Transcriptome analysis of mouse stem cells and early embryos. PLoS Biol 2003; 1:E74.

47. Nagalakshmi U, Wang Z, Waern K et al. The transcriptional landscape of the yeast genome defined by RNA sequencing. Science 2008; 320:1344-9.
48. Wilhelm BT, Marguerat S, Watt S et al. Dynamic repertoire of a eukaryotic transcriptome surveyed at single-nucleotide resolution. Nature 2008; 453:1239-43.
49. Ozsolak F, Platt AR, Jones DR et al. Direct RNA sequencing. Nature 2009; 461:814-8.
50. Boyer LA, Lee TI, Cole MF et al. Core transcriptional regulatory circuitry in human embryonic stem cells. Cell 2005; 122:947-56.
51. Kim J, Chu J, Shen X et al. An extended transcriptional network for pluripotency of embryonic stem cells. Cell 2008; 132:1049-1061.
52. Chen X, Xu H, Yuan P et al. Integration of external signaling pathways with the core transcriptional network in embryonic stem cells. Cell 2008; 133:1106-17.
53. Johnson DS, Mortazavi A, Myers RM et al. Genome-wide mapping of in vivo protein-DNA interactions. Science 2007; 316:1497-502.
54. Jothi R, Cuddapah S, Barski A et al. Genome-wide identification of in vivo protein-DNA binding sites from ChIP-Seq data. Nucleic Acids Res 2008; 36:5221-31.
55. Zhou Q, Chipperfield H, Melton DA et al. A gene regulatory network in mouse embryonic stem cells. Proc Natl Acad Sci USA 2007; 104:16438-43.
56. Walker E, Ohishi M, Davey RE et al. Prediction and testing of novel transcriptional networks regulating embryonic stem cell self-renewal and commitment. Cell Stem Cell 2007; 1:71-86.
57. Matoba R, Niwa H, Masui S et al. Dissecting Oct3/4-regulated gene networks in embryonic stem cells by expression profiling. PLoS ONE 2006; 1:e26.
58. Thanos D, Maniatis T. Virus induction of human IFN beta gene expression requires the assembly of an enhanceosome. Cell 1995; 83:1091-100.
57. Du W, Thanos D, Maniatis T. Mechanisms of transcriptional synergism between distinct virus-inducible enhancer elements. Cell 1993; 74:887-98.
60. Faiola F, Liu X, Lo S et al. Dual regulation of c-Myc by p300 via acetylation-dependent control of Myc protein turnover and coactivation of Myc-induced transcription. Mol Cell Biol 2005; 25:10220-34.
61. Ying QL, Wray J, Nichols J et al. The ground state of embryonic stem cell self-renewal. Nature 2008; 453:519-23.
62. Brons IG, Smithers LE, Trotter MW et al. Derivation of pluripotent epiblast stem cells from mammalian embryos. Nature 2007; 448:191-5.
63. Tesar PJ, Chenoweth JG, Brook FA et al. New cell lines from mouse epiblast share defining features with human embryonic stem cells. Nature 2007; 448:196-9.
64. Sato N, Meijer L, Skaltsounis L et al. Maintenance of pluripotency in human and mouse embryonic stem cells through activation of Wnt signaling by a pharmacological GSK-3-specific inhibitor. Nat Med 2004; 10:55-63.
65. Cole MF, Johnstone SE, Newman JJ et al. Tcf3 is an integral component of the core regulatory circuitry of embryonic stem cells. Genes Dev 2008; 22:746-55.
66. Pereira L, Yi F, Merrill BJ. Repression of Nanog gene transcription by Tcf3 limits embryonic stem cell self-renewal. Mol Cell Biol 2006; 26:7479-91.
67. Tam WL, Lim CY, Han J et al. T-cell factor 3 regulates embryonic stem cell pluripotency and self-renewal by the transcriptional control of multiple lineage pathways. Stem Cells 2008; 26:2019-31.
68. Efroni S, Duttagupta R, Cheng J et al. Global transcription in pluripotent embryonic stem cells. Cell Stem Cell 2008; 2:437-47.
69. Meshorer E, Yellajoshula D, George E et al. Hyperdynamic plasticity of chromatin proteins in pluripotent embryonic stem cells. Dev Cell 2006; 10:105-16.
70. Cao R, Wang L, Wang H et al. Role of histone H3 lysine 27 methylation in Polycomb-group silencing. Science 2002; 298:1039-43.
71. Endoh M, Endo TA, Endoh T et al. Polycomb group proteins Ring1A/B are functionally linked to the core transcriptional regulatory circuitry to maintain ES cell identity. Development 2008; 135:1513-24.
72. Liang J, Wan M, Zhang Y et al. Nanog and Oct4 associate with unique transcriptional repression complexes in embryonic stem cells. Nat Cell Biol 2008; 10:731-9.
73. Yeap LS, Hayashi K, Surani MA. ERG-associated protein with SET domain (ESET)-Oct4 interaction regulates pluripotency and represses the trophectoderm lineage. Epigenetics Chromatin 2009; 2:12.
74. Yuan P, Han J, Guo G et al. Eset partners with Oct4 to restrict extraembryonic trophoblast lineage potential in embryonic stem cells. Genes Dev 2009; 23:2507-20.
75. Bilodeau S, Kagey MH, Frampton GM et al. SetDB1 contributes to repression of genes encoding developmental regulators and maintenance of ES cell state. Genes Dev 2009; 23:2484-9.
76. Loh YH, Zhang W, Chen X et al. Jmjd1a and Jmjd2c histone H3 Lys 9 demethylases regulate self-renewal in embryonic stem cells. Genes Dev 2007; 21:2545-57.

CHAPTER 7

ALTERNATIVE SPLICING IN STEM CELL SELF-RENEWAL AND DIFFERENTIATION

David A. Nelles and Gene W. Yeo*

Abstract: This chapter provides a review of recent advances in understanding the importance of alternative pre-messenger RNA splicing in stem cell biology. The majority of transcribed pre-mRNAs undergo RNA splicing where introns are excised and exons are juxtaposed to form mature messenger RNA sequences. This regulated, selective removal of whole or portions of exons by alternative splicing provides avenues for control of RNA abundance and proteome diversity. We discuss several examples of key alternative splicing events in stem cell biology and provide an overview of recently developed microarray and sequencing technologies that enable systematic and genome-wide assessment of the extent of alternative splicing during stem cell differentiation.

INTRODUCTION

Stem cells are a unique resource for studying the bases of pluripotency, self-renewal and lineage specification. Embryonic stem cells remain undifferentiated in culture for long periods and are readily induced towards the three germ layers, differentiating in vitro into most if not all of the lineages that comprise a healthy organism. Thus, embryonic stem cells are a useful platform to study healthy and disease states in a multitude of lineages. Generation of cell populations enriched with a particular differentiated cell type and ongoing, detailed characterization of these cells before and after differentiation will continue to provide insight into the molecular basis of cell identity.

Gene expression studies have documented global transcriptional differences during the process of differentiation into several lineages,[1-4] but are limited in that most studies do not distinguish among alternatively spliced isoforms from the same gene locus. In

*Corresponding Author: Gene W. Yeo—Department of Cellular and Molecular Medicine, Stem Cell Program, University of California, San Diego, 9500 Gilman Drive, La Jolla, California 92037, USA. Email: geneyeo@ucsd.edu

The Cell Biology of Stem Cells, edited by Eran Meshorer and Kathrin Plath.
©2010 Landes Bioscience and Springer Science+Business Media.

this chapter, we will present several examples of alternatively spliced genes implicated in important stem cell processes and review recently available techniques that allow identification and quantitative measurement of alternative splicing events. Early work to identify splicing factors that influence alternative splicing will also be discussed.

INTRODUCTION TO ALTERNATIVE SPLICING

Transcription produces pre-messenger RNA (pre-mRNA) transcripts which are dominated by long, noncoding intronic sequences interspersed with short, 150 base exonic sequences. Intron removal, exon ligation and splice site selection is highly regulated as splicing errors can generate aberrant proteins or prevent translation of the mRNA. This process is mediated by a protein-RNA complex called the spliceosome whose stringent functional requirements are reflected in its complex makeup; it consists of five RNAs and hundreds of proteins.[5-6] This machinery interacts with cis-regulatory elements encoded in pre-mRNAs and trans-acting regulators called splicing factors.

The interactions between cis-elements, splicing factors and the spliceosome define a set of rules called the "splicing code." The components of this code determine which splice sites are chosen to generate different versions of mature mRNAs from the same pre-mRNA in the process called alternative splicing (Fig. 1). The splicing code is still largely a mystery, but its defining elements are rapidly being identified. Splicing factors include spliceosome particles, members of the serine-arginine (SR) protein family, heterogeneous nuclear ribonucleoproteins (hnRNPs) and auxiliary factors that are typically RNA binding proteins (RBPs).[7-8] Recent studies indicate that the majority of human genes are alternatively spliced, greatly increasing the protein coding potential of the genome[9-10] and genome-wide efforts to identify alternatively spliced genes in stem cells has begun.[11] The protein diversity generated by alternative splicing events is largely unexplored and provides an opportunity to better understand stem cell biology.

ALTERNATIVE SPLICING OF GENES IMPLICATED IN STEMNESS AND DIFFERENTIATION

Evidence of the influence of alternative splicing in stem cells is rapidly growing. Recent studies indicate that levels of some splice variants of stem cell-enriched genes (stemness genes) correlate with particular stages of differentiation. Some of the same studies have demonstrated disparate and reproducible phenotypes correlated with overexpression of splice variants of a stemness gene. These results hint at both the ability of splice variants from a single gene locus to differentially influence the phenotype and the importance of understanding posttranscriptional gene expression regulation afforded by alternative splicing. Here we present several examples of alternatively spliced genes important to stem cell biology.

The *POU5F1* gene is an example of a central stemness gene that is regulated by alternative splicing. This gene encodes a POU domain transcription factor, OCT4, which is a key transcriptional regulator of stem cell pluripotency. OCT4 is highly expressed in stem cells and expression of OCT4 appears to be essential for reprogramming differentiated cells to an induced pluripotent state.[12-14] The utility of OCT4 as a stemness marker was questioned after it was detected in a few somatic cell types[15-16] although finer discrimination

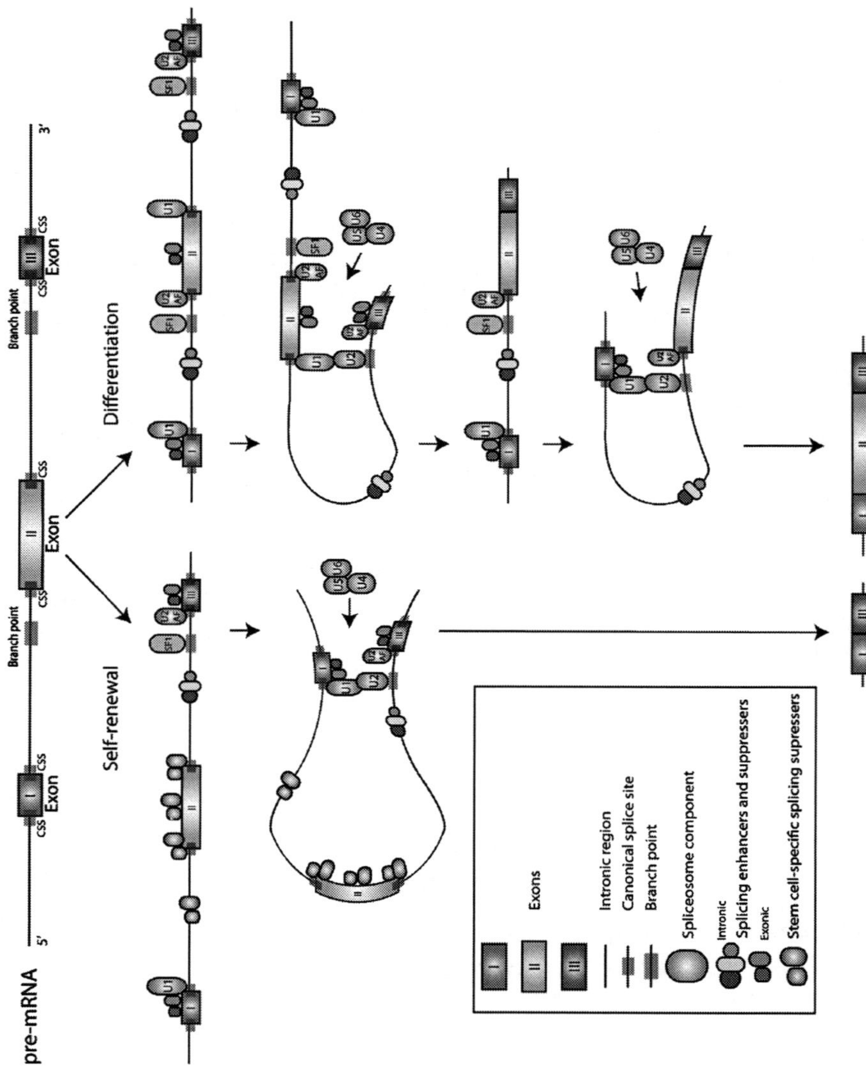

Figure 1. Legend viewed on following page.

Figure 1. Figure viewed on previous page.This diagram outlines the role of splicing factors during RNA splicing in pluripotent stem cells and differentiated cells. On the left, stem cell-specific splicing suppressers block assembly of the spliceosome near exon II and result in excision of exon II along with its flanking introns. On the right, differentiating stem cells that lack the splicing suppressor result in mRNA that includes all three exons. Splicing suppressors prevent splicing by inhibiting assembly of the spliceosome or by other mechanisms. In both cases, the core spliceosome small nuclear RNA proteins are pictured (U1, U2, U4, U5, U6) and their assembly results in catalyzed removal of a pre-mRNA region. The 5' exon end is marked by U1 while the 3' exon end is decorated with U2 auxiliary factors (U2AF). As the spliceosome assembles, the SF1-marked branch point recruits U2 and associates with U1 forming a loop. Next, U4, U5 and U6 are recruited and the intron bound to the 5' exon is cleaved, bound to the branch point and the exons are ligated. The resulting "lariat" (not pictured) and joined exons are released and the spliceosome disassembles.

among *POU5F1* gene products revealed differentially regulated splice variants. One variant called OCT4A is restricted to embryonic stem and embryonal carcinoma cells and can initiate expression at OCT4-initiated promoters.[17] OCT4B, in contrast, does not initiate expression at OCT4 promoters.[18-19] To complicate matters, a third isoform termed OCT4B1 is also highly expressed in embryonic stem and embryonal carcinoma cells, while the OCT4B isoform is expressed at low levels in many differentiated cell types.[17] The roles of these isoforms are apparently distinct but largely unknown and hint at another level of regulation of OCT4's function by alternative splicing.

Although not as well characterized as its family member OCT4, OCT2 is highly expressed in the developing central nervous system and in the adult mouse brain.[20] Its splice variants also seem to influence stem cell differentiation: the OCT2.2 variant is sufficient to induce neural phenotypes when artificially overexpressed in mouse embryonic stem cells while OCT2.4 inhibits induction of neural phenotypes even in the presence of another known inducer of neuronal differentiation.[21] Therefore, a detailed characterization of OCT2's splice variants could provide insight into neural lineage specification.

DNA methyltransferases comprise another group of genes with distinctly different alternative splicing patterns among stem cells and differentiated cells. By methylating DNA, methyltransferases epigenetically influence which genes are transcribed and provide a heritable form of expression regulation. Initial explorations revealed that DNA methyltranferase 3B (*DNMT3B*) is highly alternatively spliced; nearly 40 isoforms have been identified.[22-23] Gopalakrishnan et al recently discovered a *DNMT3B* splice variant missing exon 5 in the NH_2-terminal regulatory domain called DNMT3B3Δ5 that, in contrast to DNMT3B3, is highly expressed in embryonic stem cells and brain tissue and is down-regulated during differentiation, up-regulated in reprogrammed fibroblasts and like DNMT3B3 lacks the catalytic segment involved in methylation.[24] Importantly, DNMT3B3Δ5's increased DNA affinity compared to DNMT3B3 could indicate its role as a blocker of active forms of *DNMT3B* to prevent hypermethylation of DNA. *DNMT3B* is known to interact with a variety of other DNA methyltransferases[25] and splice variants like DNMT3BΔ5 might add a new dimension to these interactions.

The *PKCδ* gene is also highly influenced by alternative splicing and an important regulator of gene expression. *PKCδ* has long been implicated in activation of apoptotic cascades and is known to positively regulate transcription of a host of apoptotic proteins.[26] *PKCδ* is also involved in homeostatic and antiapoptotic pathways.[27-29] This dualism can be better appreciated in terms of *PKCδ*'s splice variants. PKCδI is cleavable by caspase 3, which yields a catalytic fragment known to induce apoptosis.[30] Sakurai et al discovered that another isoform, PKCδII, balances PKCδ1's activity as it is insensitive to cleavage by caspase 3.[31] Neural differentiation correlates with splicing towards the PKCδI isoform, which supports apoptosis-mediated remodeling in the developing nervous system.[32] Thus, the inclusion of intronic base pairs that distinguish PKCδI from PKCδII is sufficient to dramatically alter the regulation of apoptosis in teratocarcinoma cells before and after differentiation induction.

Alternative splicing also mediates production of splice variants with opposing functions in self-renewal. Fibroblast growth factors (FGFs) are known to positively regulate important self-renewal pathways in human embryonic stem cells.[33-34] Mayshar et al discovered that a *FGF4* splice variant referred to as FGF4 is down-regulated after human embryonic stem cell (hESC) differentiation while another splice variant called FGF4si is expressed in both pluripotent and differentiated hESCs.[35] FGF4's self-renewal potential is based upon its ability to phosphorylate ERK1/2 and activate MEK/ERK signaling and introduction of soluble FGF4si seemed to dramatically reduce phosphorylation of ERK1/2. This counteraction among splice variants demonstrates strong modulation of a pluripotency-related pathway by alternative splicing and reveals a regulatory network among splice variants from the same gene.

Adult stem cells' multipotency and self-renewal are also heavily influenced by alternative splicing. Insulin-like growth factor 1 (*IGF-1*) generates splice variants that enhance proliferation and block differentiation of muscle stem cells (IGF-IEc) or induce muscle cell growth via anabolic pathways (IGF-IEa). These splice variants are expressed in a sequential fashion in response to mechanical stress which facilitates muscle growth and repair.[36] Low levels of IGF-IEc were associated with muscle wasting in diseased patient muscle tissue,[37] which hints at the therapeutic potential of this splice variant.

Vascular endothelial growth factor A (*VEGFA*) strongly influences mesenchymal stem cells (MSCs) and is also regulated by alternative splicing. The therapeutic utility of MSCs is hinged upon their unique paracrine signaling[38-39] and splice variants of the mouse VEGF homolog affect paracrine signaling and other phenotypes in MSCs. In particular, Lin et al demonstrated that VEGF120 and VEGF188 induce expression of growth factors and immunosuppressant cytokines while VEGF164 affects expression of genes associated with remodeling and endothelial differentiation. VEGF188 also induces osteogenic phenotypes in MSCs.[40] Prospective tissue therapies rely upon VEGF to increase the regenerative potential of MSCs and only recently has the importance of choosing appropriate VEGF isoforms become apparent.

In addition to cis-acting alternative splicing, protein diversity is also amplified by trans-splicing events. Trans-splicing is the union of pre-mRNA segments from more than one gene to create a novel mRNA transcript. Few trans-spliced gene products associated with stemness have been identified, but trans-spliced mRNA from RNA binding motif protein 14 (*RBM14*) and *RBM4* indirectly affects splicing of important gene products. *RBM14* alone generates splice variants CoAA and CoAM that enhance and inhibit, respectively, cotranscriptional splicing of a variety of genes. During differentiation of embryonic stem cells via embryoid bodies, splicing switches to CoAM which inhibits the action of CoAA and induces expression of the differentiation marker SOX6.[41] When RBM4 and RBM14 are trans-spliced, splice variants and splicing regulators CoAZ and ncCoAZ generate a complex network that affects cotranscriptional splicing of the tau pre-mRNA at exon 10.[42] While the functions of these splice variants is not clear, their distinct expression profiles before and after differentiation hint at their importance to stem cell biology.

These initial studies reveal that alternative splicing regulates genes associated with almost every facet of the stem cell state (Table 1). DNA methylation is influenced by

Table 1. Summary of alternatively spliced gene products with distinct expression profiles before and after stem cell differentiation and/or gene products that mediate important stem cell processes

Gene	Isoforms	Activities	Reference
IGF-1	IGF-1Ec, IGF-1-Ea	IGF-IEc promotes proliferation and inhibits differentiation of muscle progenitors, IGF-IEa activates anabolic pathways	37
POU5F1 (OCT4)	OCT4A, OCT4B, OCT4B1	OCT4A and OCT4B1 expressed in stem cells, OCT4B in differentiated cells	17
RNA binding motif protein 4 (*RBM4*) and RNA binding motif protein 14 (*RMB14*, CoAA)*	CoAZ, ncCoAZ	CoAZ and ncCoAZ influence cotranscriptional splicing	42
DNA Methyltransferase 3 Beta (*DNMT3B*)	DNMT3B3, DNMT3B3Δ5	DNMT3B3Δ5 expressed in ES cells and functionally distinct from DNMT3B3	24
VEGFA	VEGF120, VEGF164, VEGF188	All promote MSC proliferation; some amplify paracrine signaling, osteogenic, or endothelial differentiation	40
FGF4	FGF4, FGF4si	FGF4 is important to stem cell maintenance, while FGFsi antagonizes some of FGF4's activity	35
Protein Kinase C Delta (*PKCδ*)	PKCδI, PKCδII	PKCδI and PKCδII are caspase-cleavable and incleavable, respectively	32
POU2F2 (*OCT2*)	OCT2.2, OCT2.4	OCT2.2 is sufficient to induce neural differentiation in mouse ES cells, OCT2.4 is sufficient to block neural differentiation	21
RNA binding motif protein 14 (*RMB14*, CoAA)	CoAA, CoAM	CoAA is down-regulated in favor of CoAM during early embryonic development	41

* pre-mRNAs from each are trans-spliced into a single mRNA.

highly alternatively spliced DNA methyltransferases. The activity of transcription factors both central and peripheral to stemness is also modulated by alternative splicing. Even the splicing machinery itself is regulated by trans-spliced products of *RBM4* and *RBM14*. Many of these genes' splicing patterns correlate with discrete stages of differentiation and sometimes depend on the terminal lineage of the stem cell. The majority of splice variants from stemness genes remain unidentified, but new genome-wide alternative splicing detection methods will dramatically increase the rate and reduce the cost of detecting splicing variants. Once functionally characterized, these splice variants and the underlying splicing code will reveal a previously underappreciated layer of posttranscriptional gene regulation.

GENOME-WIDE METHODS TO IDENTIFY AND DETECT ALTERNATIVE SPLICING EVENTS

Until recently, detection of alternative splicing relied upon reverse transcriptase PCR of individual mRNA fragments. Throughput was greatly increased when systematic measurement of large segments of the transcriptome was enabled by whole-genome tiling and splicing-sensitive oligonucleotide microarrays.[43-45] By designing nucleotide probes for exon sequences or exon-exon junctions for all known and predicted exons in the genome, genome-wide interrogation of alternative splicing become possible. Computational algorithms are being developed to identify differentially spliced exons from microarray data. Our group has utilized this platform to study neural differentiation of hESCs.[46] Our results revealed that alternative splicing is prevalent in groups of genes such as serine/threonine kinases and helicases. Comparative genome analysis within the intronic regions proximal to alternatively spliced exons identified putative cis-regulatory sequences that may regulate alternative splicing during neural differentiation. This approach provides a framework for comparison among other progenitor cells to identify alternative splicing-mediated pathways towards cell type specificity. For instance, a recent study comparing undifferentiated hESCs and hESC-derived progenitors revealed common and specific splicing events among cardiac and neural progenitors.[47] Further work is required to relate contrasting splicing patterns and splice variant functions to resulting phenotypes.

Unfortunately, microarray-based approaches have distinct shortcomings. Physical limitations on probe density, cross-hybridization caused by probe sequence bias and insensitivity to sparingly expressed transcripts makes detection of some alternative splicing events expensive or impossible. Additionally, arrays cannot detect un-annotated genes. Tag-based profiling methods such as massively parallel signature sequencing (MPSS), serial analysis of gene expression (SAGE), cap analysis of gene expression (CAGE) and polony multiplex analysis of gene expression (PMAGE) have much higher sensitivity and enable the discovery of novel transcripts but are cost-ineffective and time-consuming. These issues are sidestepped by next-generation sequencing technologies that produce hundreds of millions of RNA sequence reads which reveal the transcriptome's content in an inexpensive and quantitative way (Fig. 2). A comparison of mouse embryonic stem cells and embryoid bodies revealed novel alternative splicing events and demonstrated the power of this "shotgun" approach to transcriptome profiling.[48]

REGULATION OF ALTERNATIVE SPLICING BY RNA BINDING PROTEINS

A number of RNA binding proteins (RBPs) are likely to be involved in regulation of alternative splicing in pluripotent stem and differentiated cells. While relatively little is known about the splicing factors important to stem cell maintenance, several have been implicated in neural differentiation. It is thought that the majority of splicing events are regulated by splicing factors that interact directly with regions in pre-mRNA. Identifying the RNA targets of splicing factors and understanding their mechanism of action is necessary to decipher the rules of alternative splicing during stem cell differentiation.

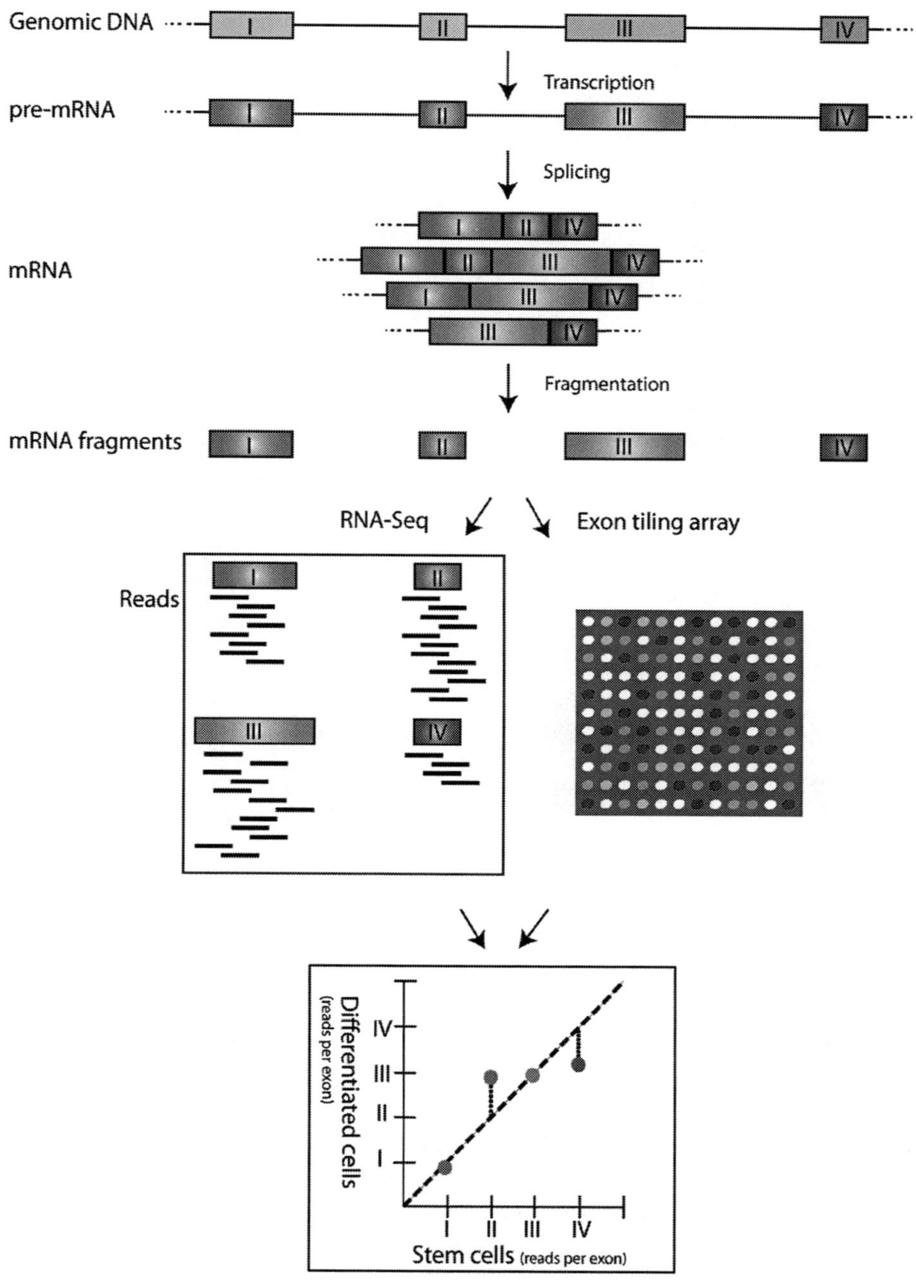

Figure 2. This schematic outlines two alternative splicing detection methods: RNA-seq and splicing-sensitive arrays. In the cell, genomic DNA is transcribed to RNA and processed into various splice variants. These splice variants are digested into short fragments and either sequenced or hybridized to a microarray. In the case of RNA-seq, short reads are aligned to the human genome and computational algorithms parse out which splice variants are expressed. Microarrays rely upon fluorophore-labeled fragments whose relative intensity on the chip reveal ratios of exon representation in mRNA. By comparing splice variant expression among differentiated and stem cells, variants enriched or depleted in stem cells can be identified.

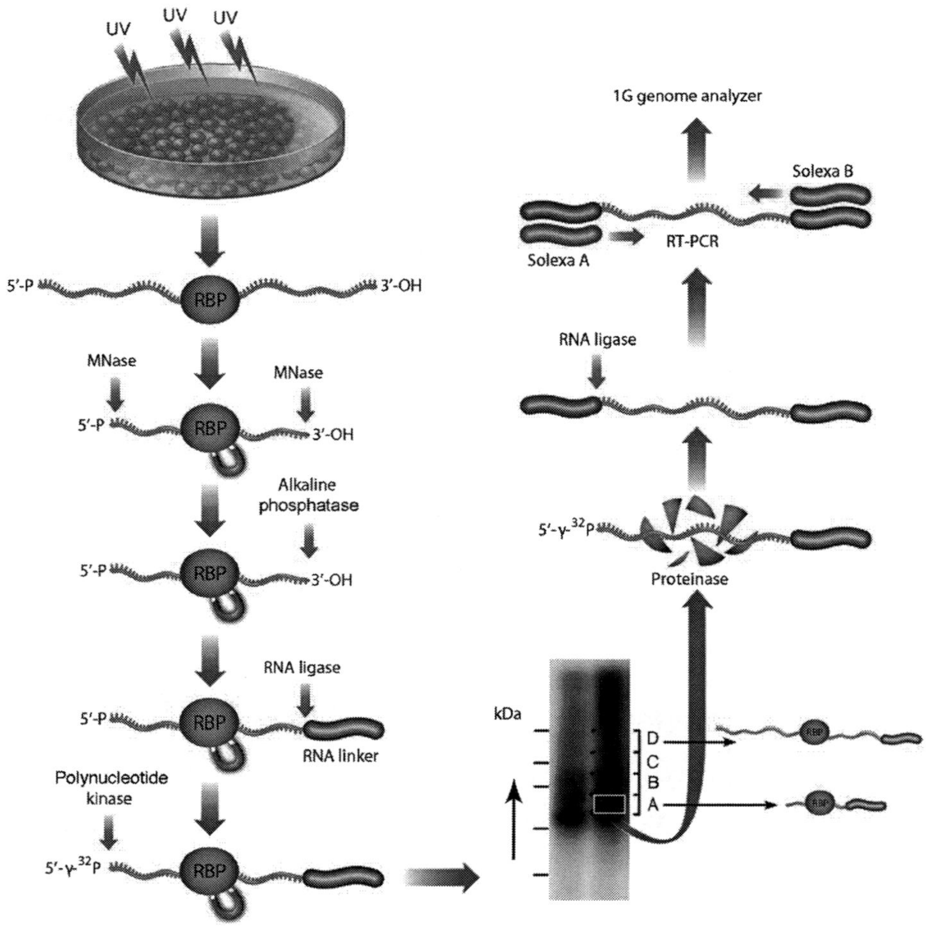

Figure 3. This schematic outlines CLIP-seq approach used with stem cells. RNA complexed with RNA binding proteins (RBPs) from UV-irradiated stem cells is enriched with an anti-RBP antibody. RNA in the complex is trimmed by MNase at two different concentrations, followed by autoradiography as illustrated. Protein-RNA covalent complexes corresponding to bands A and B are recovered following SDS-PAGE. Finally, bound RNA is amplified and then sequenced. Modified from reference 50.

In our study of neural differentiation of hESCs, we identified a conserved 'GCAUG' motif occurring near splice sites involved in differential alternative splicing.[46] This motif corresponds with the FOX1/2 splicing factor binding site, hinting at the important role of FOX splicing factors in splice site selection in hESCs.

To further investigate the role of FOX2 RBP-RNA interactions, we employed UV cross-linking and immunoprecipitation (CLIP). This technique facilitates stabilization of an RNA-protein complex in vivo via UV radiation.[49] We have utilized a modification of this technique to allow extraction of bound RNA for high-throughput sequencing in stem cells (CLIP-seq, Fig. 3). Application of CLIP-seq to isolate the RNA regions that interact with FOX2 in hESCs resulted in the identification of more than three thousand FOX2 bound regions in the human transcriptome. This RNA map of FOX2 binding sites identifies alternative splicing events regulated by FOX2.[50] FOX2 is evolutionarily conserved in mammals and is highly expressed in hESCs. Knockdown of FOX2 generated a rapid cell death phenotype in hESCs but not in neural stem cells or other cell lines. These results hint at FOX2 as an important alternative splice site selector in hESCs and further characterization may reveal its interactions with other components of the splicing machinery.

The polypyrimidine tract-binding protein (PTB) is another splicing factor that is widely expressed in the early embryo. PTB regulates alternative exon inclusion in many genes and has been implicated in aspects of mRNA regulation.[51] Several studies have shown that knockdown of the PTB protein is sufficient to trigger neuronal-specific alternative splicing in nonneuronal cells.[52-54] To probe the role of PTB in embryonic stem cells, Shibayama et al created homozygous PTB null mouse embryonic stem cells. These embryonic stem cells were viable but did not proliferate normally.[55]

SAM68 is a nucleus-localized RBP that is linked to splicing,[56] is widely expressed in multiple cell types and its overexpression inhibits neural stem cell proliferation.[57] RNAi experiments in combination with microarray analysis of altered splicing revealed exons regulated by this splicing factor in mouse neuroblastoma cells.[58] Chawla et al also demonstrated that knockdown of SAM68 prevents differentiation of mouse embryonal carcinoma cells in the presence of retinoic acid, that SAM68 is up-regulated during neural differentiation and that it affects splicing and/or regulation of genes important to neural phenotype. These signs hint at SAM68 as a powerful regulator of splicing during neural differentiation.

As more RBPs and their motifs are identified with genome-wide techniques, the rules of the RNA splicing code will become clearer. This will facilitate predictive models of RNA splicing and could afford a new level of control over stem cell fate.

CONCLUSION AND PERSPECTIVES

Until recently, studies of stem cell transcriptional regulation have focused on the mammalian genome's many transcription factors. These efforts have revealed powerful stemness-associated transcription factors that can be useful markers for stemness as well as tools for reprogramming cells to a pluripotent state. But as most cell processes are also heavily influenced by posttranscriptional gene expression regulation, refinement of our understanding of stem cell state will rely upon understanding the rules and results of alternative splicing.

The splicing field is currently in a cataloging phase to identify important splice variants and splicing factors that influence splice variant abundance. Shotgun approaches to splice variant detection allow rapid identification of more stem cell-enriched splice variants while techniques such as CLIP allow identification of splicing factors and the functional RNA elements they bind. Initial results demonstrate that differentiation correlates strongly with splicing towards particular splice variants and overexpression of some splice variants is sufficient to dramatically alter stem cell phenotype. As more lineage-specific splice variants are detected, splice variant measurements may provide a highly accurate and sensitive way to determine stem cell state and reveal new avenues for guiding stem cell lineage specification.

ACKNOWLEDGEMENTS

The authors thank Jonathan Scolnick and Stephanie Huelga for critical reading of the manuscript and Jason Nathanson for assistance with the figures. This manuscript was supported by grants to G.W.Y. from the US National Institutes of Health (HG004659 and GM084317) and the California Institute of Regenerative Medicine (RB1-01413).

REFERENCES

1. Brandenberger R, Wei H, Zhang S et al. Transcriptome characterization elucidates signaling networks that control human ES cell growth and differentiation. Nat Biotechnol 2004; 22(6):707-716.
2. Brandenberger R, Khrebtukova I, Thies RS et al. MPSS profiling of human embryonic stem cells. BMC Dev Biol 2004; 4:10.
3. Miura T, Luo Y, Khrebtukova I et al. Monitoring early differentiation events in human embryonic stem cells by massively parallel signature sequencing and expressed sequence tag scan. Stem Cells Dev 2004; 13(6):694-715.
4. Bhattacharya B, Cai J, Luo Y et al. Comparison of the gene expression profile of undifferentiated human embryonic stem cell lines and differentiating embryoid bodies. BMC Dev Biol 2005; 5:22.
5. Jurica MS, Moore MJ. Pre-mRNA splicing: awash in a sea of proteins. Mol Cell 2003; 12(1):5-14.
6. Nilsen TW. The spliceosome: the most complex macromolecular machine in the cell? Bioessays 2003; 25(12):1147-1149.
7. Matlin AJ, Clark F, Smith CW. Understanding alternative splicing: towards a cellular code. Nat Rev Mol Cell Biol 2005; 6(5):386-398.
8. Martinez-Contreras R, Cloutier P, Shkreta L et al. hnRNP proteins and splicing control. Adv Exp Med Biol 2007; 623:123-147.
9. Blencowe BJ. Alternative splicing: new insights from global analyses. Cell 2006; 126(1):37-47.
10. Wang ET, Sandberg R, Luo S et al. Alternative isoform regulation in human tissue transcriptomes. Nature 2008; 456(7221):470-476.
11. Pritsker M, Doniger TT, Kramer LC et al. Diversification of stem cell molecular repertoire by alternative splicing. Proc Natl Acad Sci USA 2005; 102(40):14290-14295.
12. Takahashi K, Tanabe K, Ohnuki M et al. Induction of pluripotent stem cells from adult human fibroblasts by defined factors. Cell 2007; 131(5):861-872.
13. Yu J, Vodyanik MA, Smuga-Otto K et al. Induced pluripotent stem cell lines derived from human somatic cells. Science 2007; 318(5858):1917-1920.
14. Lowry WE, Richter L, Yachechko R et al. Generation of human induced pluripotent stem cells from dermal fibroblasts. Proc Natl Acad Sci USA 2008; 105(8):2883-2888.
15. Tai MH, Chang CC, Kiupel M et al. Oct4 expression in adult human stem cells: evidence in support of the stem cell theory of carcinogenesis. Carcinogenesis 2005; 26(2):495-502.
16. Zangrossi S, Marabese M, Broggini M et al. Oct-4 expression in adult human differentiated cells challenges its role as a pure stem cell marker. Stem Cells 2007; 25(7):1675-1680.
17. Atlasi Y, Mowla SJ, Ziaee SA et al. OCT4 spliced variants are differentially expressed in human pluripotent and nonpluripotent cells. Stem Cells 2008; 26(12):3068-3074.
18. Cauffman G, Liebaers I, Van Steirteghem A et al. POU5F1 isoforms show different expression patterns in human embryonic stem cells and preimplantation embryos. Stem Cells 2006; 24(12):2685-2691.
19. Lee J, Kim HK, Rho JY et al. The human OCT-4 isoforms differ in their ability to confer self-renewal. J Biol Chem 2006; 281(44):33554-33565.
20. He X, Treacy MN, Simmons DM et al. Expression of a large family of POU-domain regulatory genes in mammalian brain development. Nature 1989; 340(6228):35-41.
21. Theodorou E, Dalembert G, Heffelfinger C et al. A high throughput embryonic stem cell screen identifies Oct-2 as a bifunctional regulator of neuronal differentiation. Genes Dev 2009; 23(5):575-588.
22. Wang L, Wang J, Sun S et al. A novel DNMT3B subfamily, DeltaDNMT3B, is the predominant form of DNMT3B in nonsmall cell lung cancer. Int J Oncol 2006; 29(1):201-207.
23. Ostler KR, Davis EM, Payne SL et al. Cancer cells express aberrant DNMT3B transcripts encoding truncated proteins. Oncogene 2007; 26(38):5553-5563.
24. Gopalakrishnan S, Van Emburgh BO, Shan J et al. A novel DNMT3B splice variant expressed in tumor and pluripotent cells modulates genomic DNA methylation patterns and displays altered DNA binding. Mol Cancer Res 2009; 7(10):1622-1634.

25. Margot JB, Ehrenhofer-Murray AE, Leonhardt H. Interactions within the mammalian DNA methyltransferase family. BMC Mol Biol 2003; 4:7.
26. Brodie C, Blumberg PM. Regulation of cell apoptosis by protein kinase c delta. Apoptosis 2003; 8(1):19-27.
27. Peluso JJ, Pappalardo A, Fernandez G. Basic fibroblast growth factor maintains calcium homeostasis and granulosa cell viability by stimulating calcium efflux via a PKC delta-dependent pathway. Endocrinology 2001; 142(10):4203-4211.
28. Zrachia A, Dobroslav M, Blass M et al. Infection of glioma cells with Sindbis virus induces selective activation and tyrosine phosphorylation of protein kinase C delta. Implications for Sindbis virus-induced apoptosis. J Biol Chem 2002; 277(26):23693-23701.
29. Kilpatrick LE, Lee JY, Haines KM et al. A role for PKC-delta and PI 3-kinase in TNF-alpha-mediated antiapoptotic signaling in the human neutrophil. Am J Physiol Cell Physiol 2002; 283(1):C48-57.
30. Sitailo LA, Tibudan SS, Denning MF. Bax activation and induction of apoptosis in human keratinocytes by the protein kinase C delta catalytic domain. J Invest Dermatol 2004; 123(3):434-443.
31. Sakurai Y, Onishi Y, Tanimoto Y, Kizaki H. Novel protein kinase C delta isoform insensitive to caspase-3. Biol Pharm Bull. Sep 2001;24(9):973-977.
32. Patel NA, Song SS, Cooper DR. PKCdelta alternatively spliced isoforms modulate cellular apoptosis in retinoic acid-induced differentiation of human NT2 cells and mouse embryonic stem cells. Gene Expr 2006; 13(2):73-84.
33. Dvorak P, Dvorakova D, Koskova S et al. Expression and potential role of fibroblast growth factor 2 and its receptors in human embryonic stem cells. Stem Cells 2005; 23(8):1200-1211.
34. Li J, Wang G, Wang C et al. MEK/ERK signaling contributes to the maintenance of human embryonic stem cell self-renewal. Differentiation 2007; 75(4):299-307.
35. Mayshar Y, Rom E, Chumakov I et al. Fibroblast growth factor 4 and its novel splice isoform have opposing effects on the maintenance of human embryonic stem cell self-renewal. Stem Cells 2008; 26(3):767-774.
36. Hill M, Goldspink G. Expression and splicing of the insulin-like growth factor gene in rodent muscle is associated with muscle satellite (stem) cell activation following local tissue damage. J Physiol 2003; 549(Pt 2):409-418.
37. Ates K, Yang SY, Orrell RW et al. The IGF-I splice variant MGF increases progenitor cells in ALS, dystrophic and normal muscle. FEBS Lett 2007; 581(14):2727-2732.
38. Gnecchi M, He H, Liang OD et al. Paracrine action accounts for marked protection of ischemic heart by Akt-modified mesenchymal stem cells. Nat Med 2005; 11(4):367-368.
39. Tang YL, Zhao Q, Qin X et al. Paracrine action enhances the effects of autologous mesenchymal stem cell transplantation on vascular regeneration in rat model of myocardial infarction. Ann Thorac Surg 2005; 80(1):229-236; discussion 236-227.
40. Lin H, Shabbir A, Molnar M et al. Adenoviral expression of vascular endothelial growth factor splice variants differentially regulate bone marrow-derived mesenchymal stem cells. J Cell Physiol 2008; 216(2):458-468.
41. Yang Z, Sui Y, Xiong S et al. Switched alternative splicing of oncogene CoAA during embryonal carcinoma stem cell differentiation. Nucleic Acids Res 2007; 35(6):1919-1932.
42. Brooks YS, Wang G, Yang Z et al. Functional pre-mRNA trans-splicing of coactivator CoAA and corepressor RBM4 during stem/progenitor cell differentiation. J Biol Chem 2009; 284(27):18033-18046.
43. Yeakley JM, Fan JB, Doucet D et al. Profiling alternative splicing on fiber-optic arrays. Nat Biotechnol 2002; 20(4):353-358.
44. Johnson JM, Castle J, Garrett-Engele P et al. Genome-wide survey of human alternative premRNA splicing with exon junction microarrays. Science 2003; 302(5653):2141-2144.
45. Pan Q, Shai O, Misquitta C et al. Revealing global regulatory features of mammalian alternative splicing using a quantitative microarray platform. Mol Cell 2004; 16(6):929-941.
46. Yeo GW, Xu X, Liang TY et al. Alternative splicing events identified in human embryonic stem cells and neural progenitors. PLoS Comput Biol 2007; 3(10):1951-1967.
47. Salomonis N, Nelson B, Vranizan K et al. Alternative splicing in the differentiation of human embryonic stem cells into cardiac precursors. PLoS Comput Biol 2009; 5(11):e1000553.
48. Cloonan N, Forrest AR, Kolle G et al. Stem cell transcriptome profiling via massive-scale mRNA sequencing. Nat Methods 2008; 5(7):613-619.
49. Ule J, Jensen KB, Ruggiu M et al. CLIP identifies Nova-regulated RNA networks in the brain. Science 2003; 302(5648):1212-1215.
50. Yeo GW, Coufal NG, Liang TY et al. An RNA code for the FOX2 splicing regulator revealed by mapping RNA-protein interactions in stem cells. Nat Struct Mol Biol 2009; 16(2):130-137.

51. Ashiya M, Grabowski PJ. A neuron-specific splicing switch mediated by an array of premRNA repressor sites: evidence of a regulatory role for the polypyrimidine tract binding protein and a brain-specific PTB counterpart. RNA 1997; 3(9):996-1015.
52. Boutz PL, Stoilov P, Li Q et al. A posttranscriptional regulatory switch in polypyrimidine tract-binding proteins reprograms alternative splicing in developing neurons. Genes Dev 2007; 21(13):1636-1652.
53. Makeyev EV, Zhang J, Carrasco MA et al. The MicroRNA miR-124 promotes neuronal differentiation by triggering brain-specific alternative premRNA splicing. Mol Cell 2007; 27(3):435-448.
54. Spellman R, Llorian M, Smith CW. Crossregulation and functional redundancy between the splicing regulator PTB and its paralogs nPTB and ROD1. Mol Cell 2007; 27(3):420-434.
55. Shibayama M, Ohno S, Osaka T et al. Polypyrimidine tract-binding protein is essential for early mouse development and embryonic stem cell proliferation. FEBS J 2009; 276(22):6658-6668.
56. Matter N, Herrlich P, Konig H. Signal-dependent regulation of splicing via phosphorylation of Sam68. Nature 2002; 420(6916):691-695.
57. Moritz S, Lehmann S, Faissner A et al. An induction gene trap screen in neural stem cells reveals an instructive function of the niche and identifies the splicing regulator sam68 as a tenascin-C-regulated target gene. Stem Cells 2008; 26(9):2321-2331.
58. Chawla G, Lin CH, Han A et al. Sam68 regulates a set of alternatively spliced exons during neurogenesis. Mol Cell Biol 2009; 29(1):201-213.

CHAPTER 8

MicroRNA REGULATION OF EMBRYONIC STEM CELL SELF-RENEWAL AND DIFFERENTIATION

Collin Melton and Robert Blelloch*

Abstract: Stem cell differentiation requires a complex coordination of events to transition from a self-renewing to a differentiated cell fate. Stem cells can be pluripotent (capable of giving rise to all embryonic lineages), multipotent (possessing the potential to give rise to multiple lineages) and unipotent (capable of given rise to a single cell lineage). Regardless of their potency all stem cells must silence their self-renewal program during differentiation. The self-renewal program can be defined as the integration of external and internal stimuli that enables a cell to proliferate while maintaining its potency. Two hallmarks of the self-renewal program are a self-reinforcing transcriptional network and a specialized cell-cycle profile. In this chapter we discuss the impact of various microRNAs (miRNAs) to either reinforce or inhibit the self-renewal program of stem cells and how this added regulatory layer provides robustness to cell-fate decisions. We will focus on embryonic stem cells (ESCs) describing miRNA function in self-renewal, differentiation and de-differentiation. We will compare and contrast miRNA functions in ESCs with miRNA function in lineage specific somatic stem cells and in cancer.

INTRODUCTION: THE SELF-RENEWAL PROGRAM

The stem cell self-renewal program in both embryonic and somatic stem cell populations functions to maintain potency during successive rounds of replication. The degree of potency and proliferative rate vary greatly among stem cell populations in accordance with the evolutionary pressures and biological functions of these populations.

*Corresponding Author: Robert Blelloch—The Eli and Edythe Broad Center of Regeneration Medicine and Stem Cell Research, Center for Reproductive Sciences, Biomedical Science Graduate Program and Department of Urology, University of California San Francisco, San Francisco, California 94143, USA. Email: blellochr@stemcell.ucsf.edu

The Cell Biology of Stem Cells, edited by Eran Meshorer and Kathrin Plath.
©2010 Landes Bioscience and Springer Science+Business Media.

ESCs are derived from the inner cell mass of the developing blastocyst and resemble cells of the developing epiblast. The epiblast gives rise to the embryonic endoderm, mesoderm and ectoderm, as well as the germ lineage and hence is pluripotent.[1] Epiblast cells have a rapid cell cycle. However, they eventually differentiate at which time the cell cycle extends. Like the epiblast cells, ESCs have a rapid cell cycle and are pluripotent. However, unlike epiblast cells, ESCs can self-renew indefinitely in the culture dish.

During embryonic development, the epiblast cells differentiate into specialized fetal stem cell populations that have a more limited potency. These include, among others, the fetal neural stem cells and hematopoietic stem cells. These fetal stem cells retain a high proliferative rate but possess a limited potency.[2,3] Eventually, the fetal stem cells are replaced by adult lineage specific stem cells including adult counterparts of the fetal hematopoietic and neural stem cells. The adult stem cells also have a limited potency, but unlike their fetal counterparts, typically have a slow proliferative rate. In fact, adult somatic stem cell populations are largely quiescent, although they generate transient populations of progenitor cells, which typically have a rapid proliferative rate more like that of their fetal stem cell counterparts. Quiescence in adult stem cells may have evolved to reduce the chance of harmful mutations, such as those that cause cancer.[4]

EMBRYONIC STEM CELLS

The molecular basis of the stem cell self-renewal program has been best studied in ESCs. In these cells the self-renewal program is determined by the interaction of numerous factors at the center of which is a distinct transcriptional network.[5] In ESCs, the central transcriptional network includes the transcription factors Oct4, Sox2, Nanog, Tcf3 and the Myc family of proteins (cMyc and nMyc). The coordinated actions of these transcription factors both directly and indirectly determines an epigenetic state that is poised to activate or repress upon differentiation the transcription of genes of any lineage of the three germ layers.[5] In this way the ESC transcriptional network enables its pluripotency. Additionally the ESC transcriptional network drives expression of factors that enable the cell's high proliferative rate by directly and indirectly maintaining the short ESC cell cycle.

With the induction of ESC differentiation, the many components of the self-renewal program must be shut off and a new differentiated program must be activated. Therefore, this cell fate transition is regulated by factors that both silence self-renewal and induce a lineage specific differentiation program. These factors can be classified broadly as those that influence gene expression at the level of chromatin state, transcription, transcript stability, protein translation, protein stability, or protein function.

In this chapter, we will focus on the pro-self-renewal and pro-differentiation functions of miRNAs.

miRNA BIOGENESIS AND FUNCTION

miRNAs are small noncoding RNAs which act to posttranscriptionally silence gene expression through translational inhibition and mRNA destabilization. miRNAs

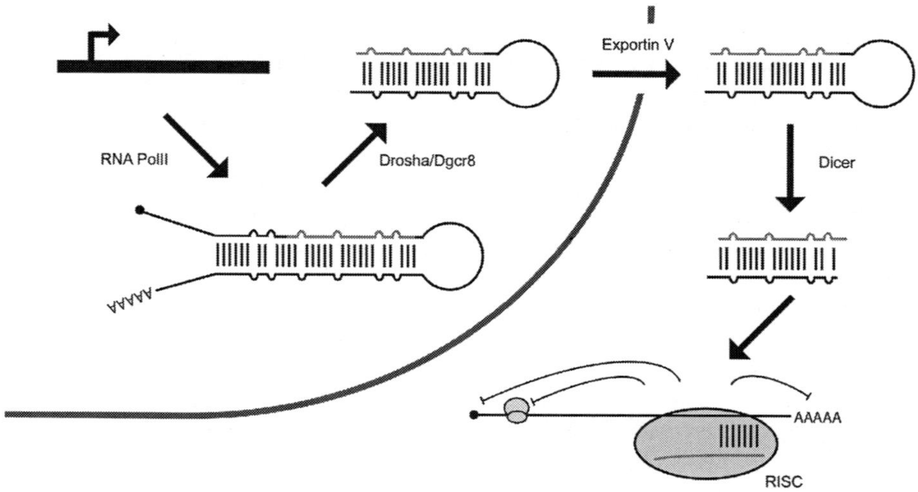

Figure 1. miRNA biogenesis. miRNAs are first transcribed as long RNA polymerase II transcripts. The hairpin structure of these transcripts is recognized by the Microprocessor complex composed of Drosha and Dgcr8 and is cleaved to form a smaller pre-miRNA hairpin. The pre-miRNA is exported from the nucleus and subsequently cleaved by Dicer to form a mature miRNA duplex. A single strand of this duplex is loaded into the RISC complex. The miRNA loaded complex destabilizes and inhibits translation of its target mRNAs.

are generated through the sequential processing of RNA transcripts (Fig. 1). miRNAs are first transcribed as long RNA polymerase II transcripts termed primary miRNAs (pri-miRNAs).[6,7] These pri-miRNAs can be either noncoding or coding. In the latter case, miRNAs will often reside within the intron of a coding gene.[8] In the nucleus, the pri-miRNA is recognized and cleaved by the microprocessor complex, which consists of the RNA binding protein DGCR8 and the RNAse III enzyme DROSHA.[9-13] This complex recognizes a stem loop structure of approximately 33 base pairs in length and posses an enzymatic activity that cleaves the loop 11 base pairs from its base leaving a characteristic 2 nucleotide 3′ overhang.[14] The processed RNA, now termed pre-miRNA, is exported from the nucleus to the cytoplasm by Exportin V where it is recognized by a second complex containing the RNAse III enzyme DICER.[15-18] This complex recognizes the pre-miRNA hairpin and cleaves it at the base of the hairpin loop again to form a 2 nucleotide 3′ overhang to generate an approximately 22 nucleotide mature miRNA duplex.[18] This mature duplex remains double-stranded until it is incorporated into the RNA-induced silencing complex (RISC). Only a single strand of the small RNA duplex is incorporated, typically the strand with the less stable 5′ end.[19]

miRNAs which are loaded into the RISC complex directly interact with their mRNA targets through base pairing to sites in the open reading frame and 3′ untranslated region. These interactions depend on base pairing of a 6-8 nucleotide seed sequence of the miRNA (nucleotides 2-8 on the 5′ end) with the mRNA target.[20] The RISC complex which is bound to target mRNAs disrupts protein production through a variety of mechanisms including disruption of ribosome initiation via interacting with the 5′ cap, prevention of ribosome elongation and promotion of RNA degradation by shortening of the polyA tail.[21]

ESCC miRNAs PROMOTE SELF-RENEWAL

Many miRNAs are co-expressed from a single transcript. One such group is the miR-290 cluster, which consists of 7 miRNAs and is highly expressed in mouse ESCs. A subgroup of the miR-290 cluster miRNAs share a common seed sequence and regulate the ESC cell cycle and, therefore, have been coined the ESCC family (ESC cell cycle promoting miRNAs).[22] Related families to the ESCC miRNAs include the miR-302 family and the mir-17/20/106 family, although the later family has a slightly different seed sequence. The ortholog of the miR-290 cluster in humans is the slightly diverged miR-370 cluster while the miR-302 clusters in mouse and human are very similar.[23,24]

The common expression of similar miRNAs in pluripotent stem cells in mouse and human suggests an important functional role in ESC self-renewal. Indeed, the first evidence for such a function was uncovered in ESC miRNA knockout models through deletion of either *Dicer* or *Dgcr8*.[25-27] These ESCs have a slowed proliferation rate and an altered cell cycle profile with an extended G1 phase.[27] These findings are particularly interesting considering that wild-type mouse ESCs are characterized by an atypical cell cycle with a abbreviated G1 phase compared to somatic cells.[28] That is, these initial findings suggested that the ESC expressed miRNAs suppress the somatic cell cycle structure.

The abbreviated G1 phase of ESCs promotes their rapid proliferation and is, at least in part, secondary to an alleviation of the G1/S restriction point.[28] In a typical somatic cell, the G1/S restriction point prevents the initiation of S phase and DNA replication. The G1/S restriction point includes a complex series of signaling events, which must reach a threshold before transitioning into S phase. Key molecular components of this reaction include, but are not limited to, the cyclins, the cyclin dependent kinases (CDKs), cdk inhibitors (CKIs), the Rb family of proteins and the E2F family of proteins.[29]

D and E type cyclins in complex with CDKs drive phosphorylation of the Rb family of proteins.[30] In mouse ESCs, CyclinE is expressed at high levels independent of cell cycle phase whereas CyclinD is not expressed.[31] CyclinE complexes with CDK2, to initiate the phosphorylation and subsequenct inactivation of the Rb family of proteins (pRb, P107 and P130). The Rb family of proteins, when in a hypophosphorylated active state, sequester activating E2Fs (E2F1-3) as well as activate repressive E2F proteins (E2F4 and 5) preventing transcription of S phase genes.[30] When Rb proteins are hyperphosphorylated and inactivated, they no longer activate the repressive E2Fs. Simultaneously, the suppression of the activating E2Fs is relieved, which allows them to drive transcription of S phase genes. Progression to S phase can be blocked by CDK inhibitors, which include members of the CIP and INK families. These inhibitors block activity of CDK/Cyclin complexes.[32] INK family inhibitors are nonfunctional in mouse ESCs as they act through CyclinD, which is not expressed at high levels. CIP family inhibitors, however, are more promiscuous in their inhibitory effects on CDK/Cyclin complexes and are able to bind and inactivate CDK2/CyclinE complexes.[32] In mouse ESCs, CIP family inhibitors are expressed at low levels, as are the Rb proteins.[28,31]

By screening miRNAs, which enhance proliferation in a *Dgcr8* knockout (–/–) ESC background, the role of ESCC miRNAs in cell cycle control was uncovered. These miRNAs not only accelerate proliferation of *Dgcr8* $^{-/-}$ ESCs, but also decrease the number of cells in the G1 phase of the cell cycle. This effect on the G1 phase is in part through direct miRNA targeting of the CIP family CDK inhibitor P21, LATS2 and some of the Rb family of proteins including pRb and P130. Through inhibition of these and other

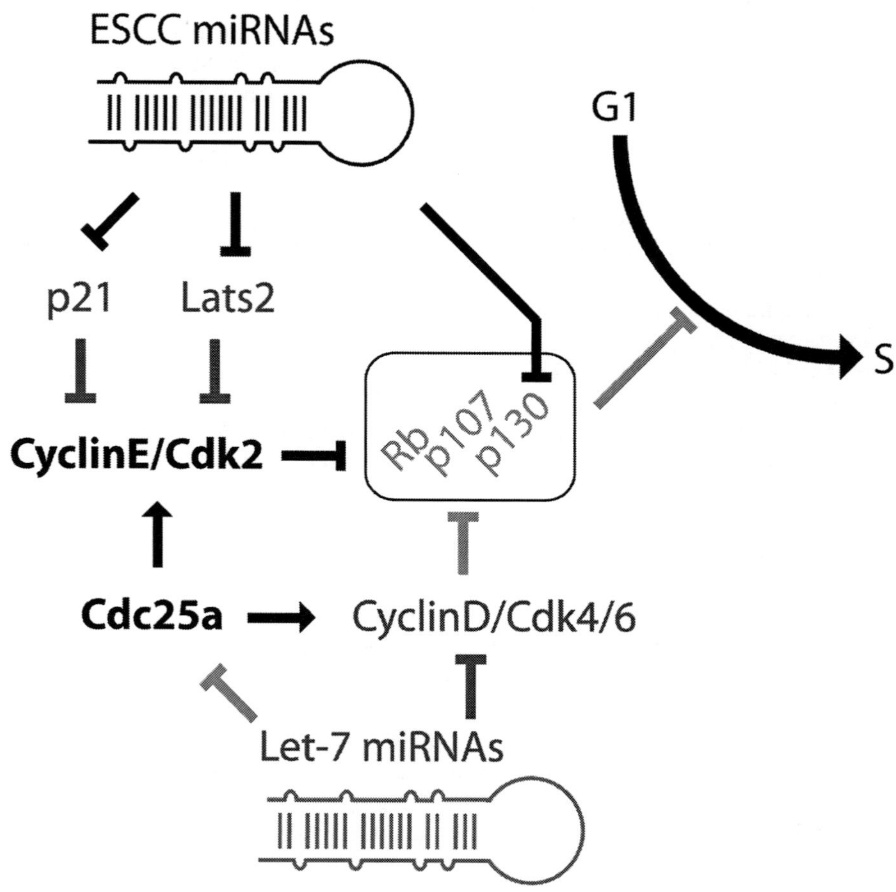

Figure 2. Let-7 and ESCC miRNAs have opposing effects on the G1-S transition. This figure represents a model of the direct inhibitory effects of the ESCC and let-7 miRNAs on factors involved in the ESC G1-S transition. As ESCs transition from a self-renewing to a differentiated state, the ESCC miRNAs are down-regulated and the let-7 miRNAs are upregulated. These changes have direct consequences on the cell cycle. Dark/bold arrows, lines and text indicate interactions, miRNAs and proteins that are up-regulated in the ESC state. Grey arrows, lines and text indicate interactions, miRNAs and proteins that are down-regulated in the ESC state. Note the interactions and functional consequences of the let-7 miRNAs on cell cycle have been tested in various somatic cell populations, but not ESCs.

predicted miRNA targets involved in the G1 phase, the ESCC miRNAs promote the ESC cell cycle (Fig. 2).[22]

Recently the impact of the ESCC miRNAs on the ESC transcriptome was analyzed in depth. It was discovered that the ESCC miRNAs indirectly activate cMyc expression.[33] Myc is a transcription factor that both promotes proliferation and is required for ESC self-renewal.[34,35] Additionally, in ESCs inhibition of Myc proteins promotes loss of ESC self-renewal, while enforced expression of cMyc prevents loss of self-renewal in the absence of LIF.[35] Lin et al recently sought to identify the mechanisms by which Myc proteins promote ESC self-renewal. In particular they found that cMyc drives transcription of numerous pro-self-renewal miRNAs including miR-141, miR-200 and miR-429. These miRNAs promote the maintenance of self-renewal in the absence of LIF although the biological mechanisms underlying these effects remain unknown.[36] Furthermore, cMyc regulates expression of the ESCC miRNAs forming a positive feedback loop as described below.

A number of other factors have been identified as indirectly upregulated by the ESCC miRNAs including the DNA methyl transferases (DNMT3a and b).[37,38] The increase in expression of these DNA methyl transferases is required to maintain appropriate DNA methylation in sub-telomeric regions, which in turn is required to prevent abnormal telomere elongation.[37] The regulation of DNMT3a and b occurs via ESCC targeting of P130—a negative regulator of DNMT3a and b transcription.[37,38] In addition to the DNA methyl transferases, a number of other pluripotency associated transcripts are indirectly upregulated by the ESCC miRNAs. These include Lin28, Trim71 and Sall4.[33] Together these numerous molecular changes induced by the ESCC miRNAs have a profound effect on promoting the cell cycle and preserving faithful maintenance of telomeres to ensure proper ESC self-renewal and maintenance of pluripotency.

miRNAs INDUCED DURING ESC DIFFERENTIATION SUPPRESS THE SELF-RENEWAL PROGRAM

As miRNAs are suited to stabilizing the self-renewing state, so are they well situated to promote the transition from self-renewal to differentiation. MicroRNAs, which silence self-renewal, can be categorized by their targets and by their expression patterns. A small number of miRNAs have been found to directly target components of the central ESC transcriptional network.[39-41] These same miRNAs are induced rapidly during ESC differentiation down specific lineages. A second class of microRNAs is induced during differentiation down a broad set of lineages and broadly suppress ESC associated genes but not the central ESC transcription factors themselves.[33] They also promote a somatic cell cycle.[42-44] These two classes of pro-differentiation miRNAs likely play distinct roles in the differentiation process. The first class of microRNAs directly suppress ESC self-renewal state, while the second class of microRNAs predominantly stabilize the differentiated state—much like the ESC microRNAs stabilize the ESC state.

MiRNAs miR-134, miR-296 and miR-470 have been discovered to directly suppress Nanog, Pou5f1 (also known as Oct4) and Sox2 in mouse ESCs.[40,41] These miRNA-target interactions occur predominantly through interactions in the open reading frame. These miRNAs are highly upregulated during retinoic acid (RA) induced differentiation, which induces predominantly neural differentiation suggesting that these miRNAs may be involved in lineage specific silencing of ESC self-renewal. In human ESCs, miR-145 was found to directly suppress ESC self-renewal via targeting Oct4, Sox2 and Klf4.[39] Understanding the biological functions and relative in vivo contributions of various direct miRNA suppressors of self-renewal will be an important area of future pursuit.

In contrast to the miRNAs which directly suppress ESC self-renewal, the let-7 family of miRNAs are stabilizers of the differentiated cell fate.[33] Mutations in let-7 were first discovered in C. elegans in a mutagenesis screen for genes that prevented terminal differentiation of seam cells in the hypodermis.[45] Since the discovery of let-7 in C. elegans, homologues of let-7 have been found in all metazoans studied.[46] In mouse and human there are 9 distinct let-7 family members with varied tissues specific expression

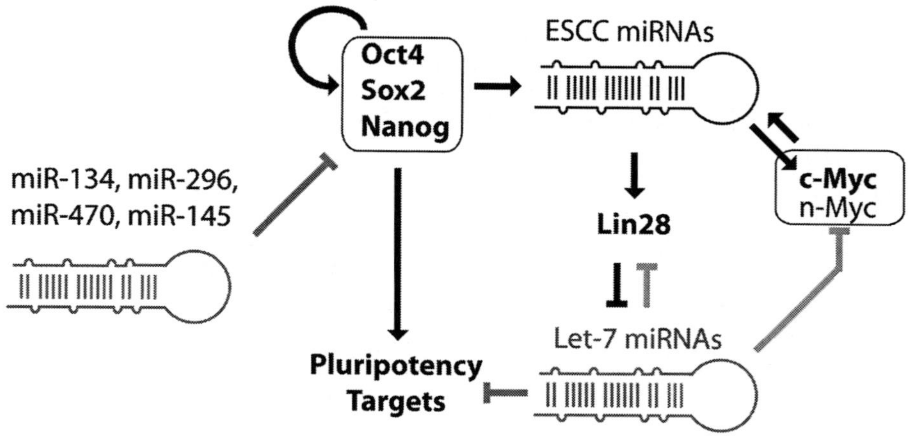

Figure 3. miRNA interactions in the ESC self-renewal network. This figure represents a model of the direct inhibitory and indirect activating effects of the ESCC, let-7 and miR-134, miR-296, miR-470 and miR-145 miRNAs. Dark/bold arrows, lines and text indicate interactions, miRNAs and proteins that are up-regulated in the ESC state. Grey arrows, lines and text indicate interactions, miRNAs and proteins that are down-regulated in the ESC state. As ESCs differentiate, the miR-134, miR-296, miR-470 and miR-145 miRNAs destabilize the Oct4/Sox2/Nanog transcriptional network to promote differentiation, whereas the let-7 miRNAs inhibit Myc and downstream targets of the Oct4/Sox2/Nanog network to stabilize the differentiated state.text.

patterns.[47-50] In ESCs an elegant mechanism exists which allows for the post-transcriptional silencing of let-7 transcripts. A complex of the RNA binding protein, Lin28 and the terminal uridyl-transferase, TUT4, binds to and induces the degradation of pre-let-7 transcripts.[51-56] Lin28 expression is quickly lost during ESC differentiation,[57,58] which allows for the rapid increase in let-7 expression.[55]

Recently, it was discovered that let-7 family members could induce silencing of self-renewal in the miRNA deficient *Dgcr8* $^{-/-}$ ESCs but not in wild-type ESCs.[33] This observation suggested that miRNAs expressed in ESCs normally prevent let-7 from silencing ESC self-renewal. Indeed, the ESCC miRNAs that predominate in ESCs, are able to prevent loss of self-renewal induced by the let-7 miRNAs. Let-7 preferentially targets transcripts that are enriched in ESCs, including many transcripts that are regulated by the pluripotency transcription factors Oct4, Sox2, Nanog and Tcf3. Additionally, a number of direct targets of let-7 are indirectly upregulated by the ESCC miRNAs, which can explain how the ESCCs antagonize let-7. Among the targets with opposing regulation by let-7 and the ESCCs are the Myc proteins, Sall4, Lin28 and Trim71.[33]

The antagonism observed between the ESCC and let-7 miRNAs and the targets which are regulated in opposing fashion by these miRNAs, suggest a network in which ESCC miRNAs and let-7 miRNAs have mutually exclusive expression and function (Fig. 3). In ESCs, the ESCC miRNAs lead to upregulation of Lin28, which directly suppresses let-7 maturation. Additionally, ESCCs indirectly upregulate cMyc and other direct let-7 targets that promote ESC self-renewal. By these mechanisms ESCC miRNAs counteract the effects of let-7. ESCC miRNA expression is promoted by Oct4, Sox2 and Nanog.[59] As ESCs differentiate, Oct4, Sox2 and Nanog expression decrease resulting in a corresponding decrease in ESCC expression. In the absence of ESCCs, Lin28 levels also decrease. In this differentiated state, let-7 is no longer inhibited and feeds back to directly target Lin28 thereby reinforcing its own expression. Furthermore, let-7 now stabilizes the differentiated state by limiting expression of factors required for the ESC fate including transcripts that were previously activated by the pluripotency transcription factors Nanog, Oct4 and Sox2.

The let-7 miRNAs in addition to suppressing the ESC transcriptional program also promote the somatic cell cycle (Fig. 2). Let-7 miRNAs target both directly and indirectly multiple activators of the G1-S transition including cdc25a, cdk6, cyclinD1 and cyclinD2.[42,44] These interactions and others contribute to the overall effect of the let-7 miRNAs on increasing the number of cells in the G1 phase of the cell cycle.[42-44] It remains unclear how or if the cell cycle directly influences ESC self-renewal. It has been postulated that in the G1 phase cells are most susceptible to pro-differentiation signaling cascades including MAPK signaling.[60] It will be important to understand in more detail the interactions between the cell cycle and the ESC transcriptional network and to understand the impact of miRNAs on these interactions.

REGULATORY NETWORKS CONTROLLING miRNA EXPRESSION

In ESCs, ESCC miRNA expression from the miR-290 cluster is controlled by the pluripotency transcription factors Nanog, Oct4, Sox2 and Tcf3 as well as by the Myc transcription factors nMyc and cMyc.[59,61] ESCC miRNAs indirectly upregulate cMyc to form a positive feedback loop which likely reinforces their own expression. When ESCs differentiate, pluripotency transcription factors are downregulated and in turn so are the ESCC miRNAs.[59]

Transcriptional control of expression of direct miRNA suppressors of ESC self-renewal remains an open area of research; however, high-throughput sequencing of chromatin immuno-precipated factors (ChIP seq) data in ESCs give us some insight into their regulation. In ESCs the miR-296 promoter is bound by Oct4, Sox2, Nanog and Tcf3; however, it is also marked by repressive H3K27 methylation and is bound by the polycomb group protein Suz12.[59] These data suggest a mechanism by which miR-296 is poised to be activated in ESCs. If upon differentiation the repressive H3K27 histone mark is rapidly lost prior to loss of Oct4, Sox2 and Nanog, these transcription factors could transiently drive transcription of miR-296. This regulation would form a negative feedback loop leading to more robust loss of ESC self-renewal. How H3K27 methylation is maintained at the miR-296 promoter in ES cells and lost with differentiation is unclear. Regulation of miR-134 and miR-470 promoters is even less well understood.

Likewise transcriptional control of let-7 expression remains relatively unclear. Different let-7 transcripts are expressed in the various differentiated tissues and thus likely diverse transcription factors are able to induce let-7 expression.[47] In ESCs, Oct4, Sox2 and Nanog drive expression of the let-7g primary transcript.[59] The primary transcripts are processed to pre-miRNAs in ESCs where they are degraded by the Lin28/Tut4 complex.[51-56] As ESCs differentiate, suppression by Lin28/Tut4 is lost and mature let-7 is produced.[55,57,58] Additional miRNAs are regulated in this way in ESCs.[53]

Recently, a new class of regulatory RNA binding proteins, the Trim-NHL proteins, has been discovered. In neural stem cells, Schwamborn et al showed that expression of Trim32 potentiates let-7 inhibition of targets and is associated with the differentiation of NSCs.[62] In ESCs, the ESCC miRNAs promote expression of Trim71 (also known as Mlin-41). Trim71 is a let-7 target essential for mouse development.[63] Rybak et al demonstrated that Trim71 acts as an ESC expressed E3 ubiquitin ligase that functions to degrade Ago2 protein, a component of the RISC complex.[64] Both Trim32 and Trim71 are members of a larger family of Trim-NHL proteins, which also include the Drosophila proteins Brat and Mei-P26. These Drosophila proteins also function to modulate the miRNA pathway through interactions with Ago1.[65] It will be important to understand if Trim71 simply functions to modulate activity of the entire miRNA pathway via degradation of Ago2 or if like Trim32 it can associate and increase the activity of specific miRNA subtypes.

miRNAs CAN PROMOTE OR INHIBIT DEDIFFERENTIATION TO IPS CELLS

ESCC miRNAs promote self-renewal in ESCs while the let-7 miRNAs promote silencing of ESC self-renewal. Reprogramming of somatic cells to induced pluripotent stem (iPS) cells can be achieved by nuclear transfer or by directed reprogramming with exogenously introduced transcription.[66] Consistent with the role of ESCC miRNAs in promoting ESC self-renewal, addition of these miRNAs to directed reprogramming assays enhances reprogramming efficiency.[61] Likewise inhibition of the let-7 miRNAs enhances reprogramming.[33] The effects of inhibiting the direct miRNA suppressors of ESC self-renewal on reprogramming remain unknown. Together, these findings demonstrate that the same mechanisms that control ESC self-renewal and differentiation also govern the dedifferentiation process.

Additionally, the ability to reprogram with cocktails of transcriptions factors with and without Myc (either Sox2, Oct4, Klf4, cMyc or Sox2, Oct4, Klf4, no cMyc) has allowed for interrogation of the function of miRNAs and miRNA inhibitors in regard to whether they function in the same or alternate pathways to each of these factors. For example, the ESCC miRNAs were shown to enhance reprogramming in the absence, but not in the presence of cMyc.[61] These findings suggest that ESCCs and Myc have redundant roles. Indeed it is now known that ESCCs induce the indirect upregulation of cMyc and that both cMyc and nMyc promote transcription of ESCC miRNAs.[33,61] Additionally, it has been discovered that inhibition of let-7 promotes reprogramming more so in the absence than in the presence of Myc.[33] This finding suggests that let-7 in somatic cells in part acts to suppress ESC self-renewal through Myc. Indeed, both cMyc and nMyc are direct targets of let-7.[33,67] It will be important and interesting to understand if there exist miRNAs, which operate in the same pathways as the other pluripotency transcription factors and whether these miRNAs can replace these transcription factors in iPS cell reprogramming.

miRNAs IN SOMATIC STEM CELLS

miRNA function in somatic stem cells remains poorly studied. Indeed, aside from ESC derived neural progenitor cells (NPCs) no detailed analysis of the miRNA repertoires of pure somatic stem cell populations has been performed. In NPCs the let-7 miRNAs are the dominant miRNA species.[59] Interestingly, recent data suggest that the let-7 miRNAs are not required for the propagation but rather the differentiation of neural stem cells in the embryonic mouse brain.[62] In this model, asymmetric divisions in neural stem cells segregates the RNA binding protein Trim32 into the daughter cell committed to differentiate further. Trim32, among other functions, increases the activity of let-7 in this cell to promote differentiation.[62]

miRNAS IN CANCER CELLS

ESCC miRNAs and the miR-17/20/106 family share a similar seed sequence. The miR-17/20/106 family has been shown to have important roles in cancer. For example, miR-93 and miR-106 miRNAs target p21 to deregulate the G1/S checkpoint and promote rapid cell proliferation in multiple tumor types.[68,69] Additionally, in vivo studies have shown important roles for these miRNAs in tumorigenesis. In particular, enforced expression of the miR-17-19b polycistron accelerates tumor formation and decreases apoptosis in an Eμ-Myc B cell lymphoma mouse.[70] The decreased apoptosis in this model is likely, at least in part, due to miR-17 family miRNAs targeting the pro-apoptotic protein Bim.[71] The miR-17/92 cluster also contributes to tumorigenesis by increasing angiogenesis in tumors.[72] The human miRNAs miR-372 and miR-373 share the ESCC seed sequence. These miRNAs cooperate with oncogenic Ras to promote tumor formation in primary human fibroblasts and are highly expressed in germ cell tumors.[73] Collectively, these data demonstrate that miRNAs that share a similar seed sequence to the ESCC miRNAs, function as potent oncogenes often by acting through similar pathways normally seen in ESCs.

In contrast to the ESCC and related miRNAs, the let-7 miRNAs act as tumor suppressors. In a model of breast cancer, a subpopulation of the cancer cells, the tumor initiating cells (TICs), can regenerate the tumor. When the TICs differentiate they are no longer capable of forming a full tumor. The let-7 miRNAs are sufficient for differentiation of these cells. In this setting, let-7 acts in part by suppressing Ras, to suppress proliferation and HMGA2, to promote differentiation of the cancer cells.[74] Likewise, in a mouse models of K-Ras induced lung cancer and in xenograft models of established cancer cell lines, addition of exogenous let-7 miRNAs suppresses while inhibition of let-7 activity promotes tumorigenesis.[75-77] Furthermore, recent evidence suggests that Lin28 through inhibition of let-7 activity can promote tumor formation.[78-81] Let-7 has been shown to target multiple oncogenes including K-Ras, N-Ras, Hmga2, cMyc, nMyc and additional factors that collectively reduce cell proliferation.[82] Together, these data strongly support a functional role for let-7 as a tumor suppressor.

CONCLUSION

The data summarized in this chapter support an important role for various miRNA species in either stabilizing the self-renewing state of stem cells or in promoting their differentiation. These miRNAs are similar to other global regulators of gene expression as different subclasses of these miRNAs can either promote or inhibit stem cell self-renewal. These impacts on self-renewal occur both through regulation of the cell cycle and the stem cell transcriptional program. As we learn more about the miRNAs that influence stem cell self-renewal it is becoming clear these miRNAs are tightly regulated in complex molecular networks. This regulation can occur at various levels both transcriptional and post-transcriptional. Furthermore, different classes of miRNAs can inhibit or activate each other's expression. Understanding the extent and function of these networks in development will greatly enhance our knowledge of both developmental and disease states.

REFERENCES

1. Surani MA, Hayashi K, Hajkova P. Genetic and epigenetic regulators of pluripotency. Cell 2007; 128(4):747-762.
2. Kriegstein A, Alvarez-Buylla A. The glial nature of embryonic and adult neural stem cells. Annu Rev Neurosci 2009; 32(1):149-184.
3. Mikkola HKA, Orkin SH. The journey of developing hematopoietic stem cells. Development 2006; 133(19):3733-3744.
4. Arai F, Suda T. Quiescent stem cells in the niche. In: StemBook, ed. The Stem Cell Research Community, StemBook, doi/10.3824/stembook.1.6.1, http://www.stembook.org. 2008.
5. Jaenisch R, Young R. Stem cells, the molecular circuitry of pluripotency and nuclear reprogramming. Cell 2008; 132(4):567-582.
6. Cai X, Hagedorn CH, Cullen BR. Human microRNAs are processed from capped, polyadenylated transcripts that can also function as mRNAs. RNA 2004; 10(12):1957-1966.
7. Lee Y, Kim M, Han J et al. MicroRNA genes are transcribed by RNA polymerase II. EMBO J 2004; 23(20):4051-4060.
8. Rodriguez A. Identification of mammalian microRNA host genes and transcription units. Genome Res 2004; 14(10a):1902-1910.
9. Basyuk E, Suavet F, Doglio A et al. Human let-7 stem-loop precursors harbor features of RNase III cleavage products. Nucleic Acids Res 2003; 31(22):6593-6597.
10. Gregory RI, Yan K, Amuthan G et al. The Microprocessor complex mediates the genesis of microRNAs. Nature 2004; 432(7014):235-240.
11. Lee Y, Ahn C, Han J et al. The nuclear RNase III Drosha initiates microRNA processing. Nature 2003; 425(6956):415-419.
12. Han J, Lee Y, Yeom K et al. The Drosha-DGCR8 complex in primary microRNA processing. Genes Dev 2004; 18(24):3016-3027.
13. Denli AM, Tops BBJ, Plasterk RHA et al. Processing of primary microRNAs by the Microprocessor complex. Nature 2004; 432(7014):231-235.
14. Han J, Lee Y, Yeom K et al. Molecular basis for the recognition of primary microRNAs by the drosha-DGCR8 complex. Cell 2006; 125(5):887-901.
15. Yi R, Qin Y, Macara IG et al. Exportin-5 mediates the nuclear export of premicroRNAs and short hairpin RNAs. Genes Dev 2003; 17(24):3011-3016.
16. Bohnsack MT, Czaplinski K, Gorlich D. Exportin 5 is a RanGTP-dependent dsRNA-binding protein that mediates nuclear export of premiRNAs. RNA 2004; 10(2):185-191.
17. Lund E, Güttinger S, Calado A et al. Nuclear export of microRNA precursors. Science 2004; 303(5654):95-98.
18. Hammond SM. Dicing and slicing: The core machinery of the RNA interference pathway. FEBS Letters 2005; 579(26):5822-5829.
19. Schwarz DS, Hutvágner G, Du T et al. Asymmetry in the assembly of the RNAi enzyme complex. Cell 2003; 115(2):199-208.
20. Bartel DP. MicroRNAs: Target recognition and regulatory functions. Cell 2009; 136(2):215-233.
21. Filipowicz W, Bhattacharyya SN, Sonenberg N. Mechanisms of posttranscriptional regulation by microRNAs: are the answers in sight? Nat Rev Genet 2008; 9(2):102-114.
22. Wang Y, Baskerville S, Shenoy A et al. Embryonic stem cell-specific microRNAs regulate the G1-S transition and promote rapid proliferation. Nat Genet 2008; 40(12):1478-1483.
23. Bar M, Wyman SK, Fritz BR et al. MicroRNA discovery and profiling in human embryonic stem cells by deep sequencing of small RNA libraries. Stem Cells 2008; 26(10):2496-2505.
24. Suh M, Lee Y, Kim JY et al. Human embryonic stem cells express a unique set of microRNAs. Dev Biol 2004; 270(2):488-498.
25. Kanellopoulou C, Muljo SA, Kung AL et al. Dicer-deficient mouse embryonic stem cells are defective in differentiation and centromeric silencing. Genes Dev 2005; 19(4):489-501.
26. Murchison EP, Partridge JF, Tam OH et al. Characterization of Dicer-deficient murine embryonic stem cells. Proc Natl Acad Sci USA 2005; 102(34):12135-12140.
27. Wang Y, Medvid R, Melton C et al. DGCR8 is essential for microRNA biogenesis and silencing of embryonic stem cell self-renewal. Nat Genet 2007; 39(3):380-385.
28. Savatier P, Huang S, Szekely L et al. Contrasting patterns of retinoblastoma protein expression in mouse embryonic stem cells and embryonic fibroblasts. Oncogene 1994; 9(3):809-818.
29. Planas-Silva MD, Weinberg RA. The restriction point and control of cell proliferation. Curr Opin Cell Biol 1997; 9(6):768-772.
30. Giacinti C, Giordano A. RB and cell cycle progression. Oncogene 2006; 25(38):5220-5227.

31. Savatier P, Lapillonne H, van Grunsven LA et al. Withdrawal of differentiation inhibitory activity/leukemia inhibitory factor up-regulates D-type cyclins and cyclin-dependent kinase inhibitors in mouse embryonic stem cells. Oncogene 1996; 12(2):309-322.
32. Mittnacht S. Control of pRB phosphorylation. Curr Opin Genet Dev 1998; 8(1):21-27.
33. Melton C, Judson R, Blelloch R. Opposing microRNA families regulate self-renewal in mouse embryonic stem cells. Nature 2010; Advance Online.
34. Singh AM, Dalton S. The cell cycle and Myc intersect with mechanisms that regulate pluripotency and reprogramming. Cell Stem Cell 2009; 5(2):141-149.
35. Cartwright P, McLean C, Sheppard A et al. LIF/STAT3 controls ES cell self-renewal and pluripotency by a Myc-dependent mechanism. Development 2005; 132(5):885-896.
36. Lin C, Jackson AL, Guo J et al. Myc-regulated microRNAs attenuate embryonic stem cell differentiation. EMBO J 2009; 28(20):3157-3170.
37. Benetti R, Gonzalo S, Jaco I et al. A mammalian microRNA cluster controls DNA methylation and telomere recombination via Rbl2-dependent regulation of DNA methyltransferases. Nat Struct Mol Biol 2008; 15(3):268-279.
38. Sinkkonen L, Hugenschmidt T, Berninger P et al. MicroRNAs control de novo DNA methylation through regulation of transcriptional repressors in mouse embryonic stem cells. Nat Struct Mol Biol 2008; 15(3):259-267.
39. Xu N, Papagiannakopoulos T, Pan G et al. MicroRNA-145 regulates OCT4, SOX2 and KLF4 and represses pluripotency in human embryonic stem cells. Cell 2009; 137(4):647-658.
40. Tay YM, Tam W, Ang Y et al. MicroRNA-134 modulates the differentiation of mouse embryonic stem cells, where it causes posttranscriptional attenuation of Nanog and LRH1. Stem Cells 2008; 26(1):17-29.
41. Tay Y, Zhang J, Thomson AM et al. MicroRNAs to Nanog, Oct4 and Sox2 coding regions modulate embryonic stem cell differentiation. Nature 2008; 455(7216):1124-1128.
42. Johnson CD, Esquela-Kerscher A, Stefani G et al. The let-7 microRNA represses cell proliferation pathways in human cells. Cancer Res 2007; 67(16):7713-7722.
43. Kumar MS, Erkeland SJ, Pester RE et al. Suppression of nonsmall cell lung tumor development by the let-7 microRNA family. Proc Natl Acad Sci USA 2008; 105(10):3903-3908.
44. Schultz J, Lorenz P, Gross G et al. MicroRNA let-7b targets important cell cycle molecules in malignant melanoma cells and interferes with anchorage-independent growth. Cell Res 2008; 18(5):549-557.
45. Reinhart BJ, Slack FJ, Basson M et al. The 21-nucleotide let-7 RNA regulates developmental timing in Caenorhabditis elegans. Nature 2000; 403(6772):901-906.
46. Pasquinelli AE, Reinhart BJ, Slack F et al. Conservation of the sequence and temporal expression of let-7 heterochronic regulatory RNA. Nature 2000; 408(6808):86-89.
47. Landgraf P, Rusu M, Sheridan R et al. A mammalian microRNA expression atlas based on small RNA library sequencing. Cell 2007; 129(7):1401-1414.
48. Griffiths-Jones S. The microRNA Registry. Nucl Acids Res 2004; 32(suppl_1):D109-111.
49. Griffiths-Jones S, Grocock RJ, van Dongen S et al. miRBase: microRNA sequences, targets and gene nomenclature. Nucl Acids Res 2006; 1;34(Database issue):D140-4
50. Griffiths-Jones S, Saini HK, van Dongen S et al. miRBase: tools for microRNA genomics. Nucl Acids Res 2008; 36(suppl_1):D154-158.
51. Hagan JP, Piskounova E, Gregory RI. Lin28 recruits the TUTase Zcchc11 to inhibit let-7 maturation in mouse embryonic stem cells. Nat Struct Mol Biol 2009; 16(10):1021-1025.
52. Heo I, Joo C, Cho J et al. Lin28 mediates the terminal uridylation of let-7 precursor MicroRNA. Mol Cell 2008; 32(2):276-84.
53. Heo I, Joo C, Kim Y et al. TUT4 in concert with Lin28 suppresses microRNA biogenesis through premicroRNA uridylation. Cell 2009; 138(4):696-708.
54. Rybak A, Fuchs H, Smirnova L et al. A feedback loop comprising lin-28 and let-7 controls prelet-7 maturation during neural stem-cell commitment. Nat Cell Biol 2008; 10(8):987-93.
55. Thomson JM, Newman M, Parker JS et al. Extensive posttranscriptional regulation of microRNAs and its implications for cancer. Genes Dev 2006; 20(16):2202-7.
56. Viswanathan SR, Daley GQ, Gregory RI. Selective blockade of microRNA processing by Lin28. Science 2008; 320(5872):97-100.
57. Wu L, Belasco JG. Micro-RNA regulation of the mammalian lin-28 gene during neuronal differentiation of embryonal carcinoma cells. Mol Cell Biol 2005; 25(21):9198-9208.
58. Yang D, Moss EG. Temporally regulated expression of Lin-28 in diverse tissues of the developing mouse. Gene Expression Patterns 2003; 3(6):719-726.
59. Marson A, Levine SS, Cole MF et al. Connecting microRNA genes to the core transcriptional regulatory circuitry of embryonic stem cells. Cell 2008; 134(3):521-533.
60. Burdon T, Smith A, Savatier P. Signalling, cell cycle and pluripotency in embryonic stem cells. Trends Cell Biol 2002; 12(9):432-438.

61. Judson RL, Babiarz JE, Venere M et al. Embryonic stem cell-specific microRNAs promote induced pluripotency. Nat Biotechnol 2009; 27(5):459-461.
62. Schwamborn JC, Berezikov E, Knoblich JA. The TRIM-NHL Protein TRIM32 activates microRNAs and prevents self-renewal in mouse neural progenitors. Cell 2009; 136(5):913-925.
63. Maller Schulman BR, Liang X, Stahlhut C et al. The let-7 microRNA target gene, Mlin41/Trim71 is required for mouse embryonic survival and neural tube closure. Cell Cycle 2008; 7(24):3935-3942.
64. Rybak A, Fuchs H, Hadian K et al. The let-7 target gene mouse lin-41 is a stem cell specific E3 ubiquitin ligase for the miRNA pathway protein Ago2. Nat Cell Biol 2009; 11(12):1411-1420.
65. Neumuller RA, Betschinger J, Fischer A et al. Mei-P26 regulates microRNAs and cell growth in the Drosophila ovarian stem cell lineage. Nature 2008; 454(7201):241-245.
66. Hochedlinger K, Plath K. Epigenetic reprogramming and induced pluripotency. Development 2009; 136(4):509-523.
67. Kumar MS, Lu J, Mercer KL et al. Impaired microRNA processing enhances cellular transformation and tumorigenesis. Nat Genet 2007; 39(5):673-677.
68. Ivanovska I, Ball AS, Diaz RL et al. MicroRNAs in the miR-106b Family Regulate p21/CDKN1A and Promote Cell Cycle Progression. Mol Cell Biol 2008; 28(7):2167-2174.
69. Petrocca F, Visone R, Onelli MR et al. E2F1-Regulated microRNAs impair TGF[beta]-dependent cell-cycle arrest and apoptosis in gastric cancer. Cancer Cell 2008; 13(3):272-286.
70. He L, Thomson JM, Hemann MT et al. A microRNA polycistron as a potential human oncogene. Nature 2005; 435(7043):828-833.
71. Mendell JT. miRiad roles for the miR-17-92 cluster in development and disease. Cell 2008; 133(2):217-222.
72. Dews M, Homayouni A, Yu D et al. Augmentation of tumor angiogenesis by a Myc-activated microRNA cluster. Nat Genet 2006; 38(9):1060-1065.
73. Voorhoeve PM, le Sage C, Schrier M et al. A genetic screen implicates miRNA-372 and miRNA-373 as oncogenes in testicular germ cell tumors. Cell 2006; 124(6):1169-1181.
74. Yu F, Yao H, Zhu P et al. let-7 regulates self renewal and tumorigenicity of breast cancer cells. Cell 2007; 131:1109-1123.
75. Trang P, Medina PP, Wiggins JF et al. Regression of murine lung tumors by the let-7 microRNA. Oncogene [Internet] 2009 [cited 2009 Dec 16]; Available from: http://www.ncbi.nlm.nih.gov/pubmed/19966857.
76. Kumar MS, Erkeland SJ, Pester RE et al. Suppression of nonsmall cell lung tumor development by the let-7 microRNA family. Proc Natl Acad Sci USA 2008; 105(10):3903-3908.
77. Esquela-Kerscher A, Trang P, Wiggins JF et al. The let-7 microRNA reduces tumor growth in mouse models of lung cancer. Cell Cycle 2008; 7(6):759-764.
78. Chang T, Zeitels LR, Hwang H et al. Lin-28B transactivation is necessary for Myc-mediated let-7 repression and proliferation. Proc Natl Acad Sci USA 2009; 106(9):3384-3389.
79. Iliopoulos D, Hirsch HA, Struhl K. An epigenetic switch involving NF-kappaB, Lin28, Let-7 MicroRNA and IL6 links inflammation to cell transformation. Cell 2009; 139(4):693-706.
80. Viswanathan SR, Powers JT, Einhorn W et al. Lin28 promotes transformation and is associated with advanced human malignancies. Nat Genet 2009; 41(7):843-848.
81. Dangi-Garimella S, Yun J, Eves EM et al. Raf kinase inhibitory protein suppresses a metastasis signalling cascade involving LIN28 and let-7. EMBO J 2009; 28(4):347-358.
82. Büssing I, Slack FJ, Großhans H. let-7 microRNAs in development, stem cells and cancer. Trends Mol Med 2008; 14(9):400-409.

CHAPTER 9

TELOMERES AND TELOMERASE IN ADULT STEM CELLS AND PLURIPOTENT EMBRYONIC STEM CELLS

Rosa M. Marión and Maria A. Blasco*

Abstract: Telomerase expression is silenced in most adult somatic tissues with the exception of adult stem cell (SC) compartments, which have the property of having the longest telomeres within a given tissue. Adult SC compartments suffer from telomere shortening associated with organismal aging until telomeres reach a critically short length, which is sufficient to impair SC mobilization and tissue regeneration. p53 is essential to prevent that adult SC carrying telomere damage contribute to tissue regeneration, indicating a novel role for p53 in SC behavior and therefore in the maintenance of tissue fitness and tumor protection. Reprogramming of adult differentiated cells to a more pluripotent state has been achieved by various means, including somatic cell nuclear transfer and, more recently, by over expression of specific transcription factors to generate the so-called induced pluripotent stem (iPS) cells. Recent work has demonstrated that telomeric chromatin is remodeled and telomeres are elongated by telomerase during nuclear reprogramming. These findings suggest that the structure of telomeric chromatin is dynamic and controlled by epigenetic programs associated with the differentiation potential of cells, which are reversed by reprogramming. This chapter will focus on the current knowledge of the role of telomeres and telomerase in adult SC, as well as during nuclear reprograming to generate pluripotent embryonic-like stem cells from adult differentiated cells.

*Corresponding Author: Maria A. Blasco—Telomeres and Telomerase Group, Molecular Oncology Program, Spanish National Cancer Centre (CNIO), Melchor Fernández Almagro 3, Madrid, E-28029, Spain. Email: mblasco@cnio.es

The Cell Biology of Stem Cells, edited by Eran Meshorer and Kathrin Plath.
©2010 Landes Bioscience and Springer Science+Business Media.

INTRODUCTION

One of the best-known cell-intrinsic events associated with aging is the progressive shortening of telomeres, the natural ends of chromosomes. The speed at which telomeres shorten with aging seems to vary between men and women and can be influenced by factors considered to accelerate aging and to be a risk of premature death, such as perceived stress, smoking and obesity, all of which have been proposed to negatively impact on telomere length.[1-4] Telomere shortening is also accelerated in various human diseases associated with aging, such as cardiovascular disease, dementia and infections, among others,[5-11] as well as in human syndromes caused by mutations in telomere maintenance pathways, including germ line mutations in telomerase components.[12-15] A current model suggests that short telomeres represent a chronic type of DNA damage which is propagated to daughter cells and that can cause aging by impairing the ability of adult stem cells (SC) to regenerate and repair tissues, as well as that leads to cell loss via induction of cell senescence and apoptosis.[16-19]

During development cells gradually lose their self-renewal and differentiation capacities, however, differentiated somatic cells can be reverted to a more pluripotent state.[20] This reversal to a developmentally more-primitive state is termed nuclear reprogramming. The mechanisms underlying nuclear reprogramming are thought to involve genome-wide changes in chromatin structure and gene expression,[21-23] which are dictated by epigenetic states correlated with the differentiation potential of cells.

Nuclear reprogramming has been achieved by using somatic cell nuclear transfer (SCNT),[24] although other methods, including fusion of differentiated cells with embryonic stem (ES) cells, have been also described.[25] Nuclear reprogramming by SCNT is achieved by insertion of the nucleus of a differentiated somatic cell into an enucleated, unfertilized egg cell (the oocyte) of the same species (Fig. 1).[24] Along this process, the adult differentiated nucleus "goes back" in development to a zygotic state with the potential to generate an entire organism, genetically identical to the donor of the somatic cell. SCNT has been successfully performed in various mammalian species, including mice, sheep, cattle, pigs, rabbits and cats.[24,26-30] These studies demonstrated that the ability to de-differentiate and reprogram nuclei to acquire totipotency inheres in mammalian oocytes. Moreover, the success of the technique gave promise to applications such as gene manipulation, species preservation, livestock propagation, as well as the generation of patient-specific pluripotent stem cells that could be used for the study and treatment of human diseases.

The recent discovery that mouse and human somatic cells can be reprogrammed to the so-called induced pluripotent stem (iPS) cells by over expression of four or less transcription factors,[31,32] has generated much excitement. iPS cells represent a new source of patient-specific stem cells. Importantly, iPS cell generation bypasses the technical difficulties and ethical controversies associated with SCNT (Fig. 1), which involve the generation and destruction of human embryos. Full characterization of iPS cells is still in progress, but initial studies indicate that iPS cells have the same properties as ES cells, including similar global gene expression and epigenetic patterns, as well as the ability to contribute to mouse embryonic development and form teratomas when injected into nude mice.[33]

The successful use of reprogrammed cells in cell therapy requires that these cells have sufficient proliferative potential and are able to maintain genomic integrity to ensure long-term functionality. In this regard, a proper telomere length and function

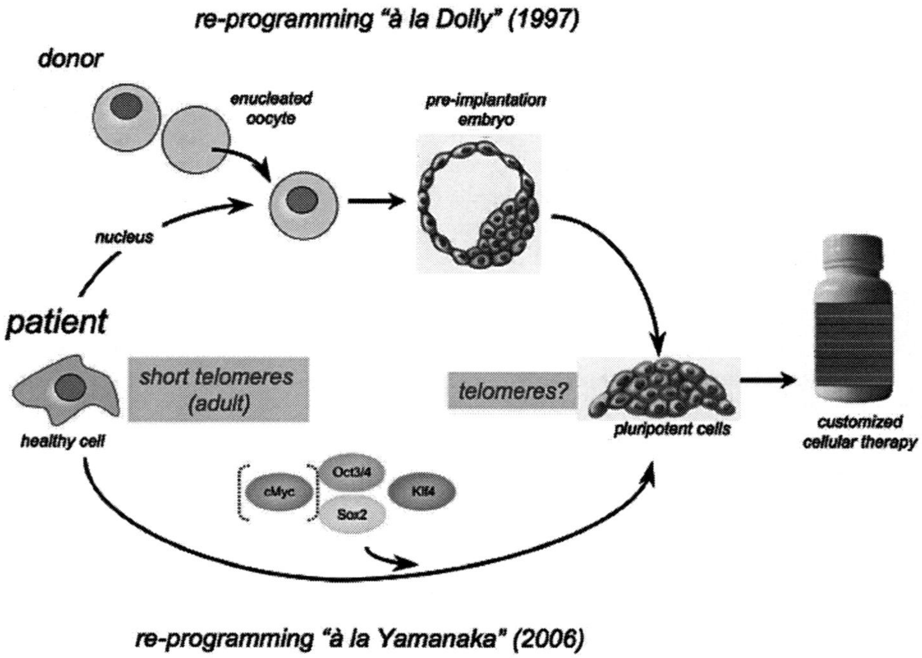

Figure 1. Strategies to induce nuclear reprogramming. Generation of pluripotent stem cells from adult differentiated somatic cells by different nuclear reprogramming procedures. Nuclear reprogramming by somatic cell nuclear transfer (top) is achieved by transplantation of the nucleus of the differentiated donor cell into an enucleated oocyte. Pluripotent stem cells are then obtained from the pre-implantation embryo generated. An alternative mechanism (bottom), which does not involve the generation and destruction of embryos, is the generation of induced pluripotent stem (iPS) cells by overexpression of a set of "stemness" transcription factors. In any case, donor adult cells present shortened telomeres, raising the question whether telomeres are rejuvenated to a stem cell-like state during nuclear reprogramming.

is essential to ensure chromosomal stability of the resulting reprogrammed cells. Since telomeres shorten with each cell cycle and telomere length is tightly linked to cellular age, it is of relevance to determine whether the shortened telomeres of a differentiated donor cell are fully restored to an ES cell-like length during reprogramming.

REGULATION OF TELOMERES AND TELOMERASE

Telomeres are ribonucleoprotein heterochromatic structures at the ends of chromosomes that protect them from degradation and from being detected as double-strand DNA breaks.[34,35] Mammalian telomeres consist of tandem repeats of the TTAGGG sequence bound by a six subunit -protein complex known as shelterin.[35-37] Telomere function depends on a minimal length of telomeric repeats and the binding of the shelterin complex.[37] Incomplete DNA replication of chromosome ends (the so-called "end-replication problem") results in progressive shortening of telomeres, a defect which is propagated into daughter cells and can eventually lead to critically short/uncapped telomeres and cell cycle arrest/senescence.[38,39] Progressive telomere shortening is proposed to be one of the mechanisms underlying organismal aging.[40] Telomere length is maintained by telomerase, a reverse transcriptase encoded by the *Tert* (telomerase reverse transcriptase) and *Terc* (telomerase RNA component) genes, which is able to add telomeric repeats de novo after each round of cell division onto chromosome ends.[41,42] Alternative ways to maintain telomere length exist, such as ALT (alternative lengthening of telomeres), which relays on homologous recombination between telomeric or subtelomeric sequences.[43] Current evidence suggests that telomere-elongation mechanisms are regulated by the epigenetic status of telomeric chromatin and by the telomere-binding proteins.[44,45] In particular, both telomeric and subtelomeric regions are enriched in histone marks characteristic of repressed heterochromatin domains, such as trimethylation of H3K9 and H4K20 and binding of heterochromatin protein 1 (HP1).[46-48] Also, subtelomeric DNA is heavily methylated.[47] Loss of these heterochromatic marks is concomitant with excessive telomere elongation. In particular, abnormally long telomeres are observed upon loss of H3K9m3, HP1 and H4K20m3 marks in cells deficient for the Suv39h or Suv4-20h histone methyl transferases,[46,48] as well as upon loss of subtelomeric DNA methylation in cells deficient for Dicer or the DNMT1 and DNMT3ab DNA methyl transferases.[47,49]

In addition, telomeres are actively transcribed in yeast, zebra fish and mammals generating long, noncoding RNAs known as TelRNAs or TERRAs.[44,45,50,51] TERRAs can associate with the telomeric chromatin, where they are proposed to act as negative regulators of telomere length based on their ability to act as potent inhibitors of telomerase in vitro. [44,45,50,51]In line with this, TERRAs are down-regulated in association with cancer progression, a situation that requires efficient elongation of short telomeres by telomerase. Finally, telomere-binding proteins such as TRF1, TRF2, Tin2, TANK1 and TANK2 have been also shown to influence telomere length.[52-55] Altogether, these observations point to a higher order structure at telomeres that is epigenetically regulated and important for telomere length control (Fig. 2).

Telomerase expression is restricted to embryonic development as well as to adult stem cell compartments.[16,56-58] However, telomerase activity in these tissues is not sufficient to prevent telomere shortening with age.[39,59] Mutations in telomerase components present in patients suffering from dyskeratosis congenita, aplastic anemia and idiopathic pulmonary fibrosis[12-15] further accelerate telomere shortening and lead to premature loss of tissue regeneration and disease, indicating that telomerase levels in the adult organism are rate-limiting for organism fitness. Moreover, telomerase-deficient mice show premature aging and a decreased proliferative potential of adult stem cell populations.[60-64] This role of telomerase and telomere length in organ homeostasis and adult stem cell biology highlights the importance of understanding the effects of nuclear reprogramming on telomerase activity, telomere length and telomeric chromatin.

ROLE OF TELOMERES AND TELOMERASE IN ADULT SC COMPARTMENTS

Cancer and aging, two biological processes in which telomerase activity has been implicated, are increasingly seen as SC diseases.[40,57,65] In particular, cancer may often originate from the transformation of normal SC, while aging has been associated with a progressive decline in the number and/or functionality of certain SC.[57,65]

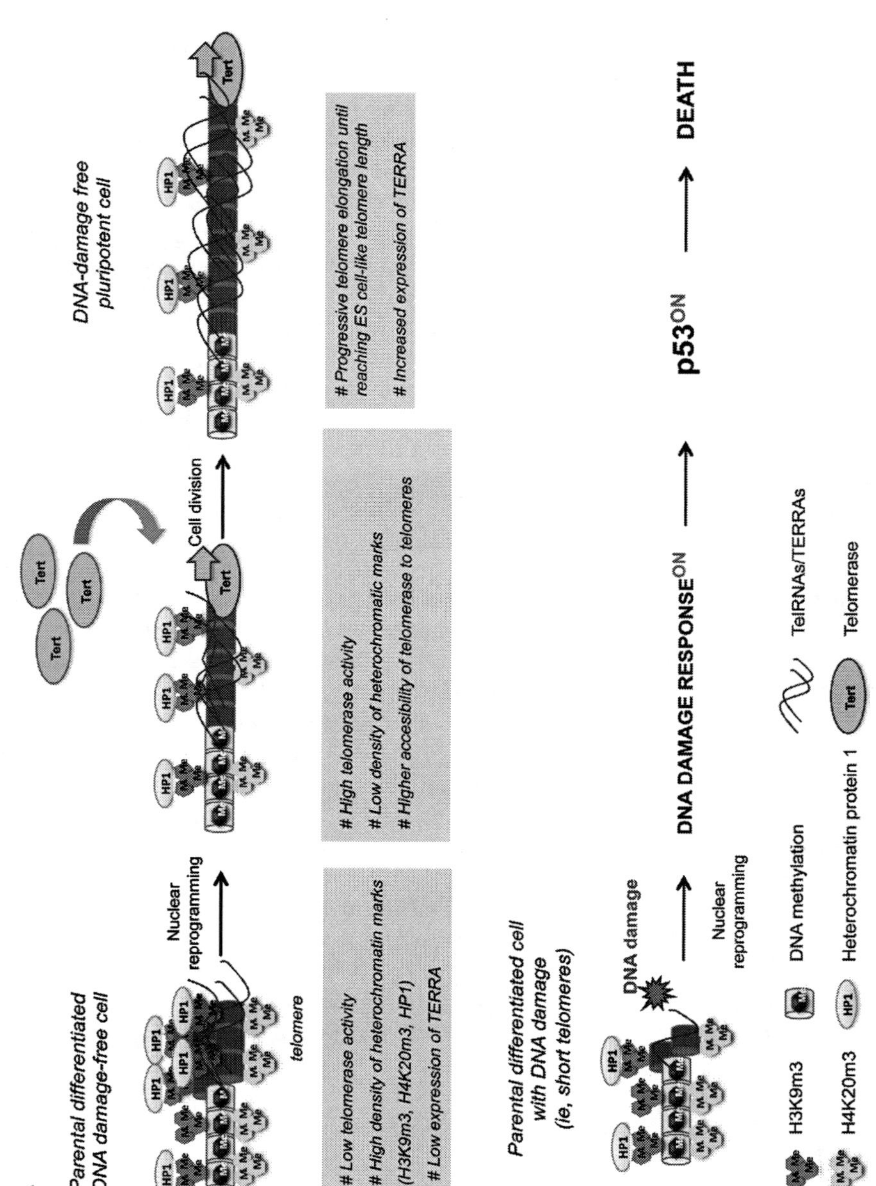

Figure 2. Legend viewed on following page.

Figure 2. Figure viewed on previous page. Reprogramming of telomeres upon induction of pluripotency. A) Telomeres in adult differentiated cells are organized into a highly compact chromatin with high levels of heterochromatic marks and low telomeric RNAs TERRAs/TelRNA expression. Nuclear reprogramming results in a reduction of H3K9m3, H4K20m3 and HP1 heterochromatic marks that transform the telomeric chromatin into a less compacted and more accessible structure, concomitant with a dramatic upregulation of telomerase enzyme. Telomerase efficiently elongates telomeres and levels of TERRAs/TelRNA expression are upregulated. Telomere elongation continues postreprogramming until the natural limit of telomere length of pluripotent cells has been reached. B) Suboptimal cells, such as those with critically short telomeres, are eliminated during the reprogramming process. Critically short telomeres and other types of DNA damage, turn on the DNA damage response at the onset of reprogramming and induce p53-dependent apoptosis, preventing the generation of pluripotent cells from suboptimal damaged cells.

The fact that telomerase activity is largely restricted to SC, suggests that telomerase levels in these cells may be determinant for organism fitness. Evidence for a rate-limiting role of telomerase in human aging and life span, has come from the study of human diseases associated with mutations in telomerase components. As discussed above, mutations in the telomerase core components, Tert and Terc, are present in patients that suffer from aplastic anemia, idiopathic pulmonary fibrosis and dyskeratosis congenita. These lethal diseases are adult-onset and are characterized by a premature loss of tissue regeneration associated with telomere shortening.[12-15,66-68] In an analogous manner, telomerase deficiency in mice results in decreased median and maximum life span already within the first mouse generation, also indicating that telomerase is rate-limiting for aging in mice.[69]

During the last years, the specific role of telomerase in different stem cells compartments has been elucidated, including hematopoietic stem cells (HSC),[70-73] epidermal stem cells (ESC)[56,57] and neural stem cells (NSC).[64] The use of loss-of-function and gain-of-function mouse models for telomerase, has served to establish the role of telomere length and telomerase activity on ESC behavior.[56,57,74] Telomere shortening in the context of Terc-deficient mice has been shown to result in decreased functionality of their skin ESC compartment.[56] In particular, mobilization (proliferation and migration) of ESC out of the hair follicle niche upon mitogen-induced proliferation is partially inhibited in mice with a slight reduction in telomere length (early generation *Terc*$^{-/-}$ mice) and strongly inhibited in mice with critically short telomeres (late generation *Terc*$^{-/-}$ mice).[56] The immediate consequences of such mobilization defect are lower rates of proliferation in the hair follicle stem cell niche as well as in the adjacent transient-amplifying compartment, resulting in defective hair growth and a stunted hyperplasic response.[56] This defective stem cell function could be fully rescued by telomerase re-introduction and elongation of short telomeres,[75] as well as by p53 abrogation in the absence of telomere elongation.[76] These results highlight short telomeres as causative of SC dysfunction in a p53-dependent manner. Besides the skin, other tissues with a high cell turnover such as bone marrow, intestine and testis, show atrophies in Terc-deficient mice with critically short telomeres,[61,62] supporting the notion that telomere length is a determinant for tissue fitness in the wide context of the organism.

Finally, it is important to note that the effects of telomere length and telomerase activity on different stem cell compartments (ESC, HSC and adult NSC) are cell autonomous, as demonstrated using in vitro clonogenicity assays.[56,63] In addition, it has been recently shown that short telomeres in the context of the Terc-deficient mouse model may also limit the ability of stem cell microenvironments to sustain the proper functioning of transplanted wild-type stem cells.[77] All together, these results suggest that telomere shortening with aging is not only an intrinsic factor leading to aberrant stem cell functioning but also may affect the viability of the stem cell environment further aggravating stem cell dysfunction with aging. This is relevant for designing potential therapeutic strategies based on telomerase reactivation, since it indicates that the effects of telomerase and telomere length on stem cell behavior are dependent on both the stem cells and the physiological niche micro-environments. These findings on the negative impact of short telomeres on the regenerative capacity of tissues also lead to the provocative idea that boosted telomerase expresion in adult tissues may have antiaging effects by extending tissue fitness and longevity. Importantly, telomerase over-expression in the context of mice engeneered to be cancer resistant, significantly delayed aging and resulted in a 40% extension of median longevity,[78] first demonstrating anti-aging activity of telomerase in the context of the organism.

TELOMERES AND TELOMERASE REGULATION DURING REPROGRAMMING BY SCNT

When Dolly the sheep was reported as the first adult mammal generated using SCNT,[24] important questions were raised regarding the "age" of her cells. Since Dolly was cloned from an adult cell, it was intriguing to determine whether Dolly's cells maintained the telomere length corresponding to the age of the donor cell or whether they were "rejuvenated" to the length of pluripotent ES cells during reprogramming. These were relevant questions that raised concerns not only regarding possible developmental problems of the cloned animals, but also regarding the quality of the resulting ES cells as well as their potential use in regenerative medicine therapies.

The analysis of Dolly's telomeres, cloned from a cultured mammary cell from a 6-year-old animal, revealed that they were shorter, by approximately 20%, when compared with age-matched controls.[79] Similarly, sheep cloned by nuclear transfer of cultured cells from embryonic or fetal tissue showed a 10-15% telomere shortening comparing with age-matched controls.[79] These disappointing observations suggested that the telomeric cellular clock had not been reset to zero during reprogramming. However, follow-up analyses across a variety of SCNT-derived animals gave rather different results, with the majority of cloned animals presenting a normal telomere length compared to age-matched controls. In particular, cloned mice by means of SCNT showed telomeres that were stable for six generations.[80] Similarly, cloned cattle derived from fibroblasts cells,[81,82] cumulus cells[82] and granulosa cells,[83] all showed normal telomere length. Finally, cloned sheep derived from fibroblast cells also showed restored telomere length.[84] Interestingly, even in the instance when senescent cells with very short telomeres were used as nuclear donors, telomere length was restored and even enhanced by the cloning process.[81] These findings suggested that the-already shortened telomeres of somatic donor cells could be re-elongated during reprogramming, although the degree of elongation was quite variable, underscoring the complexity of telomere length control in the clones. To date, it is still not clear why Dolly's telomeres were so unusually short, but it has been suggested that this variability may be the consequence of differences in donor cell type, in nuclear transfer procedures and the species.[85]

The demonstration of telomere elongation during SCNT opened new questions, such as when during reprogramming are telomeres being elongated and whether this elongation is telomerase-dependent. The latter is a relevant question, as telomeres are known to lengthen during the early embryo cleave cycles following fertilization not by telomerase but through a recombination-based mechanism.[58] In contrast, telomeres are elongated in a telomerase-dependent manner at the transition from the morula to the blastocyst stage both in mice and cattle embryos.[86] When SCNT-derived embryos were studied at the morula to blastocysts transition, telomeres were comparable to those of the donor cells at the morula stage, but restored to normal length at the blastocyst stage, regardless of the telomere length of donor nuclei.[86] These data suggested that telomerase activation could be taking place during the SCNT process. Indeed, telomerase reactivation was observed in cloned embryos obtained from nuclear transfer of donor nuclei that showed low or undetectable levels of telomerase.[81-83,87] The temporal profile of the telomerase activity during development was similar in cloned embryos and fertilized embryos, with the highest level in the blastocyst stage[82,83,87] and correlates with the moment of telomere elongation.[86]

These results indicate that restoration of telomere length in SCNT cloned animals occurs during embryogenesis, very likely through an increase in telomerase activity at the morula to blastocysts transition.

TELOMERES AND TELOMERASE REGULATION DURING iPS CELL GENERATION

When iPS cells were first described,[31,32] it became of immediate interest to determine whether telomeres acquired the characteristics of ES cells telomeres, including a much longer telomere length than that of the differentiated parental cells. In this regard, two well-defined characteristics of ES cells, such as high levels of TERT and high telomerase activity,[88] were readily described for iPS cells.[31,32,89] As the presence of active telomerase does not necessarily mean net telomere elongation,[90] it remained unclear whether telomeres were being elongated or not during iPS cell generation, a process that occurs in an in vitro cell culture setting in contrast to SCNT. A recent study demonstrated that telomerase-dependent telomere elongation occurs in iPS cells derived from mouse embryonic fibroblasts (MEFs), which continues postreprogramming until reaching ES cell telomere length.[91] Even when fibroblasts were derived from old donor mice, which showed much shorter telomeres than MEFs, telomeres were efficiently elongated in the resulting iPS cells. These results suggested that iPS cells derived from donors with a limited telomere reserve, such as elderly individuals or patients with diseases characterized by short telomeres, will rejuvenate their telomeres as long as they carry a functional telomerase pathway.

Interestingly, the resulting iPS cells showed a decreased density of H3K9m3 and H4K20m3 heterochromatic marks at telomeres compared to the parental MEF, reaching similarly low levels to those of ES cell telomeres.[91] These results are in line with a higher plasticity in the chromatin of pluripotent ES cells compared to that of differentiated cells.[92] Also in agreement with an opening of the telomeric chromatin associated with nuclear reprogramming, we observed similarly elevated telomere recombination frequencies both in ES cells and in iPS cells when compared to parental MEF, which showed much lower frequencies of recombination.[91] Finally and in agreement with their longer telomeres, TERRA levels are efficiently increased in iPS compared to the MEF. This accumulation of TERRA in turn may serve as a mechanism to negatively regulate telomere elongation by telomerase once the iPS cells reach the ES cell telomere length.[91] Since it has been described that histone H3.3 and ATRX (alpha thalassemia/mental retardation syndrome X-Linked) are present at telomeres of mouse ES cells,[93,94] it is likely that they could also be found at the telomeres of iPS cells, although this has not been proven yet. Together, these observations demonstrate that generation of iPS cells involves a change in the epigenetic status of telomeres towards a more open chromatin conformation with a lower density of heterochromatic histone marks, which is coincidental with increased TERRA transcription, increased telomere recombination and continuous telomere elongation until reaching ES cell telomere length. Since TERRA has been proposed to negatively regulate telomerase activity,[51] increased expression of TERRA in iPS may serve as a counting mechanism of telomere length that would inhibit telomerase activity once the iPS cells reach the ES cell telomere length. These results prove that telomeric chromatin is dynamic and reprogrammable depending of the differentiation stage of cells (Fig. 2).

TELOMERASE ACTIVATION IS ESSENTIAL FOR THE "GOOD" QUALITY OF THE RESULTING iPS CELLS

Reprogramming efficiency of cells derived from increasing generations of telomerase-deficient mice, which present a higher frequency of critically short telomeres and chromosome end-to-end fusions, was dramatically reduced, indicating that a minimum telomere length is required for efficient reprogramming.[91] Crucially, reintroduction of telomerase reduced the frequency of short telomeres and largely restored iPS cell generation efficiency. These results suggested that damaged/uncapped telomeres are responsible for their failure to reprogram and highlighted the existence of 'reprogramming barriers' that abort the reprogramming of cells with uncapped telomeres. Since p53 has a key involvement in preventing the propagation of DNA-damaged cells, including those containing short telomeres, its possible involvement as a reprogramming barrier was readily tested.[95] Abrogation of p53 allowed efficient reprogramming of cells with critically short telomeres and other types of DNA damage,[95] demonstrating that p53 is critical in preventing the generation of human and mouse pluripotent cells from suboptimal parental cells, including those with critically short telomeres. In line with these results, other studies have also shown that p53 limits the production of iPS cells.[96-99] Overall, these findings demonstrate that telomere length and telomeric chromatin are rejuvenated during in vitro reprogramming and highlight the important role of telomere biology and dynamics in iPS cell generation and functionality (Fig. 2).

REGULATION OF TELOMERE REPROGRAMMING

Despite the emerging details describing the profound changes on telomeres during nuclear reprogramming, how telomere reprogramming is regulated is still largely unknown. Due to the importance of telomerase in this process, it would be important to determine how telomerase activity is upregulated. The proto-oncogene c-Myc, one of the four factors involved in the reprogramming by defined factors, transcriptionally regulates Tert,[100] suggesting that could be responsible for the telomerase activation observed in iPS cells. However, iPS cells generated with or without c-Myc showed similar levels of telomerase activity.[91] Thus, the mechanism for telomerase activation during iPS generation remains unknown. However, telomerase activity is not the only factor determining telomere length. Telomerase-mediated telomere elongation and maintenance depend on telomere structure, which is regulated by epigenetic modifications at telomeres and by telomere binding proteins.[44,45] Nuclear reprogramming requires the removal of the tissue-specific epigenetic pattern imposed on the chromatin during cellular differentiation and division, which is accomplished through large scale reorganization of chromatin structure and functions, as global changes of DNA methylation or histone modifications that lead to a more open state of the chromatin.[23] These epigenetic alterations have been shown to also alter telomere chromatin during reprogramming.[91] The reprogramming of telomeric chromatin into a more open conformation observed in iPS cells may be required to allow telomerase access to the end of the telomere and posterior telomere lengthening (Fig. 2), although this has not been formally proven. In line with this, aberrant epigenetic modifications in cloned embryos obtained by SCNT may explain the differences in telomere lengths observed as the result of differences in the accessibility of telomerase to the telomere. In this regard, it is interesting to note that cloned embryos showed abnormal methylation patterns.[22] On the other hand, epigenetic changes occurring during reprogramming could also have a direct impact in telomerase expression.

Finally, telomere binding proteins are also mediators of telomere length that may inhibit or facilitate the binding of telomerase to telomeric DNA. A possibility exist that the expression or function of these proteins is regulated during reprogramming to contribute to telomere rejuvenation, although data supporting this hypothesis is not available. Thus, future studies defining the role of chromatin modifying activities and telomere binding proteins on telomere reprogramming would be of great interest.

CONCLUSION

Shortening of telomeres to a critically short length in adult SC, something that is associated with organismal aging, is sufficient to impair SC mobilization and tissue regeneration and is proposed to be a key determinant of organismal longevity. In turn, nuclear reprogramming, achieved either by SCNT or by expression of defined transcription factors, increases telomerase activity and restores telomere length, leading to a rejuvenated status of telomeres. Moreover, telomeric chromatin is also reprogrammed to an ES cell-like state, with a more open conformation and increased expression of TERRAs, at least in induced pluripotent stem cells obtained by in vitro reprogramming. All together the data available show that a complex regulation of telomeres occurs during differentiation and reprogramming and demonstrate that telomeric chromatin structure is dynamic and defined by cell-type specific epigenetic programs that can be reversed by reprogramming.

ACKNOWLEDGEMENTS

R.M.M. is a “Ramon y Cajal” senior scientist. M.A. Blasco´s laboratory is funded by the Spanish Ministry of Innovation and Science, the European Union, the European Reseach Council (ERC Advance Grants), the Spanish Association Against Cancer (AECC) and the Körber European Science Award to M.A. Blasco.

REFERENCES

1. Epel ES, Blackburn EH, Lin J et al. Accelerated telomere shortening in response to life stress. Proc Natl Acad Sci USA 2004; 101:17312-17315.
2. Valdes AM, Andrew T, Gardner JP et al. Obesity, cigarette smoking and telomere length in women. Lancet 2005; 366:662-664.
3. Canela A, Vera E, Klatt P et al. High-thoughput telomere length quantification by FISH and its application to human population studies. Proc Natl Acad Sci USA 2007; 104:5300-5305.
4. Cherkas LF, Aviv A, Valdes AM et al. The effects of social status on biological aging as measured by white-blood-cell telomere length. Aging Cell 2006; 5:361-365.
5. Oh H, Wang SC, Prahash A et al. Telomere attrition and Chk2 activation in human heart failure. Proc Natl Acad Sci USA 2003; 100:5378-5383.
6. O'Sullivan JN, Bronner MP, Brentnall TA et al. Chromosomal instability in ulcerative colitis is related to telomere shortening. Nat Genet 2002; 32:280-284.
7. Wiemann SU, Satyanarayana A, Tsahuridu M et al. Hepatocyte telomer shortening and senescence are general markers of human liver cirrhosis. FASEB J 2002; 16:935-942.
8. Samani NJ, Boultby R, Butler R et al. Telomere shortening in atherosclerosis. Lancet 2001; 358:472-473.
9. Wolthers KC, Bea G, Wisman A et al. T-cell telomere length in HIV-1 infection: no evidence for increased CD4+ T-cell turnover. Science 1996; 274:1543-1547.
10. Cawthon RM, Smith KR, O'Brien E et al. Association between telomere length in blood and mortality in people aged 60 years or older. Lancet 2003; 361:393-395.
11. Honig LS, Schupf N, Lee JH et al. Shorter telomeres are associated with mortality in those with APOE epsilon4 and dementia. Ann Neurol 2006; 60:181-187.
12. Mitchell JR, Wood E, Collins K. A telomerase component is defective in the human disease dyskeratosis congenital. Nature 1999; 402:551-555.
13. Yamaguchi H, Calado RT, Ly H et al. Mutations in TERT, the gene for telomerase reverse transcriptase, in aplastic anemia. N Engl J Med 2005; 352:1413-1424.
14. Tsakiri KD, Cronkhite JT, Kuan PJ et al. Adult-onset pulmonary fibrosis caused by mutations in telomerase. Proc Natl Acad Sci USA 2007; 104:7552-7557.
15. Armanios MY, Chen JJ, Cogan JD et al. Telomerase mutations in families with idiopathic pulmonary fibrosis. N Engl J Med 2007; 356:1317-1326.
16. Blasco MA. Telomere length, stem cells and aging. Nat Chem Biol 2007; 3:640-649.
17. Serrano M, Blasco MA. Cancer and ageing: convergent and divergent mechanisms. Nat Rev Mol Cell Biol 2007; 9:715-722.
18. Finkel T, Serrano M, Blasco MA. The common biology of cancer and ageing. Nature 2007; 448:767-774.
19. Collado M, Blasco MA, Serrano M. Cellular senescence in cancer and aging. Cell 2007; 130:223-233.
20. Gurdon JB, Melton DA. Nuclear reprogramming in cells. Science 2008; 322:1811-1815.
21. Rideout WM 3rd, Eggan K, Jaenisch R. Nuclear cloning and epigenetic reprogramming of the genome. Science 2001; 293:1093-1098.
22. Yang X, Smith SL, Tian XC et al. Nuclear reprogramming of cloned embryos and its implications for therapeutic cloning. Nat Genet 2007; 39:295-302.
23. Hochedlinger K, Plath K. Epigenetic reprogramming and induced pluripotency. Development 2009; 136:509-523.
24. Wilmut I, Schnieke AE, McWhir J et al. Viable offspring derived from fetal and adult mammalian cells. Nature 1997; 385:810-813.
25. Tada M, TakahamaY, Abe K et al. Nuclear reprogramming of somatic cells by in vitro hybridization with ES cells. Curr Biol 2001; 11:1553-1558.
26. Wakayama T, Perry AC, Zuccotti M et al. Full-term development of mice from enucleated oocytes injected with cumulus cell nuclei. Nature 1998; 394:369-374.
27. Kato Y, Tani T, Sotomaru Y et al. Eight calves cloned from somatic cells of a single adult. Science 1998; 282:2095-2098.
28. Polejaeva IA, Chen SH, Vaught TD et al. Cloned pigs produced by nuclear transfer from adult somatic cells. Nature 2000; 407:86-90.
29. Chesné P, Adenot PG, Viglietta C et al. Cloned rabbits produced by nuclear transfer from adult somatic cells. Nat Biotechnol 2002; 20:366-369.
30. Shin T, Kraemer D, Pryor J et al. A cat cloned by nuclear transplantation. Nature 2002; 415:859.
31. Takahashi K, Yamanaka S. Induction of pluripotent stem cells from mouse embryonic and adult fibroblast cultures by defined factors. Cell 2006; 126:663-676.
32. Takahashi K, Tanabe K, Ohnuki M et al. Induction of pluripotent stem cells from adult human fibroblasts by defined factors. Cell 2007; 131:861-872.

33. Amabile G, Meissner A. Induced pluripotent stem cells: current progress and potential for regenerative medicine. Trends Mol Med 2009; 15:59-68.
34. Chan SR, Blackburn EH. Telomeres and telomerase. Philos Trans R Soc Lond B Biol Sci 2004; 359:109-121.
35. Palm W, de Lange T. How shelterin protects mammalian telomeres. Annu Rev Genet 2008; 42:301-334.
36. Blackburn EH. Switching and signaling at the telomere. Cell 2001; 106:661-673.
37. de Lange T. Shelterin: the protein complex that shapes and safeguards human telomeres. Genes Dev 2005; 19:2100-2110.
38. d'Adda di Fagagna F, Reaper PM, Clay-Farrace L et al. A DNA damage checkpoint response in telomere-initiated senescence. Nature 2003; 426:194-198.
39. Harley CB, Futcher AB, Greider CW. Telomeres shorten during ageing of human fibroblasts. Nature 1990; 345:458-460.
40. Blasco MA. Telomeres and human disease: ageing, cancer and beyond. Nat Rev Genet 2005; 6:611-622.
41. Chan SW, Blackburn EH. New ways not to make ends meet: telomerase, DNA damage proteins and heterochromatin. Oncogene 2002; 21:553-563.
42. Collins K, Mitchell JR. Telomerase in the human organism. Oncogene 2002; 21:564-579.
43. Dunham MA, Neumann AA, Fasching CL et al. Telomere maintenance by recombination in human cells. Nature Genet 2000; 26:447-450.
44. Schoeftner S, Blasco MA. A 'higher order' of telomere regulation: telomere heterochromatin and telomeric RNAs. EMBO J 2009; 28:2323-2336.
45. Schoeftner S, Blasco MA. Chromatin regulation and noncoding RNAs at mammalian telomeres. Semin Cell Dev Biol 2009; doi:10.1016/j.semcdb.2009.09.015.
46. García-Cao M, O'Sullivan R, Peters AH et al. Epigenetic regulation of telomere length in mammalian cells by the Suv39h1 and Suv39h2 histone methyltransferases. Nat Genet 2004; 36:94-99.
47. Gonzalo S, Jaco I, Fraga MF et al. DNA methyltransferases control telomere length and telomere recombination in mammalian cells. Nat Cell Biol 2006; 8:416-424.
48. Benetti R, Gonzalo S, Jaco I et al. Suv4-20h deficiency results in telomere elongation and derepression of telomere recombination. J Cell Biol 2007; 178:925-936.
49. Benetti R, Gonzalo S, Jaco I et al. A mammalian microRNA cluster controls DNA methylation and telomere recombination via Rbl2-dependent regulation of DNA methyltransferases. Nat Struct Mol Biol 2008; 15:268-279.
50. Azzalin CM, Reichenbach P, Khoriauli L et al. Telomeric repeat containing RNA and RNA surveillance factors at mammalian chromosome ends. Science 2007; 318:798-801.
51. Schoeftner S, Blasco MA. Developmentally regulated transcription of mammalian telomeres by DNA-dependent RNA polymerase II. Nat Cell Biol 2008; 10:228-236.
52. van Steensel B, de Lange T. Control of telomere length by the human telomeric protein TRF1. Nature 1997; 385:740-743.
53. Muñoz P, Blanco R, Flores JM et al. XPF nuclease-dependent telomere loss and increased DNA damage in mice overexpressing TRF2 result in premature aging and cancer. Nat Genet 2005; 37:1063-1071.
54. Ye JZ, de Lange T. TIN2 is a tankyrase 1 PARP modulator in the TRF1 telomere length control complex. Nat Genet 2004; 36:618-623.
55. Smith S, de Lange T. Tankyrase promotes telomere elongation in human cells. Curr Biol 2000; 10:1299-1302.
56. Flores I, Cayuela ML, Blasco MA. Effects of telomerase and telomere length on epidermal stem cell behavior. Science 2005; 309:1253-1256.
57. Flores I, Benetti R, Blasco MA. Telomerase regulation and stem cell behavior. Curr Opin Cell Biol 2006; 18:254-260.
58. Liu L, Bailey SM, Okuka M et al. Telomere lengthening early in development. Nat Cell Biol 2007; 9:1436-1441.
59. Flores I, Canela A, Vera E et al. The longest telomeres: A general signature of adult stem cell compartments. Genes Dev 2008; 22:654-667.
60. Blasco MA, Lee HW, Hande MP et al. Telomere shortening and tumor formation by mouse cells lacking telomerase RNA. Cell 1997; 91:25-34.
61. Lee HW, Blasco MA, Gottlieb GJ et al. Essential role of mouse telomerase in highly proliferative organs. Nature 1998; 392:569-574.
62. Herrera E, Samper E, Martín-Caballero J et al. Disease states associated with telomerase deficiency appear earlier in mice with short telomeres. EMBO J 1999; 18:2950-2960.
63. Samper E, Fernández P, Eguía R et al. Long-term repopulating ability of telomerase-deficient murine hematopoietic stem cells. Blood 2002; 99:2767-2775.
64. Ferrón S, Mira H, Franco S et al. Telomere shortening and chromosomal instability abrogates proliferation of adult but not embryonic neural stem cells. Development 2004; 131:4059-4070.

65. Bell DR, Van Zant G. Stem cells, aging and cancer: inevitabilities and outcomes. Oncogene 2004; 23:7290-7296.
66. Vulliamy T, Marrone A, Goldman F et al. The RNA component of telomerase is mutated in autosomal dominant dyskeratosis congenita. Nature 2001; 413:432-435.
67. Vulliamy T, Marrone A, Szydlo R et al. Disease anticipation is associated with progressive telomere shortening in families with dyskeratosis congenita due to mutations in TERC. Nat Genet 2004; 36:447-449.
68. Marrone A, Stevens D, Vulliamy T et al. Heterozygous telomerase RNA mutations found in dyskeratosis congenita and aplastic anemia reduce telomerase activity via haploinsufficiency. Blood 2004; 104:3936-3942.
69. García-Cao I, García-Cao M, Tomás-Loba A et al. Increased p53 activity does not accelerate telomere-driven ageing. EMBO Rep 2006; 7:546-552.
70. Vaziri H, Dragowska W, Allsopp RC et al. Evidence for a mitotic clock in human hematopoietic stem cells: loss of telomeric DNA with age. Proc Natl Acad Sci USA 1994; 91:9857-9860.
71. Allsopp RC, Cheshier S, Weissman IL. Telomere shortening accompanies increased cell cycle activity during serial transplantation of hematopoietic stem cells. J Exp Med 2001; 193:917-924.
72. Allsopp RC, Morin GB, DePinho R et al. Telomerase is required to slow telomere shortening and extend replicative lifespan of HSCs during serial transplantation. Blood 2003; 102:517-520.
73. Allsopp RC, Morin GB, Horner JW et al. Effect of TERT over-expression on the long-term transplantation capacity of hematopoietic stem cells. Nat Med 2003; 9:369-371.
74. Sarin KY, Cheung P, Gilison D et al. Conditional telomerase induction causes proliferation of hair follicle stem cells. Nature 2005; 436:1048-1052.
75. Siegl-Cachedenier I, Flores I, Klatt P et al. Telomerase reverses epidermal hair follicle stem cell defects and loss of long-term survival associated with critically short telomeres. J Cell Biol 2007; 179:277-290.
76. Flores I, Blasco MA. A p53-dependent response limits epidermal stem cell functionality and organismal size in mice with short telomeres. PLoS One 2009; 4(3):e4934. doi:10.1371/journal.pone.0004934
77. Ju Z, Jiang H, Jaworski M et al. Telomere dysfunction induces environmental alterations limiting hematopoietic stem cell function and engraftment. Nat Med 2007; 3:742-747.
78. Tomás-Loba A, Flores I, Fernández-Marcos PJ et al. Telomerase reverse transcriptase delays aging in cancer-resistant mice. Cell 2008; 135:609-622.
79. Shiels PG, Kind AJ, Campbell KH et al. Analysis of telomere lengths in cloned sheep. Nature 1999; 399:316-317.
80. Wakayama T, Shinkai Y, Tamashiro KL et al. Cloning of mice to six generations. Nature 2000; 407:318-319.
81. Lanza RP, Cibelli JB, Blackwell C et al. Extension of cell life-span and telomere length in animals cloned from senescent somatic cells. Science 2000; 288:665-669.
82. Tian XC, Xu J, Yang X. Normal telomere lengths found in cloned cattle. Nat Genet 2000; 26:272-273.
83. Betts D, Bordignon V, Hill J et al. Reprogramming of telomerase activity and rebuilding of telomere length in cloned cattle. Proc Natl Acad Sci USA 2001; 98:1077-1082.
84. Clark AJ, Ferrier P, Aslam S et al. Proliferative lifespan is conserved after nuclear transfer. Nat Cell Biol 2003; 5:535-538.
85. Miyashita N, Shiga K, Yonai M et al. Remarkable differences in telomere lengths among cloned cattle derived from different cell types. Biol Reprod 2002; 66:1649-1655.
86. Schaetzlein S, Lucas-Hahn A, Lemme E et al. Telomere length is reset during early mammalian embryogenesis. Proc Natl Acad Sci USA 2004; 101:8034-8038.
87. Xu J, Yang X. Telomerase activity in early bovine embryos derived from parthenogenetic activation and nuclear transfer. Biol Reprod 2001; 64:770-774.
88. Thomson JA, Itskovitz-Eldor J, Shapiro SS et al. Embryonic stem cell lines derived from human blastocysts. Science 1998; 282:1145-1147.
89. Stadtfeld M, Maherali N, Breault DT et al. Defining molecular cornerstones during fibroblast to iPS cell reprogramming in mouse. Cell Stem Cell 2008; 2:230-240.
90. Zhu J, Wang H, Bishop JM et al. Telomerase extends the lifespan of virus-transformed human cells without net telomere lengthening. Proc Natl Acad Sci USA 1999; 96:3723-3728.
91. Marion RM, Strati K, Li H et al. Telomeres acquire embryonic stem cell characteristics in induced pluripotent stem cells. Cell Stem Cell 2009; 4:141-154.
92. Meshorer E, Yellajoshula D, George E et al. Hyperdynamic plasticity of chromatin proteins in pluripotent embryonic stem cells. Dev Cell 2006; 10:105-116.
93. Wong LH, Ren H, Williams E et al. Histone H3.3 incorporation provides a unique and functionally essential telomeric chromatin in embryonic stem cells. Genome Res 2009; 19:404-414.
94. Wong LH, McGhie JD, Sim M. ATRX interacts with H3.3 in maintaining telomere structural integrity in pluripotent embryonic stem cells. Genome Res 2010; doi 10.1101/gr.101477.109.

95. Marión RM, Strati K, Li H et al. A p53-mediated DNA damage response limits reprogramming to ensure iPS cell genomic integrity. Nature 2009; 460:1149-1153.
96. Li H, Collado M, Villasante A et al. The Ink4/Arf locus is a barrier for iPS cell reprogramming. Nature 2009; 460:1136-1139.
97. Hong H, Takahashi K, Ichisaka T et al. Suppression of induced pluripotent stem cell generation by the p53-p21 pathway. Nature 2009; 460:1132-1135.
98. Kawamura T, Suzuki J, Wang YV et al. Linking the p53 tumour suppressor pathway to somatic cell reprogramming. Nature 2009; 460:1140-1144.
99. Utikal J, Polo JM, Stadtfeld M et al. Immortalization eliminates a roadblock during cellular reprogramming into iPS cells. Nature 2009; 460:1145-1148.
100. Wang J, Xie LY, Allan S et al. Myc activates telomerase. Genes Dev 1998; 12:1769-1774.

CHAPTER 10

X CHROMOSOME INACTIVATION AND EMBRYONIC STEM CELLS

Tahsin Stefan Barakat and Joost Gribnau*

Abstract: X chromosome inactivation (XCI) is a process required to equalize the dosage of X-encoded genes between female and male cells. XCI is initiated very early during female embryonic development or upon differentiation of female embryonic stem (ES) cells and results in inactivation of one X chromosome in every female somatic cell. The regulation of XCI involves factors that also play a crucial role in ES cell maintenance and differentiation and the XCI process therefore provides a beautiful paradigm to study ES cell biology. In this chapter we describe the important *cis* and *trans* acting regulators of XCI and introduce the models that have been postulated to explain initiation of XCI in female cells only. We also discuss the proteins involved in the establishment of the inactive X chromosome and describe the different chromatin modifications associated with the inactivation process. Finally, we describe the potential of mouse and human ES and induced pluripotent stem (iPS) cells as model systems to study the XCI process.

INTRODUCTION

In many species, the sex of an individual is genetically determined by genes located on sex chromosomes.[1] Mammals are heterogametic, in which the female nucleus contains two X chromosomes and a male nucleus one X chromosome and one Y chromosome. Sex chromosomes originated from a pair of autosomes and divergence of these autosomes in proto X and proto Y chromosomes was initiated by the emergence of the key male sex determining gene SRY, which evolved from the ancestral SOX3 gene on the proto Y chromosome.[2-5] In subsequent steps the proto Y chromosome acquired genes involved in male fertility, resulting in a genomic region which was nonhomologous with the X chromosome. It is thought that the absence of homology initiated degeneration of the Y

*Corresponding Author: Joost Gribnau—Department of Reproduction and Development, Room Ee 09-71, Erasmus MC, 3015 GE Rotterdam, The Netherlands. Email: j.gribnau@erasmusmc.nl

The Cell Biology of Stem Cells, edited by Eran Meshorer and Kathrin Plath.
©2010 Landes Bioscience and Springer Science+Business Media.

chromosome. Nowadays the Y chromosome contains only a limited number of genes, most of them involved in male fertility and sex determination.[6,7] The X chromosome is still able to recombine in the female germ line, which prevented degradation and therefore resulted in the maintenance of a large chromosome containing more then a thousand genes, involved in a plethora of biological functions, varying from brain development to metabolism and fertility.[1,8]

Since both sexes contain an equal number of autosomal chromosomes, an equal dosage of X linked genes is needed to accomplish functional cell physiology. In placental mammals, dosage compensation of X linked genes between both sexes is achieved by inactivation of one of the two X chromosomes in females, in a process called X chromosome inactivation (XCI).[9-11] XCI occurs during early female development and results in functional hetero-chromatinization and silencing of the X chromosome, which is maintained during subsequent cell divisions throughout life.[12] Therefore in both sexes, only one X chromosome is functionally active. Expression of X encoded genes from this chromosome and the single X in male cells is twofold up regulated compared to autosomes, thereby further contributing to proper dosage compensation of X-encoded genes.[13-16] In mouse, X chromosome inactivation (XCI) is present in two forms. In the extra-embryonic tissues XCI is imprinted, with the paternal X chromosome (Xp) being inactivated in all cells.[17] This process is initiated very early during development, around the two- to eight-cell stage[18,19] and is maintained in the developing extra-embryonic tissues of the embryo, including the fetal placenta. In contrast, the Xp is reactivated in

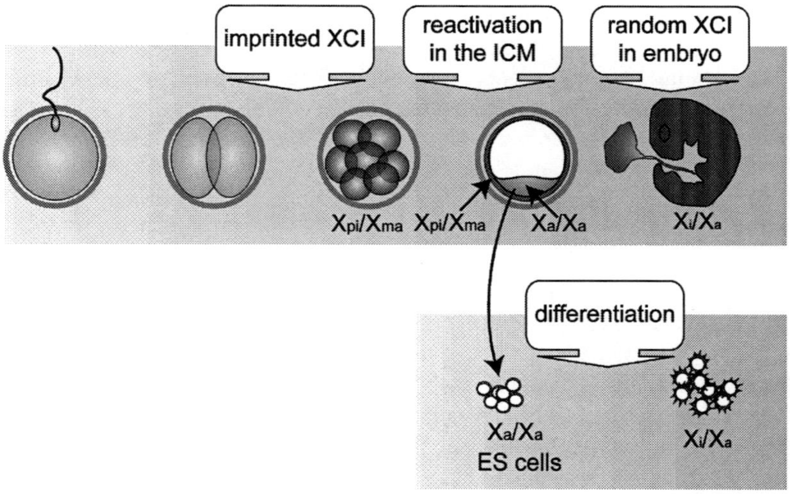

Figure 1. X chromosome inactivation during female mouse development. During early mouse development, at the 2 to 8 cell stage, imprinted XCI is initiated, in which the paternal X chromosome is always inactivated (Xpi, red cells). Imprinted XCI is maintained in the extraembryonic trophectoderm (the future placenta) and the primitive endoderm. In the inner cell mass of the blastocyst, imprinted XCI is reversed and both X chromosomes again become transcriptionally active (XaXa, grey). Around the time of implantation of the embryo in the uterus, random XCI starts in the fetal precursor cells which arise from the ICM of the early embryo (XiXa, blue cells paternal X active, red cells maternal X active). XCI is thus coupled to development and cell differentiation. Therefore embryonic stem (ES) cells, which are isolated from the ICM of the blastocyst, can be used to study XCI in vitro, as random XCI is initiated in female ES cells upon differentiation. A color version of this figure is available at www.landesbioscience.com/curie.

the inner cell mass (ICM), which gives rise to the embryo proper, after which random X inactivation is initiated around day 5.5 of development (Fig. 1).

Embryonic stem (ES) cells are derived from the inner cell mass (ICM) of a blastocyst. They are characterized by the ability of self renewal and pluripotency with the capacity to form all cell types of the embryo proper and adult organism upon differentiation.[20,21] Besides potential applications for regenerative medicine, ES cells are an ideal study system for early mammalian development from the pre-implantation period onwards.[22,23] Female mouse ES cells retain two active X chromosomes (Xa) and upon differentiation these cells initiate random XCI, making them the prevailing model system to study XCI.[24,25] Besides simulating early development, XCI in ES cells itself is important to accomplish proper cell function and developmental potential. The recent discovery of induced pluripotent stem (iPS) cells, which have ES cell characteristics and are derived from somatic cells by reprogramming with defined pluripotency factors,[26-31] has reemphasized the importance of XCI. The active status of X chromosomes in ES and iPS cells can be used as a pluripotency marker, as during iPS reprogramming the inactivated X chromosome (Xi) from somatic cells becomes reactivated in iPS cells.[32,33]

In this chapter, we first describe the genetic key players involved in XCI, followed by a discussion of different models proposed to explain the counting and choice of X chromosomes during XCI and the establishment and maintenance of the inactive X during subsequent cell divisions. Finally, the differences between mouse and human ES cells regarding XCI are discussed.

CIS ACTING FACTORS IN XCI

Transcriptional silencing of a whole chromosome during development has fascinated biologists for decades and in recent years a considerable amount of knowledge has been acquired contributing to our understanding of the molecular mechanisms involved in XCI. Genetic studies in mice and humans with X-to-autosome translocations have revealed that a major X-linked control locus, the X inactivation center (Xic in mice and XIC in humans) is necessary for XCI to occur.[34-37] The Xic, which encompasses more then 1 Mb on the mouse X chromosome,[38] has been shown to contain at least 4 genes which are involved in the process of XCI (Fig. 2). Three of these genes, *Xist*, *Tsix* and *Xite*, are noncoding and represent the master switch locus involved in silencing of the X chromosome in *cis*. The fourth gene, *Rnf12* which will be discussed in later sections, encodes an E3 ubiquitin ligase involved in regulation of XCI in *trans*.

The first gene which has been found to map to the XIC/Xic is the X-inactive specific transcript (*Xist* in mice, *XIST* in humans).[37,39,40] *Xist* is the only known gene which is specifically expressed from the Xi. *Xist* is a noncoding gene, consisting of 7 exons in the mouse and 8 exons in humans, producing a poly-adenylated RNA molecule (17 kb in human and 15 kb in mouse), which is subject to alternative splicing.[41,42] *Xist* RNA is tightly associated with the Xi[43,44] and it is required for XCI to occur in *cis*, as knockout studies in female ES cells and mice have shown that X chromosomes bearing a deletion of the *Xist* gene are unable to inactivate the mutated X.[45-47] During XCI, expression of *Xist* is up regulated from the future Xi.[48,49] *Xist* RNA molecules spread from the Xic across the X chromosome in *cis* and cover the Xi completely. By spreading over the Xi, *Xist* RNA induces heterochromatinization of the X chromosome by attracting chromatin modifiers, involved in gene silencing.[50-52]

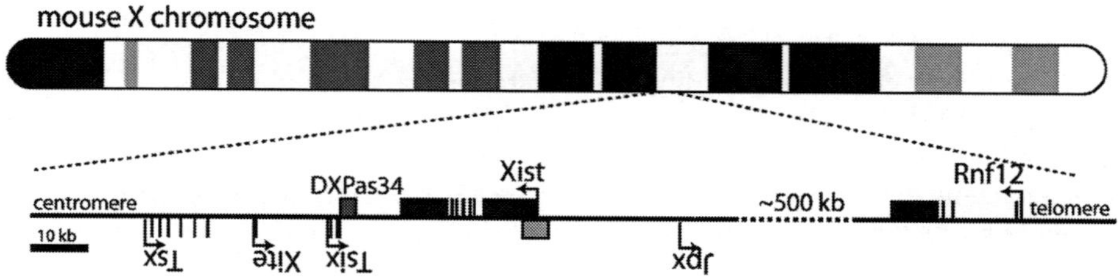

Figure 2. The X inactivation center on the X chromosome. Schematic representation of the mouse X chromosome and the location of part of the X inactivation center including the genes, *Xist*, *Tsix*, *Xite* and *Rnf12* which are involved in regulating XCI.

Another noncoding gene located within the Xic is *Tsix*, which is transcribed antisense to *Xist*.[53] *Tsix* contains four exons and at least two transcriptional start sites, producing a 40 kb transcript, which only localizes to the Xic, as determined by RNA-FISH experiments. In mice the *Tsix* gene fully overlaps with *Xist* and *Tsix* has been shown to negatively regulate expression of *Xist*, as a deletion of *Tsix* leads to up regulation of *Xist* transcription and exclusive inactivation of the mutated X chromosome in female cells.[54,55] Prior to XCI, expression of *Tsix* is from both X chromosomes in a 10 to 100 times excess compared to *Xist* and after initiation of XCI expression is continued transiently on the future Xa.[56] On the Xi, transcriptional shutdown of *Tsix* is accompanied by chromatin changes at the *Tsix* promoter.[57] The exact mechanisms involved in *Tsix*-mediated silencing of *Xist* are unknown. Since expression of *Tsix* transcripts is found in a gradient along the *Xist* gene, with more transcripts in the 5' portion of *Tsix* relative to the 3' portion of the gene, a role for transcriptional interference as a mechanism to suppress *Xist* has been proposed.[49,56,58] Another possible mechanism by which *Tsix* might suppress *Xist* transcription is via RNA-mediated silencing. It has been shown that *Tsix* regulates the methylation status and thus the activity of the *Xist* promoter, via de novo methyltransferase 3a (DNMT3A).[49] Also, active chromatin marks are more abundant at the *Xist* promoter in cells with a deficient *Tsix* gene in *cis*, whereas marks of repressed chromatin are reduced.[59,60] Antisense transcription through the *Xist* promoter itself seems to be crucial for the establishment of repressive chromatin marks, as a truncation of *Tsix* to 93% of its normal length failed to induce *Xist* silencing.[61] Also deleting the DXPas34 element, which is a CpG island located downstream to the major *Tsix* promoter and also initiates antisense transcription, abrogates *Xist* silencing in *cis*, thereby further emphasizing the importance of antisense transcription in *Tsix*-mediated silencing of *Xist*.[60,62,63] Furthermore, the methylation status of this CpG island coincides perfectly with antisense transcription through *Xist*.[64,65] *Xist* and *Tsix* transcripts are partially overlapping and therefore a possible role for an RNAi-mediated mechanism regulating XCI can not be excluded.[66] Small xiRNAs, ranging in size from 25 to 42 nucleotides have indeed been detected from different regions within the *Xist* gene and a mutation of the endonuclease *Dicer* resulted in a loss of xiRNA formation and decreased methylation of *Xist*, implicating a role for *Dicer* in XCI. This is disputed by others, who found that *Dicer* null-ES cells show normal XCI and that the effects on the Xa are mediated by a decreased activity of de novo methyltransferases rather than a direct effect of *Dicer*.[67,68] Therefore, at present the exact role of small RNAs in XCI initiation is unclear. Also, over expression in *cis* of *Tsix* cDNA in a cell line with abrogated endogenous *Tsix* transcription did not result in restoration of *Tsix*-mediated *Xist* silencing, which argues against an RNAi-mediated process.[69]

The third noncoding gene involved in XCI, *Xite*, for X chromosome intergenic transcript element,[70] is located approximately 10 kb upstream of *Tsix* and its expression and methylation pattern during XCI is similar to that of *Tsix*. *Xite* is believed to be the positive regulator of *Tsix*.[71] Deletion of *Xite* results in a reduced antisense transcription through the *Xist* locus, implying a similar role for *Xite* in inhibiting *Xist* expression as for *Tsix* and DXPas34.[65,70]

TRANS ACTING FACTORS IN XCI

How are *Xist*, *Tsix* and *Xite* regulated? Among the proteins involved in *Tsix* regulation are the insulator protein CTCF and the transcription factor yin yang 1 (YY1), for which several tandemly organized binding sites have been identified in the *DXpas34* region and the *Xite* promoter.[72] Knockout studies involving *Yy1*, or partial ablation of *Yy1* and *Ctcf* through RNAi mediated repression, revealed down-regulation of *Tsix* expression and concomitant up-regulation of *Xist* expression, supporting a role for YY1 and CTCF in *Tsix* expression.[73] Recently, it has been shown that the pluripotency transcription factor OCT4 is able to bind to the DXpas34 element and the *Xite* promoter and may cooperate with CTCF in the regulation of *Tsix* transcription.[74] Another study showed that the pluripotency factors NANOG, OCT4 and SOX2[75-78] can repress *Xist* expression by binding to a region within *Xist* intron 1.[79] *Nanog*-deficient ES cells show up-regulation of *Xist* and this preceded down-regulation of *Tsix*, suggesting a *Tsix*-independent mechanism for the pluripotency factors in *Xist* repression. Interestingly, Nanog expression is also required for reactivation of the inactive paternal X chromosome in the ICM, most likely through suppression of *Xist*.[80]Therefore it seems likely that the well characterized key players of pluripotency are involved in repressing XCI prior to differentiation through different pathways.

A transgenesis screen in ES cells using BAC transgenes covering part of the Xic region, resulted in the identification of X-encoded *Rnf12* as an activator of XCI.[81] *Rnf12* is located 500 kb telomeric from *Xist*, within the region that originally delineated the Xic and encodes an E3 ubiquitin ligase which has previously shown to be involved in the regulation of LIM-homeodomain transcription factors, estrogen dependent transcription activation and telomere length homeostasis.[82-84] Additional copies of *Rnf12* can induce ectopic XCI in male ES cells on the single X chromosome and in a significant proportion of female cells on both X chromosomes.[81] Initiation of XCI was reduced in heterozygous *Rnf12* female ES cells compared with wild type female ES cells. These findings indicate an important dose-dependent role for RNF12 in activation of XCI, although the molecular mechanism by which RNF12 mediates XCI remains elusive so far. Also, more unidentified activators of XCI must be present to explain initiation of XCI in female *Rnf12* heterozygous knockout ES cells.

COUNTING AND CHOICE

A long standing question in XCI research involved the counting and choice mechanism regulating initiation of XCI and required to inactivate the proper number of X chromosomes.

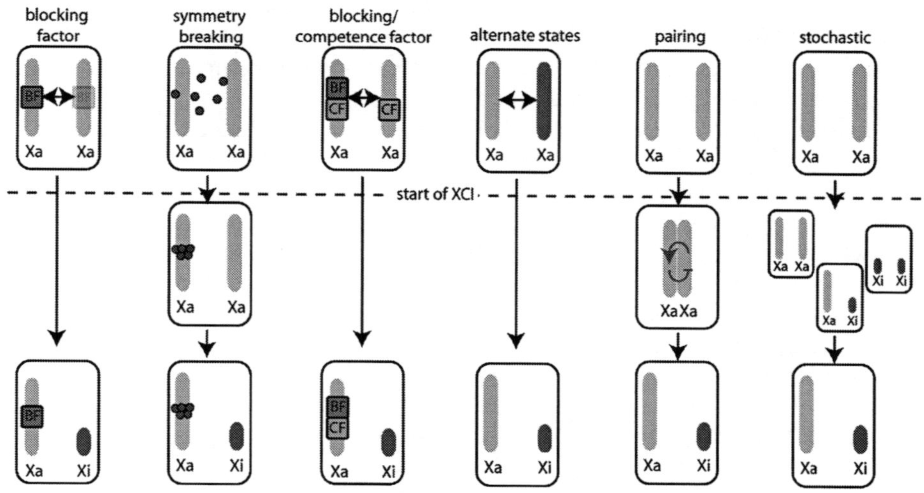

Figure 3. Models explaining counting and choice in X chromosome inactivation. Several models have been postulated to explain initiation of XCI in female cells only. The BF model presumes the presence of a single autosomally encoded protein or entity, the blocking factor (BF), in the nucleus that prevents XCI on one X. The symmetry breaking model proposes that the BF is composed of many proteins that self-assemble on the future Xa. An additional X-encoded competence factor (CF) is added in the BF/CF model. One CF copy is required for initiation of XCI on the unprotected X, whereas the other CF copy is titrated away by the BF on the future Xa. In the alternative states model XCI is predetermined and dependent on differences in the chromatin state of both X chromosomes. Female specific initiation of XCI has also been explained by a pairing model, where both Xic's in female cells come in close proximity, facilitating a cross communication process that regulates initiation of XCI. Finally, a stochastic model has been postulated to explain XCI. In this model, X chromosomes have a probability to initiate XCI and only cells with the right number of active X chromosomes survive.

How does a cell sense the number of X chromosomes present in a nucleus and how many of them need to be inactivated? A number of clinical observations in patients with an aberrant number of X chromosomes has shed some light on this question. In patients with a supernumerary number of X chromosomes, like so-called 46,XXX super females or 46, XXY Klinefelter patients, all but one X chromosome become inactivated.[85-87] In Turner syndrome, the single X chromosome present in these females does not undergo XCI, whereas in tetraploid female embryos two inactivated X chromosomes are found.[88,89] From this, the general rule has been deduced that XCI results in one Xa per diploid genome.[90]

Several models, with increasing complexity and partially overlapping mechanisms, have tried to explain these observations (Fig. 3). The blocking factor (BF) model predicts the presence of an autosomally encoded factor which is present in one entity in a diploid nucleus.[35,36] The BF is thought to act through binding to a DNA element, called counting element and only interacts with one X chromosome, thereby rescuing this X from XCI in *cis*. Since there is only enough BF in a diploid nucleus to prevent XCI on one X chromosome, all additional X chromosomes will become silenced. The related symmetry breaking model states that the BF is not a single entity, but consists of several autosomally encoded molecules which assemble on the future Xa, thereby preventing XCI.[91,92] Since the Xic is necessary for XCI to occur, as XCI is initiated only in the presence of more than one Xic in a diploid nucleus, it seems likely that the postulated counting element must be located within the Xic. Several studies made use of transgenes and deletions, to identify the counting element, reasoning that when an additional counting element would be introduced into male ES cells, this should be able to titrate away the limiting BF and induce counting. Therefore, XCI should also occur on the endogenous X chromosome, which is now no longer protected by the BF.

Indeed, several transgenic male ES cell lines have been generated with autosomally integrated transgenes covering *Xist* and *Tsix*, or *Xist* alone, in which ectopic XCI occurred on the single endogenous X chromosome.[93-98] Autosomal *Xist* expression and silencing of autosomal genes adjacent to the integration site was also observed. The transgenes used varied from large YACs carrying >500 kb to small cosmids of only 35 kb covering only *Xist* and flanking regions.[94] These studies therefore indicated that factors involved in counting may be located within the sequences covered by the transgenes. However, other related studies did not show induction of counting using similar transgenes,[81,99,100] or showed that only multicopy transgenes are able to induce counting.[101] Interestingly, studies involving a deletion of *Xist* on one X chromosome in female ES cells, which had shown that *Xist* is necessary for XCI to occur in *cis*, also showed that *Xist* transcription and the deleted part of the *Xist* gene are not involved in counting, as XCI is normally initiated on the wild type X chromosome.[45-47,102-104] In contrast, different male *Tsix* mutant ES cells, some already generated prior to the identification of *Tsix*, displayed initiation of XCI on the single X chromosome, which suggested a role for the deleted sequences in the counting process.[55,58,60,63,105] The same mutations did however not disturb the counting process in female cells, but resulted in preferential inactivation of the mutated allele, showing that *Tsix* mediates silencing of *Xist* in *cis*.[54] These findings could be explained through a mechanism whereby the respective mutations disrupted the counting element, preventing BF binding. However, a heterozygous deletion of a region including *Xist*, *Tsix* and Xite (ΔXTX), in female ES cells and mice did not result in a disturbed counting process, as the wild type X chromosome was normally inactivated.[106] This finding indicated that *Xist*, *Tsix* and Xite are not required for the counting process and locate the counting element outside the deleted region, although studies with *Xist* transgenes suggest that overlapping sequences may be involved which play a redundant role in counting and initiation of XCI.

In contrast to all previously reported *Tsix* mutations, one described *Tsix* mutation (ΔCpG), which involves a deletion of the DXPas34 region, did not result in aberrant XCI in male cells.[107] Interestingly, female cells with a homozygous ΔCpG *Tsix* mutation showed chaotic XCI, with many cells initiating XCI on both X chromosomes. Based on this finding an X-encoded competence factor (CF) was hypothesized, involved in activation of XCI. One model comprising the combined action of a BF and CF postulates that the abundantly present CF inactivates all X chromosomes but not the one to which BF is bound.[103] Another hypothesis states that also the X-encoded CF is limiting and is titrated away by one 'copy' of the autosomally-encoded BF, which corresponds to a single X chromosome.[107] When more than one X chromosome is present in a diploid background, the extra copies of CF will not be titrated by the BF and will inactivate the remaining unprotected X chromosome(s)

The model of alternate states proposes that the two X chromosomes in an XX cell are already different prior to XCI.[108] This model is supported by the fact that cohesion of sister chromatids has been shown to be differentially regulated between two X chromosomes in undifferentiated cells. Also differences in methylation and the chromatin state may play a role. Therefore an inherent epigenetic difference between two genetically identical chromosomes may exist prior to the initiation of XCI. However further experimental validations for these observation are needed.

A different model explains counting and choice in XCI by transient transvection or pairing events of the two Xic's present in a female diploid nucleus.[109,110] This model is supported by observations that in early differentiating ES cells, there is a nonrandom spatial distribution of the Xic's in the nucleus, at which the Xic's move closer to each other prior to the onset of XCI. This transient pairing event may therefore play a role in the regulation of counting and choice. Pairing is facilitated by *Tsix* and *Xite* sequences and it seems to be dependent on the action of CTCF.[111] The stem cell transcription factor *Oct4* and transcription mediated by RNA polymerase II are also crucial for the pairing events.[74,111] A genomic region containing part of the *Slc16A2* gene, located 250 to 350 kb telomeric of *Xist* has been identified which also mediates pairing of the X chromosomes at the onset of XCI[112] and was proposed to play a role in the activation of XCI. At present, it is unclear whether pairing has a functional role in XCI, or is a consequence of the transcriptional activation, which may result in relocation of the Xic in the nucleus. Interestingly, pairing appears not to be required for initiation of XCI, as XCI is initiated in $XX^{\Delta 65kb}$ ES cells with a deletion distal to *Xist* removing *Tsix* and *Xite* sequences, which abolish the XCI pairing event.[109,110] Therefore additional studies are needed to clarify the role of pairing in XCI.

Many of the above discussed models assume that the XCI process is deterministic and mutually exclusive, in which always the correct number of X chromosomes are inactivated in female cells. However, in vitro studies with diploid and tetraploid ES and ICM cells revealed a significant percentage of cells with too many or too few Xi's,[89,106,113] suggesting a stochastic mechanism directing the XCI process, with an independent probability for every X chromosome to initiate XCI.[106] Comparison of the relative number of cells that initiated XCI between different diploid, triploid and tetraploid ES cells indicated that the X to autosome ratio determines the probability for an X chromosome to be inactivated.[114] The probability is the resultant of different factors: X-encoded XCI-activators and autosomally encoded XCI-inhibitors that promote or repress *Xist* accumulation, respectively. Upon development or differentiation, the concentration of the XCI-activators will rise and/or the concentration of the XCI-inhibitors will decrease and in female cells this will be sufficient to generate a specific probability in time for enough *Xist* to accumulate and start to spread in *cis* (Fig. 4). XCI-inhibitors are involved in setting up a threshold that has to be overcome by *Xist* to accumulate. Because the XCI-activator gene is X-linked, spreading of *Xist* will down-regulate the XCI-activator gene in *cis*, preventing the second X chromosome from inactivation. In this model, initiation of spreading is a stochastic event, so that the chance for silencing of the XCI-activator gene on either X is equal. In male cells the concentration of the XCI-activator will not be sufficient to break the threshold and initiate XCI. Therefore, female specific initiation of XCI is obtained through a sex-dependent dosage difference in X-encoded XCI-activators that promote *Xist* accumulation. Cell lines and mice that harbor *Xist* or *Tsix* mutations that affect the expression of one of the genes indicate that *Xist* and *Tsix* are the major players in setting up the probability and that the XCI-activators and XCI-inhibitors are likely to act through these genes.

Several XCI-inhibitors have been identified, some of them acting through activation of *Tsix* expression, like *Yy1*, *Ctcf* and *Oct4*, or by direct suppression of *Xist*, like *Nanog*, *Oct4* and *Sox2*.[72-74,79] *Rnf12* was identified as an X-encoded activator of XCI.[81] The observed ectopic XCI induced by *Rnf12* transgenes correlates with expression of transgenic

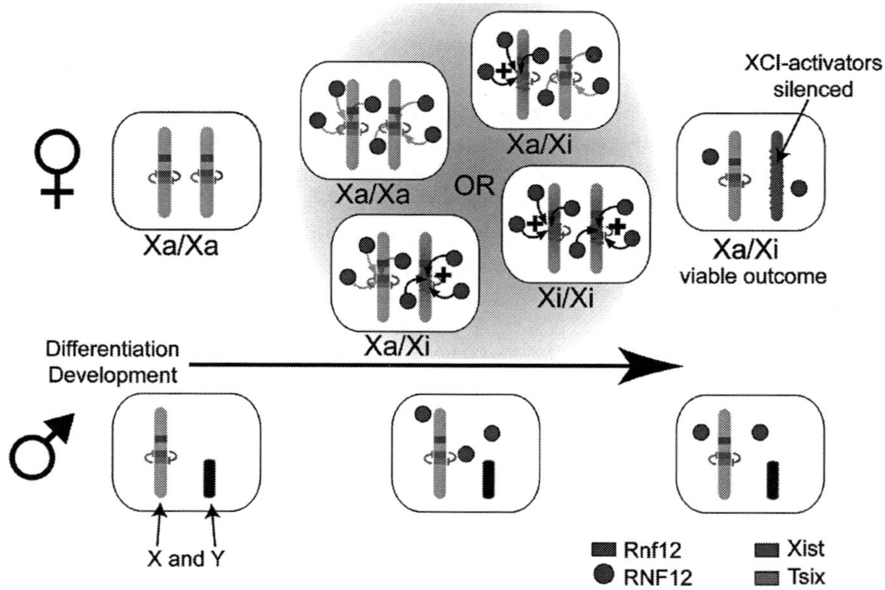

Figure 4. A stochastic model for XCI. A stochastic model for XCI postulates that initiation of XCI is dependent on X-encoded activators of XCI and autosomally encoded inhibitors of XCI. Upon differentiation or during development the nuclear concentration of the XCI-activator RNF12 (blue) and putative other XCI activators increases, whereas the concentration of XCI-inhibitors such as NANOG, OCT4 and SOX2 decreases (not shown), or remains constant in time (YY1 and CTCF, not shown). XCI-activators are X-encoded and therefore only in female cells the balance in the nuclear concentration of these factors is sufficient to generate an independent and continuous probability to initiate XCI (grey and black arrows) and results in cells that do not initiate XCI, or initiate XCI on one or both X chromosomes (+). Initiation of XCI results in spreading of Xist and silencing of XCI-activator genes in cis, resulting in down-regulation of the XCI-activator concentration, which prevents XCI on the remaining Xa. A color version of this figure is available at www.landesbioscience.com/curie.

Rnf12/RNF12, providing evidence for a dose-dependent role of RNF12 in activation of XCI. In agreement with this, heterozygous *Rnf12* knockout female ES cells only initiate XCI in a significant reduced number of cells. RNF12 expression in wild type ES cells is up-regulated around the time XCI is initiated and prior to XCI, a twofold higher dosage of RNF12 protein has been detected in female compared to male ES cells. *Rnf12* itself is subject to rapid silencing upon XCI induction, which will facilitate inhibition of initiation of XCI on the second X chromosome in female cells. How *Rnf12* regulates XCI is unknown and future studies have to address the molecular mechanisms involved in *Rnf12*-mediated activation of XCI.

The recent identification of inhibitors and activators of XCI support a stochastic model for XCI. This indicates that the XCI process does not involve a choice mechanism, but is dictated by an initiation and feedback process, which is regulated through an intricate balance between inhibiting factors and activators of XCI.

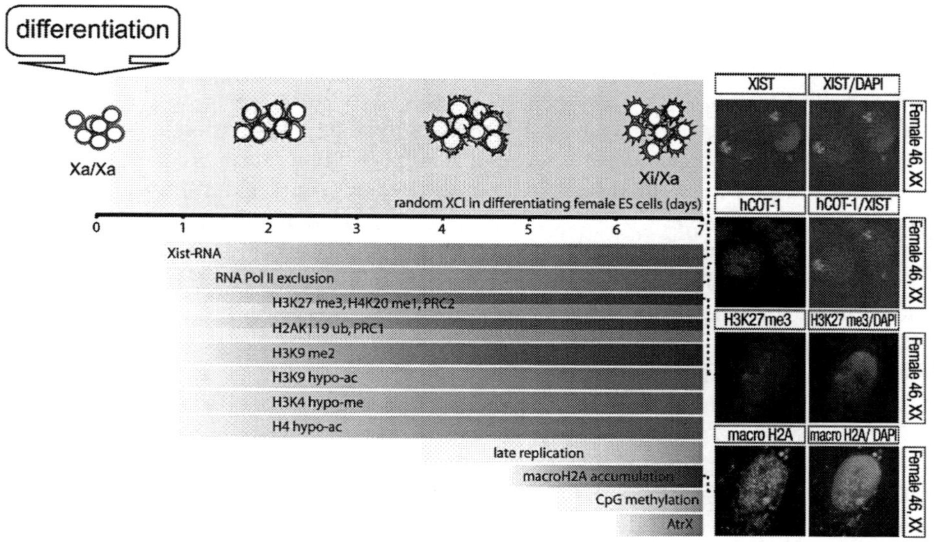

Figure 5. Epigenetic changes on the Xi. Initiation of random XCI is induced upon differentiation of female ES cells. The first change on the future Xi is spreading of *Xist* RNA in *cis* (red), followed by exclusion of RNA polymerase II (brown), the loss of active chromatin marks (blue) and gain of chromatin marks or incorporation of histone variants specific for inactive chromatin (purple). These chromatin changes are accompanied by accumulation of proteins and protein complexes (green) and other epigenetic changes including a shift to late replication in S phase and CpG methylation (orange). The right panels show female human fibroblast subjected to *XIST* RNA FISH, RNA FISH with a Cot probe detecting RNA Polymerase II transcribed repetitive regions, an immuno staining with anti H3K27 me3 antibody and an immuno staining with anti macroH2A antibody. Each cell contains one *XIST* signal coating the Xi, which corresponds with a Cot negative area, accumulation of H3K27 me3 and incorporation of macroH2A. A color version of this figure is available at www.landesbioscience.com/curie.

SILENCING AND MAINTENANCE OF SILENCING

Once XCI is initiated, a series of events takes place which changes the euchromatin of the active X chromosome into the tightly packed inactive heterochromatin of the Xi, which can be recognized as the Barr body in female somatic cells.[115,116] The first step in this cascade of events is the transcriptional up regulation and spreading of *Xist* on the future Xi[37,39] (Fig. 5). The *Xist* transcript contains several repeats, of which the A repeat, located in the 5'portion of *Xist*, is involved in the silencing process.[117] Coating of the future Xi with *Xist* RNA leads to a rapid deprivation of RNA polymerase II and associated transcription factors, which leads to an immediate reduction of gene transcription on this chromosome and the creation of a silent nuclear compartment.[118,119] One of the earliest events after *Xist* accumulation are specific chromatin changes, including the loss of active chromatin marks like H3K9 acetylation and H3K4 mono- and di-methylation,[19,120,121] as well as a gain of silent chromatin marks like H3K27 tri-methylation (H3K27-me3),[50,51,57,122] H3K9 di-methylation (H3K9-me2)[121,123-126] and H4K20-mono-methylation (H4K20-me1),[127] followed by the incorporation of histone variants including macroH2A[128] and changes in DNA methylation and replication timing.[129-131]

To initiate the establishment of chromatin changes and silencing, *Xist* needs to accumulate and spread along the X chromosome. The exact mechanism involved in *Xist* spreading is not clear. Studies on X-to-autosome translocations and autosomally integrated *Xist* transgenes have indicated that *Xist* spreading on autosomal regions is not complete, which indicated that X-chromosomal sequences are needed for efficient spreading.[35,95,132–135] Booster elements or way stations have been hypothesized to facilitate the *Xist* transmission along the X chromosome[136] and LINE-1 retrotransposons were postulated to be the most likely candidate for being these booster elements,[137–140] as they are one of the most common repetitive sequences in mammalian genomes and are enriched on the X chromosome compared to autosomes.[141-144] Regions on the X chromosome that undergo XCI have a higher LINE-1 density than regions surrounding genes that escape inactivation[8,142,143,145-147] and the density of LINE-1's on the fused autosome in X-to-autosome translocations seems to correlate with the amount of *Xist* spreading.[133-135,148-153] However, also other sequences may be important for the *Xist* spreading, as others did not find a correlation between LINE-1 density and spreading, or suggested a less critical role.[57,146,154-156] How *Xist* interacts with these sequences during the spreading process and whether these interactions are direct or mediated by other molecules is at present not clear.

After *Xist* accumulation, one of the first histone modifications detected on the Xi is H3K27-me3. This modification is dependent on the Polycomb repressive complex PRC2, which accumulates on the Xi and consists of the proteins EED, RbAp46/48, SUZ12 and EZH2, of which EZH2 is a methyl transferase implicated in H3K27 tri-methylation.[51,157,158] Recently, a direct interaction between EZH2 and the A repeat of *Xist* has been reported and *Ezh2* knockdown studies in female ES cells indicated a role for PRC2 in establishment of the Xi in random XCI.[52] Interestingly, a knockout of EED resulting in a non-functional PRC2 complex and the absence of H3K27-me3 only reveals a defect in imprinted XCI in mice,[159] with re-activation of the Xi in extra-embryonic tissues. Despite defects in embryonic development due to the absence of EED, random XCI was not affected in the embryo,[160] which contrasts with results obtained after knock down of *Ezh2* in vitro.[52] A different study, employing autosomally integrated inducible *Xist* transgenes, indicated that in the absence of EED some components of the PRC1 complex (MPH1 and MPH2) do not localize to the *Xist* coated autosome. However, recruitment of RING1B, which is part of the PRC1 complex and *Xist* mediated silencing of autosomal sequences were not affected.[161] Accumulation of H2AK119ub1 (mono-ubiquitylation of histone H2A on lysine 119) on the Xi is dependent on RING1B,[162,163] but also a homozygous *Ring1b* mutation does not affect *Xist* mediated silencing of an autosome with a transgenic insertion of *Xist*.[161] These findings indicate that PRC1 and PRC2 are dispensable for random XCI, which does not exclude the possibility that these complexes are involved in redundant mechanisms in the establishment and maintenance of the Xi.

Also other proteins have been implicated in establishment and maintenance of the Xi. An E3 ubiquitin ligase complex consisting of the speckle-type POZ protein SPOP and CULLIN3 has been shown to be involved in the regulation of macroH2A deposition on the Xi.[164] SPOP and CULLIN3 ubiquitinate both BMI1, a component of the PRC1 complex and macroH2A. Ubiquitination of macroH2A appears crucial for the recruitment of macroH2A to the Xi and RNAi-mediated knock-down of either SPOP or CULLIN3 results in diminished macroH2A staining on the Xi. RNAi-mediated knock-down of either macroH2A or SPOP/CULLIN3 in combination with demethylation and deacetylation inhibitor treatment, resulted in reactivation of an Xi-linked reporter gene. Reactivation was not found with a demethylation and deacetylation inhibitor treatment alone, indicating a role for macroH2A in the maintenance of the silent

Xi state. The silencing function of macroH2A might be indirectly established by recruitment of the Poly (ADP-ribose) polymerase 1, PARP-1.[165] PARP-1 is a nuclear enzyme involved in modulating chromatin structure, sensing DNA damage and regulation of gene expression.[166-168] MacroH2A is able to recruit PARP-1 to the Xi and to inhibit the catalytic activity of PARP-1. Enzymatically inactive PARP-1 is able to bind nucleosomes and inhibit transcription.[169] Depletion of PARP-1 in combination with demethylation and deacetylation inhibitor treatment leads to reactivation of an Xi-linked reporter gene. Hence, macroH2A might collaborate in gene silencing by modulating the enzymatic activity and thus the ability to silence, of PARP-1. The nuclear scaffold protein SAF-A is also enriched on the Xi. SAF-A has been characterized as a DNA/RNA binding protein and is a putative component of the nuclear scaffold involved in regulation of gene expression and DNA replication. Enrichment of SAF-A is dependent on the RNA binding domain of SAF-A, which might be involved in *Xist* mediated recruitment to the Xi.[170] Recruitment of *Xist* to the Xi may be facilitated by BRCA1, a protein involved in many pathways including checkpoint activation and DNA repair. Association of BRCA1 with the Xi was found in a small percentage of cells and reconstitution experiments indicated a role for BRCA1 in *Xist/XIST* localization.[171,172] Nonetheless, this claim is disputed by others and the exact role of BRCA1 in XCI remains to be determined.[173,174] SMCHD1 also associates with the Xi and is involved in the methylation of CpG islands of genes subject to XCI.[175] SMCHD1 contains a SMC-hinge domain which is found in proteins involved in cohesion and chromosome condensation. *SmcHD1* knockout mice show defects in maintenance of the Xi in embryonic and extra-embryonic tissues. Analysis of *Atrx* heterozygous female knockout mice implicated a role for ATRX (a chromatin remodeling protein named alpha-thalassaemia and mental retardation on the X chromosome) in imprinted XCI.[176] Female mice inheriting a mutated allele through the female germ line fail to inactivate the paternal X in extra-embryonic tissues, suggesting a failure in imprinted XCI. Interestingly, at later stages of ES cell differentiation, ATRX also associates with the Xi, supporting a role in maintenance of the Xi.[177] *SatB1* and *SatB2* have also been implicated in the establishment of the Xi. These cancer associated genes encode nuclear proteins, which act as genome organizers and gene regulators.[178] Knockdown studies of *SatB1* and *SatB2* showed a partial defect in silencing of the X chromosome in differentiating female ES cells, suggesting a direct role for these proteins in heterochromatinization of the Xi.[179]

Interestingly, once the Xi is established, expression of *Xist* is no longer necessary for the maintenance, as a conditional deletion of *Xist* does not lead to reactivation of silenced genes.[47] Multiple redundant proteins and epigenetic layers are thus involved in the establishment and maintenance of the Xi. Together, they ensure that the Xi is stably silenced, leading to persistent dosage compensation. Only a limited number of X-linked genes is able to escape XCI mediated silencing in the mouse. In humans, more escapees have been identified.[180-183] Therefore, in different species, the robustness of silencing might differ, which might be caused by a differential evolutionary need for dosage compensation between species. The genes which escape dosage compensation and do not have Y-chromosomal homologues, are thus differentially expressed between males and females and may play a role in establishing differences between both sexes.

In summary, many different redundant epigenetic layers and only partly understood complex interactions are involved in the silencing and maintenance of the silent state of the X chromosome. Although many investigations have focussed on these mechanisms, only the first steps have been made towards a complete understanding of the silencing and maintenance of XCI.

XCI AND HUMAN ES CELLS

Studying X chromosome inactivation in humans is challenging. Due to ethical reasons, the use of early human embryos for research purposes has been widely restricted. Therefore, most of the obtained knowledge on human XCI has come from studies which made use of different model systems, including mouse-human cell hybrids, human embryonal carcinoma and tumour cell lines[184,185] and human transgenes integrated in mouse ES cells.[81,97,186-188] The derivation of human ES (hES) cells promised the availability of a potent study model for human XCI, comparable to mouse ES cells. Different studies explored XCI in hES cells, with varying and conflicting results.[189-194] Most of the female hES cell lines display an inactivated X chromosome already in the undifferentiated state characterized by *XIST* expression, *XIST* coating and accumulated markers of heterochromatin on the Xi. Other undifferentiated cell lines have active X chromosomes and have the potential to inactivate during differentiation, comparable to mouse ES cells (Table 1). Interestingly, certain cell lines behave different in distinct laboratories, with some sub clones showing random XCI upon differentiation, whereas others show XCI hallmarks prior to differentiation. In a survey of 11 characterized hES cell lines, Silva et al identified three different classes of hES cells with regard to XCI.[191] The first class only displays XCI characteristics upon differentiation. The second and third classes have an inactive X chromosome in the undifferentiated state, but in the third group *XIST* expression is lost. Although this last category does no longer express *XIST*, other XCI markers are still present, like the exclusion of Cot-1 RNA from the X chromosome. Interestingly, in these cells H3K27 tri-methylation is also lost, which indicates that H3K27 tri-methylation is dispensable for the maintenance of XCI. These results support previous findings that in mice and human recruitment of H3K27 tri-methylation is *XIST* dependent.[50]

Table 1. Characteristics of mouse and human pluripotent stem cells

	Mouse			Human	
	ES Cells	iPS Cells	EpiSC	ES Cells	iPS
XaXa	+	+	–	+/–	–[1]
XaXi	–	–	+	+/–	+[1]
Xist/XIST	–	–	+	+/–	+[1]
H3K27 me3, Cot exclusion	–	–	+	+/–	+[1]
bFGF, Activin/Nodal sign.	–	–	+	+	+
Lif/Stat3 signalling	+	+	–	–	–
diff. in extra embr. tissue	–	–	+	+	?

Comparison of XCI characteristics and growth conditions of different reported human and mouse pluripotent female stem cells (1, only one female iPS line was tested). Indicated are the presence (+) or absence (-) of an Xi and epigenetic changes associated with XCI in the different pluripotent stem cells described. Also shown are the tissue culture conditions required to maintain these stem cells (+), and the potential to differentiate in extra-embryonic tissues (+).

How can the differences between hES and mouse ES cells regarding XCI be explained? In mice, cells of the ICM show two Xa's prior to differentiation. A species specific difference in the time window in which XCI occurs could explain why hES cells, which are also derived from the ICM, display XCI characteristics prior to differentiation. Does random XCI in human embryos already occur at an earlier stage and do hES cells therefore display an Xi? Alternatively, reactivation of the Xi established during the early cleavage stages might occur later in humans compared to mouse. In mice, imprinted XCI occurs prior to implantation and causes inactivation of the Xp.[17-19,195,196] Imprinted XCI is maintained in the extra-embryonic tissues, but is reversed in the cells of the ICM (Fig. 1), by reactivating the Xp, most likely initiated by *Nanog* expression in cells of the epiblast.[80] Although in human XCI is also initiated during the early cleavage divisions,[197] it is unclear whether human XCI is imprinted.[198-207] In addition it is unclear at what stage the Xi is reactivated, if reactivated at all and hence, XCI in undifferentiated hES cells could be due to persistence of XCI from the early pre-implantation embryo, rather then precocious initiation of random XCI. In support of this, a highly skewed XCI pattern has been observed in female hES cells with a preference for one of the two X chromosomes to be inactivated.[190,208] Unfortunately, the paternal origin of the inactivated X chromosome was not determined.

Beside variations in the XCI timing between species, the observed differences in XCI between undifferentiated human and mouse ES cells could also be explained by fundamental differences between the two cell types.[209,210] Human and mouse ES cells differ significantly in morphology, clonogenicity, molecular profile and culture requirements. For example, hES cells need bFGF and Activin/Nodal signalling for their self renewal, whereas mouse ES cells depend on LIF/Stat3 signalling[211,212] and both cell types are characterized by a genome wide difference in pluripotency factor promoter occupancy.[213] Although both cell types have the potential to differentiate in vitro into cell types of all germ layers and to form teratomas in vivo, only for mouse ES cells the capability to contribute to the germ line and generate an entire animal has been tested. For hES cells the generation of chimaeras is not possible due to ethical reasons. Therefore at present it is unclear whether hES cells are indeed the proper human equivalent of mouse ES cells. In mice, an additional pluripotent cell population has been isolated from the postimplantation mouse epiblast using culture conditions including bFGF and Activin, in the absence of LIF.[214,215] These so-called EpiSCs (postimplantation epiblast derived stem cells) express the pluripotency factors *Oct4*, *Sox2* and *Nanog*, but differ in morphology and in the expression of certain transcripts from mouse ES cells, making them more comparable to hES cells.[216] Although these cells can differentiate in vitro into cells of all germ layers and extra-embryonic trophectoderm, they are extremely inefficient in contributing to chimaeras and germ line transmission has not been reported. Interestingly, female EpiSCs have undergone XCI, leaving one X chromosome active. Therefore, it is possible that XCI observed in hES cells is a reflection of the difference between human and mouse ES cells, with hES cells being in fact the human counterpart of mouse EpiSCs.[217] In agreement with this, mouse EpiSCs and hES cells share the ability to differentiate in vitro into extra-embryonic tissues,[218] whereas mouse ES cells can only differentiate into cells of the three germ layers. It has been shown that EpiSCs can be differentiated from mouse ES cells in culture, which indicates that EpiSCs are a more restricted derivative from ES cells.[219] Also reprogramming of EpiSCs to ES cells by over expression of the pluripotency factor *Klf4*,[219] *Nanog*[80] or extended culture in LIF,[220] is accompanied by the reactivation of the inactive X chromosome. How can human ES cells be the human equivalent of a more differentiated cell type, despite the fact that human ES cells are also derived from the ICM, like mouse ES cells? It is possible that the culture conditions used for the derivation of hES cells allow the differentiation of ground state hES cells to hES cells with EpiSCs characteristics and prevent in vitro stabilization of the pluripotent state of the human ICM.[221] Also differences in early development between rodents and humans, such as the presence of an egg cylinder or the presence of a temporary arrest in embryonic development called diapause specific for the mouse, might explain why in humans a shorter time window may exist to derive ground state ES cells, comparable to mice.[222] Therefore it is possible that during human ES cell derivation in fact only more differentiated cells are derived, which might therefore explain XCI in these cells.

The presence of an inactive X in undifferentiated hES cells could also be explained by a selective pressure against two active X chromosomes. In fact, during development, only a short time window exists in which cells in the early embryo have two active X chromosomes. Since these cells are not programmed for an infinitive state in the ICM, these cells might adapt to culture conditions during ES cell derivation by initiating XCI, which might confer them with a growth or survival advantage. Selection against two active X chromosomes is also seen in some mouse ES cell lines and female mouse ES cells show genome wide hypomethylation,[223] possibly resulting in genome instability. This may explain why many inbred mouse ES cells loose one of the two X chromosomes during expansion and only ES cells from hybrid crosses stably propagate two X chromosomes. Since hES cells are not characterized by a tendency to X chromosomal loss, they might employ XCI to prevent activity of two X chromosomes.

Taken together, the finding that XCI is already initiated in undifferentiated hES cells could have many different reasons and most likely a combination of these mechanisms will turn out to explain the epigenetic fluidity observed in hES cells. Future studies are needed to determine the origin of this finding, since concerns about the epigenetic stability of hES cells and differentiation potential of these cells, should be clarified before the introduction of hES cells for any clinical application.[191,224-226]

Could the recent emergence of human iPS cells provide us with a better model to study XCI? For mouse iPS cells, reactivation of the silent X chromosome of somatic cells during reprogramming has been shown[32,33] and these cells initiate XCI upon differentiation. Although until today no studies have addressed the XCI status in human iPS cells, preliminary results from our laboratory indicate that also in human iPS cells precocious *XIST* accumulation is observed (T.S.B and J.G unpublished observations). Whether this is caused by culture artefacts, fundamental differences in the pluripotent state of human ES and iPS cells compared to mouse, or failure of complete reprogramming needs to be clarified in future investigations.

CONCLUSION

Mouse ES cells provide a powerful model system to study XCI, as XCI is initiated in female ES cells upon differentiation. Recent findings indicate that XCI is suppressed in female ES through the action of *Nanog*, *Sox2* and *Oct4*, which also play a key role in maintaining the pluripotent ground state of ES cells as part of a complex transcription factor network. Activation of XCI requires down regulation of these factors and up-regulation of X-encoded activators of XCI, including *Rnf12* which is also involved in differentiation processes. This indicates that XCI is regulated through the same factors that are also involved in pluripotency and differentiation. Nevertheless, many factors regulating XCI and the downstream targets of these factors still need to be identified. The tight link between pluripotency and XCI underscores that the epigenetic status of the X chromosome provides a potent readout to study pluripotency and nuclear reprogramming of female mouse cells. Unfortunately, for human female ES cells the picture appears to be more complex and it remains unclear whether female human ES cells provide a true model system to study the XCI process. Clearly more studies have to be performed to solve this issue.

ACKNOWLEDGEMENTS

We would like to thank all members of the laboratory for helpful and stimulating discussions. This work was supported by a NWO TOP grant to J.G. and a grant from the Dutch government (BSIK programme 03038, SCDD).

REFERENCES

1. Ohno S. Sex chromosomes and Sex-linked genes. 1967, Berlin: Springer.
2. Charlesworth B. The evolution of sex chromosomes. Science 1991; 251(4997):1030-3.
3. Graves JA. The origin and function of the mammalian Y chromosome and Y-borne genes—an evolving understanding. Bioessays 1995; 17(4):311-20.
4. Graves JA, Koina E, Sankovic N. How the gene content of human sex chromosomes evolved. Curr Opin Genet Dev 2006; 16(3):219-24.
5. Graves JA. Sex chromosome specialization and degeneration in mammals. Cell 2006; 124(5):901-14.
6. Skaletsky H, Kuroda-Kawaguchi T, Minx PJ et al. The male-specific region of the human Y chromosome is a mosaic of discrete sequence classes. Nature 2003; 423(6942):825-37.
7. Koopman P, Gubbay J, Vivian N et al. Male development of chromosomally female mice transgenic for Sry. Nature 1991; 351(6322):117-21.
8. Ross MT, Grafham DV, Coffey AJ et al. The DNA sequence of the human X chromosome. Nature 2005; 434(7031):325-37.
9. Lyon MF. Gene action in the X-chromosome of the mouse (Mus musculus L.). Nature 1961; 190:372-3.
10. Lyon MF, Phillips RJ, Searle AG. A test for mutagenicity of caffeine in mice. Z Vererbungsl 1962; 93:7-13.
11. Lyon MF. Sex chromatin and gene action in the mammalian X-chromosome. Am J Hum Genet 1962; 14: 135-48.
12. Davidson RG, Nitowsky HM, Childs B. Demonstration of two populations of cells in the human female heterozygous for glucose-6-phosphate dehydrogenase variants. Proc Natl Acad Sci USA 1963; 50: 481-5.
13. Adler DA, Rugarli EI, Lingenfelter PA et al. Evidence of evolutionary up-regulation of the single active X chromosome in mammals based on Clc4 expression levels in Mus spretus and Mus musculus. Proc Natl Acad Sci USA 1997; 94(17):9244-8.
14. Birchler JA, Fernandez HR, Kavi HH. Commonalities in compensation. Bioessays 2006; 28(6):565-8.
15. Lin H, Gupta V, Vermilyea MD et al. Dosage compensation in the mouse balances up-regulation and silencing of X-linked genes. PLoS Biol 2007; 5(12):e326.
16. Nguyen DK. Disteche CM. Dosage compensation of the active X chromosome in mammals. Nat Genet 2006; 38(1):47-53.
17. Takagi N, Sasaki M. Preferential inactivation of the paternally derived X chromosome in the extraembryonic membranes of the mouse. Nature 1975; 256(5519):640-2.
18. Huynh KD, Lee JT. Inheritance of a pre-inactivated paternal X chromosome in early mouse embryos. Nature 2003; 426(6968):857-62.
19. Okamoto I, Otte AP, Allis CD et al. Epigenetic dynamics of imprinted X inactivation during early mouse development. Science 2004; 303(5658):644-9.
20. Evans MJ, Kaufman MH. Establishment in culture of pluripotential cells from mouse embryos. Nature 1981; 292(5819):154-6.

21. Martin GR. Isolation of a pluripotent cell line from early mouse embryos cultured in medium conditioned by teratocarcinoma stem cells. Proc Natl Acad Sci USA 1981; 78(12):7634-8.
22. Thomson JA, Itskovitz-Eldor J, Shapiro SS et al. Embryonic stem cell lines derived from human blastocysts. Science 1998; 282(5391):1145-7.
23. Keller G. Embryonic stem cell differentiation: emergence of a new era in biology and medicine. Genes Dev 2005; 19(10):1129-55.
24. Leahy A, Xiong JW, Kuhnert F et al. Use of developmental marker genes to define temporal and spatial patterns of differentiation during embryoid body formation. J Exp Zool 1999; 284(1):67-81.
25. Monk M. A stem-line model for cellular and chromosomal differentiation in early mouse-development. Differentiation 1981; 19(2):71-6.
26. Takahashi K, Yamanaka S. Induction of pluripotent stem cells from mouse embryonic and adult fibroblast cultures by defined factors. Cell 2006; 126(4):663-76.
27. Nakagawa M, Koyanagi M, Tanabe K et al. Generation of induced pluripotent stem cells without Myc from mouse and human fibroblasts. Nat Biotechnol 2008; 26(1):101-6.
28. Wernig M, Meissner A, Foreman R et al. In vitro reprogramming of fibroblasts into a pluripotent ES-cell-like state. Nature 2007; 448(7151):318-24.
29. Yu J, Vodyanik MA, Smuga-Otto K et al. Induced pluripotent stem cell lines derived from human somatic cells. Science 2007; 318(5858):1917-20.
30. Takahashi K, Tanabe K, Ohnuki M et al. Induction of pluripotent stem cells from adult human fibroblasts by defined factors. Cell 2007; 131(5):861-72.
31. Park IH, Zhao R, West JA et al. Reprogramming of human somatic cells to pluripotency with defined factors. Nature 2008; 451(7175):141-6.
32. Maherali N, Sridharan R, Xie W et al. Directly reprogrammed fibroblasts show global epigenetic remodeling and widespread tissue contribution. Cell Stem Cell 2007; 1(1):55-70.
33. Stadtfeld M, Maherali N, Breault DT et al. Defining molecular cornerstones during fibroblast to iPS cell reprogramming in mouse. Cell Stem Cell 2008; 2(3):230-40.
34. Russell LB. Mammalian X-chromosome action: inactivation limited in spread and region of origin. Science 1963; 140:976-8.
35. Rastan S. Non-random X-chromosome inactivation in mouse X-autosome translocation embryos—location of the inactivation centre. J Embryol Exp Morphol 1983; 78:1-22.
36. Rastan S, Robertson EJ. X-chromosome deletions in embryo-derived (EK) cell lines associated with lack of X-chromosome inactivation. J Embryol Exp Morphol 1985; 90:379-88.
37. Brown CJ, Ballabio A, Rupert JL et al. A gene from the region of the human X inactivation centre is expressed exclusively from the inactive X chromosome. Nature 1991; 349(6304):38-44.
38. Heard E, Avner P. Role play in X-inactivation. Hum Mol Genet 1994; 3 Spec No:1481-5.
39. Borsani G, Tonlorenzi R, Simmler MC et al. Characterization of a murine gene expressed from the inactive X chromosome. Nature 1991; 351(6324):325-9.
40. Brockdorff N, Ashworth A, Kay GF et al. Conservation of position and exclusive expression of mouse Xist from the inactive X chromosome. Nature 1991; 351(6324):329-31.
41. Brockdorff N, Ashworth A, Kay GF et al. The product of the mouse Xist gene is a 15 kb inactive X-specific transcript containing no conserved ORF and located in the nucleus. Cell 1992; 71(3):515-26.
42. Brown CJ, Hendrich BD, Rupert JL et al. The human XIST gene: analysis of a 17 kb inactive X-specific RNA that contains conserved repeats and is highly localized within the nucleus. Cell 1992; 71(3):527-42.
43. Clemson CM, McNeil JA, Willard HF et al. XIST RNA paints the inactive X chromosome at interphase: evidence for a novel RNA involved in nuclear/chromosome structure. J Cell Biol 1996; 132(3):259-75.
44. Jonkers I, Monkhorst K, Rentmeester E et al. Xist RNA is confined to the nuclear territory of the silenced X chromosome throughout the cell cycle. Mol Cell Biol 2008; 28(18):5583-94.
45. Penny GD, Kay GF, Sheardown SA et al. Requirement for Xist in X chromosome inactivation. Nature 1996; 379(6561):131-7.
46. Marahrens Y, Panning B, Dausman J et al. Xist-deficient mice are defective in dosage compensation but not spermatogenesis. Genes Dev 1997; 11(2):156-66.
47. Csankovszki G, Panning B, Bates B et al. Conditional deletion of Xist disrupts histone macroH2A localization but not maintenance of X inactivation. Nat Genet 1999; 22(4):323-4.
48. Kay GF, Penny GD, Patel D et al. Expression of Xist during mouse development suggests a role in the initiation of X chromosome inactivation. Cell 1993; 72(2):171-82.
49. Sun BK, Deaton AM, Lee JT. A transient heterochromatic state in Xist preempts X inactivation choice without RNA stabilization. Mol Cell 2006; 21(5):617-28.
50. Plath K, Fang J, Mlynarczyk-Evans SK et al. Role of histone H3 lysine 27 methylation in X inactivation. Science 2003; 300(5616):131-5.
51. Silva J, Mak W, Zvetkova I et al. Establishment of histone h3 methylation on the inactive X chromosome requires transient recruitment of Eed-Enx1 polycomb group complexes. Dev Cell 2003; 4(4):481-95.

52. Zhao J, Sun BK, Erwin JA et al. Polycomb proteins targeted by a short repeat RNA to the mouse X chromosome. Science 2008; 322(5902):750-6.
53. Lee JT, Davidow LS, Warshawsky D. Tsix, a gene antisense to Xist at the X-inactivation centre. Nat Genet 1999; 21(4):400-4.
54. Lee JT, Lu N. Targeted mutagenesis of Tsix leads to nonrandom X inactivation. Cell 1999; 99(1):47-57.
55. Clerc P, Avner P. Role of the region 3' to Xist exon 6 in the counting process of X-chromosome inactivation. Nat Genet 1998; 19(3):249-53.
56. Shibata S, Lee JT. Characterization and quantitation of differential Tsix transcripts: implications for Tsix function. Hum Mol Genet 2003; 12(2):125-36.
57. Marks H, Chow JC, Denissov S et al. High-resolution analysis of epigenetic changes associated with X inactivation. Genome Res 2009; 19(8):1361-73.
58. Luikenhuis S, Wutz and A, Jaenisch R. Antisense transcription through the Xist locus mediates Tsix function in embryonic stem cells. Mol Cell Biol 2001; 21(24):8512-20.
59. Navarro P, Pichard S, Ciaudo C et al. Tsix transcription across the Xist gene alters chromatin conformation without affecting Xist transcription: implications for X-chromosome inactivation. Genes Dev 2005; 19(12):1474-84.
60. Vigneau S, Augui S, Navarro P et al. An essential role for the DXPas34 tandem repeat and Tsix transcription in the counting process of X chromosome inactivation. Proc Natl Acad Sci USA 2006; 103(19):7390-5.
61. Ohhata T, Hoki Y, Sasaki H et al. Crucial role of antisense transcription across the Xist promoter in Tsix-mediated Xist chromatin modification. Development 2008; 135(2):227-35.
62. Cohen DE, Davidow LS, Erwin JA et al. The DXPas34 repeat regulates random and imprinted X inactivation. Dev Cell 2007; 12(1):57-71.
63. Debrand E, Chureau C, Arnaud D et al. Functional analysis of the DXPas34 locus, a 3' regulator of Xist expression. Mol Cell Biol 1999; 19(12):8513-25.
64. Prissette M, El-Maarri O, Arnaud D et al. Methylation profiles of DXPas34 during the onset of X-inactivation. Hum Mol Genet 2001; 10(1):31-8.
65. Boumil RM, Ogawa Y, Sun BK et al. Differential methylation of Xite and CTCF sites in Tsix mirrors the pattern of X-inactivation choice in mice. Mol Cell Biol 2006; 26(6):2109-17.
66. Ogawa Y, Sun BK, Lee JT. Intersection of the RNA interference and X-inactivation pathways. Science 2008; 320(5881):1336-41.
67. Nesterova TB, Popova BC, Cobb BS et al. Dicer regulates Xist promoter methylation in ES cells indirectly through transcriptional control of Dnmt3a. Epigenetics Chromatin 2008; 1(1):2.
68. Kanellopoulou C, Muljo SA, Dimitrov SD et al. X chromosome inactivation in the absence of Dicer. Proc Natl Acad Sci USA 2009; 106(4):1122-7.
69. Shibata S Lee JT. Tsix transcription- versus RNA-based mechanisms in Xist repression and epigenetic choice. Curr Biol 2004; 14(19):1747-54.
70. Ogawa Y, Lee JT. Xite, X-inactivation intergenic transcription elements that regulate the probability of choice. Mol Cell 2003; 11(3):731-43.
71. Stavropoulos N, Rowntree RK, Lee JT. Identification of developmentally specific enhancers for Tsix in the regulation of X chromosome inactivation. Mol Cell Biol 2005; 25(7):2757-69.
72. Chao W, Huynh KD, Spencer RJ et al. CTCF, a candidate trans-acting factor for X-inactivation choice. Science 2002; 295(5553):345-7.
73. Donohoe ME, Zhang LF, Xu N et al. Identification of a Ctcf cofactor, Yy1, for the X chromosome binary switch. Mol Cell 2007; 25(1):43-56.
74. Donohoe ME, Silva SS, Pinter SF et al. The pluripotency factor Oct4 interacts with Ctcf and also controls X-chromosome pairing and counting. Nature 2009; 460(7251):128-32.
75. Chambers I, Colby D, Robertson M et al. Functional expression cloning of Nanog, a pluripotency sustaining factor in embryonic stem cells. Cell 2003; 113(5):643-55.
76. Mitsui K, Tokuzawa Y, Itoh H et al. The homeoprotein Nanog is required for maintenance of pluripotency in mouse epiblast and ES cells. Cell 2003; 113(5):631-42.
77. Nichols J, Zevnik B, Anastassiadis K et al. Formation of pluripotent stem cells in the mammalian embryo depends on the POU transcription factor Oct4. Cell 1998; 95(3):379-91.
78. Avilion AA, Nicolis SK, Pevny LH et al. Multipotent cell lineages in early mouse development depend on SOX2 function. Genes Dev 2003; 17(1):126-40.
79. Navarro P, Chambers I, Karwacki-Neisius V et al. Molecular coupling of Xist regulation and pluripotency. Science 2008; 321(5896):1693-5.
80. Silva J, Nichols J, Theunissen TW et al. Nanog is the gateway to the pluripotent ground state. Cell 2009; 138(4):722-37.
81. Jonkers I, Barakat TS, Achame EM et al. RNF12 is an X-Encoded dose-dependent activator of X chromosome inactivation. Cell 2009; 139(5):999-1011.

82. Bach I, Rodriguez-Esteban C, Carrière C et al. RLIM inhibits functional activity of LIM homeodomain transcription factors via recruitment of the histone deacetylase complex. Nat Genet 1999; 22(4):394-9.
83. Her YR. Chung IK. Ubiquitin Ligase RLIM Modulates Telomere Length Homeostasis through a Proteolysis of TRF1. J Biol Chem 2009; 284(13):8557-66.
84. JJohnsen SA, Güngör C, Prenzel T et al. Regulation of estrogen-dependent transcription by the LIM cofactors CLIM and RLIM in breast cancer. Cancer Res 2009; 69(1):128-36.
85. Jacobs Pa, Baikie Ag, Brown Wm et al. Evidence for the existence of the human "super female". Lancet 1959; 2(7100):423-5.
86. Maclean N, Mitchell JM. A survey of sex-chromosome abnormalities among 4514 mental defectives. Lancet 1962; 1(7224):293-6.
87. Grumbach MM, Morishima A, Taylor JH. Human Sex Chromosome Abnormalities in Relation to DNA Replication and Heterochromatinization. Proc Natl Acad Sci USA 1963; 49(5):581-9.
88. Carr DH. Chromosome studies in selected spontaneous abortions. 1. Conception after oral contraceptives. Can Med Assoc J 1970; 103(4):343-8.
89. Webb S, de Vries TJ, Kaufman MH. The differential staining pattern of the X chromosome in the embryonic and extraembryonic tissues of postimplantation homozygous tetraploid mouse embryos. Genet Res 1992; 59(3):205-14.
90. Harnden DG. Nuclear sex in triploid XXY human cells. Lancet 1961; 2(7200):488.
91. Nicodemi M, Prisco A. Symmetry-breaking model for X-chromosome inactivation. Phys Rev Lett 2007; 98(10):108104.
92. Nicodemi M, Prisco A. Self-assembly and DNA binding of the blocking factor in x chromosome inactivation. PLoS Comput Biol 2007; 3(11):e210.
93. Lee JT, Strauss WM, Dausman JA et al. A 450 kb transgene displays properties of the mammalian X-inactivation center. Cell 1996; 86(1):83-94.
94. Herzing LB, Romer JT, Horn JM et al. Xist has properties of the X-chromosome inactivation centre. Nature 1997; 386(6622):272-5.
95. Lee JT, Jaenisch R. Long-range cis effects of ectopic X-inactivation centres on a mouse autosome. Nature 1997; 386(6622):275-9.
96. Lee JT, Lu N, Han Y. Genetic analysis of the mouse X inactivation center defines an 80-kb multifunction domain. Proc Natl Acad Sci USA 1999; 96(7):3836-41.
97. Migeon BR, Kazi E, Haisley-Roystor C et al. Human X inactivation center induces random X chromosome inactivation in male transgenic mice. Genomics 1999; 59(2):113-21.
98. Migeon BR, Winter H, Kazi E et al. Low-copy-number human transgene is recognized as an X inactivation center in mouse ES cells, but fails to induce cis-inactivation in chimeric mice. Genomics 2001; 71(2):156-62.
99. Heard E, Kress C, Mongelard F et al. Transgenic mice carrying an Xist-containing YAC. Hum Mol Genet 1996; 5(4):441-50.
100. Matsuura S, Episkopou V, Hamvas R et al. Xist expression from an Xist YAC transgene carried on the mouse Y chromosome. Hum Mol Genet 1996; 5(4):451-9.
101. Heard E, Mongelard F, Arnaud D et al. Xist yeast artificial chromosome transgenes function as X-inactivation centers only in multicopy arrays and not as single copies. Mol Cell Biol 1999; 19(4):3156-66.
102. Gribnau J et al. X chromosome choice occurs independently of asynchronous replication timing. J Cell Biol 2005; 168(3):365-73.
103. MarahrensY, Loring J, Jaenisch R. Role of the Xist gene in X chromosome choosing. Cell 1998; 92(5):657-64.
104. Nesterova TB, Johnston CM, Appanah R et al. Skewing X chromosome choice by modulating sense transcription across the Xist locus. Genes Dev 2003; 17(17):2177-90.
105. Sado T, Li E, Sasaki H. Effect of TSIX disruption on XIST expression in male ES cells. Cytogenet Genome Res 2002; 99(1-4):115-8.
106. Monkhorst K, Jonkers I, Rentmeester E et al. X inactivation counting and choice is a stochastic process: evidence for involvement of an X-linked activator. Cell 2008; 132(3):410-21.
107. Lee JT. Homozygous Tsix mutant mice reveal a sex-ratio distortion and revert to random X-inactivation. Nat Genet 2002; 32(1):195-200.
108. Mlynarczyk-Evans S, Royce-Tolland M, Alexander MK et al. X chromosomes alternate between two states prior to random X-inactivation. PLoS Biol 2006; 4(6):e159.
109. Bacher CP, Guggiari M, Brors B et al. Transient colocalization of X-inactivation centres accompanies the initiation of X inactivation. Nat Cell Biol 2006; 8(3):293-9.
110. Xu N, Tsai CL, Lee JT. Transient homologous chromosome pairing marks the onset of X inactivation. Science 2006; 311(5764):1149-52.
111. Xu N, Donohoe ME, Silva SS et al. Evidence that homologous X-chromosome pairing requires transcription and Ctcf protein. Nat Genet 2007; 39(11):1390-6.

112. Augui S, Filion GJ, Huart S et al. Sensing X chromosome pairs before X inactivation via a novel X-pairing region of the Xic. Science 2007; 318(5856):1632-6.
113. Takagi N. Variable X chromosome inactivation patterns in near-tetraploid murine EC x somatic cell hybrid cells differentiated in vitro. Genetica 1993; 88(2-3):107-17.
114. Monkhorst K, de Hoon B, Jonkers I et al. The probability to initiate X chromosome inactivation is determined by the X to autosomal ratio and X chromosome specific allelic properties. PLoS One 2009; 4(5):e5616.
115. Barr ML, Bertram EG. A morphological distinction between neurones of the male and female and the behaviour of the nucleolar satellite during accelerated nucleoprotein synthesis. Nature 1949; 163(4148):676.
116. Ohno S, Hauschka TS. Allocycly of the X-chromosome in tumors and normal tissues. Cancer Res 1960; 20:541-5.
117. Wutz A, Rasmussen TP, Jaenisch R. Chromosomal silencing and localization are mediated by different domains of Xist RNA. Nat Genet 2002; 30(2):167-74.
118. Chaumeil J, Le Baccon P, Wutz A et al. A novel role for Xist RNA in the formation of a repressive nuclear compartment into which genes are recruited when silenced. Genes Dev 2006; 20(16):2223-37.
119. Clemson CM, Hall LL, Byron M et al. The X chromosome is organized into a gene-rich outer rim and an internal core containing silenced nongenic sequences. Proc Natl Acad Sci USA 2006; 103(20):7688-93.
120. Goto Y, Gomez M, Brockdorff N et al. Differential patterns of histone methylation and acetylation distinguish active and repressed alleles at X-linked genes. Cytogenet Genome Res 2002; 99(1-4):66-74.
121. Heard E, Rougeulle C, Arnaud D et al. Methylation of histone H3 at Lys-9 is an early mark on the X chromosome during X inactivation. Cell 2001; 107(6):727-38.
122. Mak W, Baxter J, Silva J et al. Mitotically stable association of polycomb group proteins eed and enx1 with the inactive x chromosome in trophoblast stem cells. Curr Biol 2002; 12(12):1016-20.
123. Boggs BA, Cheung P, Heard E et al. Differentially methylated forms of histone H3 show unique association patterns with inactive human X chromosomes. Nat Genet 2002; 30(1):73-6.
124. Mermoud JE, Popova B, Peters AH et al. Histone H3 lysine 9 methylation occurs rapidly at the onset of random X chromosome inactivation. Curr Biol 2002; 12(3):247-51.
125. Peters AH, Mermoud JE, O'Carroll D et al. Histone H3 lysine 9 methylation is an epigenetic imprint of facultative heterochromatin. Nat Genet 2002; 30(1):77-80.
126. Rougeulle C, Chaumeil J, Sarma K et al. Differential histone H3 Lys-9 and Lys-27 methylation profiles on the X chromosome. Mol Cell Biol 2004; 24(12):5475-84.
127. Kohlmaier A, Savarese F, Lachner M et al. A chromosomal memory triggered by Xist regulates histone methylation in X inactivation. PLoS Biol 2004; 2(7):E171.
128. Costanzi C, Pehrson JR. Histone macroH2A1 is concentrated in the inactive X chromosome of female mammals. Nature 1998; 393(6685):599-601.
129. Lock LF, Takagi N, Martin GR. Methylation of the Hprt gene on the inactive X occurs after chromosome inactivation. Cell 1987; 48(1):39-46.
130. Norris DP, Brockdorff N, Rastan S. Methylation status of CpG-rich islands on active and inactive mouse X chromosomes. Mamm Genome 1991; 1(2):78-83.
131. Mohandas T, Sparkes RS, Shapiro LJ. Reactivation of an inactive human X chromosome: evidence for X inactivation by DNA methylation. Science 1981; 211(4480):393-6.
132. Russell LB, Montgomery CS. Comparative studies on X-autosome translocations in the mouse. II. Inactivation of autosomal loci, segregation and mapping of autosomal breakpoints in five T (X;1) S. Genetics 1970; 64(2):281-312.
133. Duthie SM, Nesterova TB, Formstone EJ et al. Xist RNA exhibits a banded localization on the inactive X chromosome and is excluded from autosomal material in cis. Hum Mol Genet 1999; 8(2):195-204.
134. Keohane AM et al. H4 acetylation, XIST RNA and replication timing are coincident and define x; autosome boundaries in two abnormal X chromosomes. Hum Mol Genet 1999; 8(2):377-83.
135. Popova BC, Tada T, Takagi N et al. Attenuated spread of X-inactivation in an X; autosome translocation. Proc Natl Acad Sci USA 2006; 103(20):7706-11.
136. Gartler SM, Riggs AD. Mammalian X-chromosome inactivation. Annu Rev Genet 1983; 17:155-90.
137. Lyon MF. X-chromosome inactivation: a repeat hypothesis. Cytogenet Cell Genet 1998; 80(1-4):133-7.
138. Lyon MF. LINE-1 elements and X chromosome inactivation: a function for "junk" DNA? Proc Natl Acad Sci USA 2000; 97(12):6248-9.
139. Lyon MF. Do LINEs Have a Role in X-Chromosome Inactivation? J Biomed Biotechnol 2006; 2006(1):59746.
140. Lyon MF. No longer 'all-or-none'. Eur J Hum Genet 2005; 13(7):796-7.
141. Furano AV. The biological properties and evolutionary dynamics of mammalian LINE-1 retrotransposons. Prog Nucleic Acid Res Mol Biol 2000; 64:255-94.
142. Waters PD, Dobigny G, Pardini AT et al. LINE-1 distribution in Afrotheria and Xenarthra: implications for understanding the evolution of LINE-1 in eutherian genomes. Chromosoma 2004; 113(3):137-44.

143. Parish DA, Vise P, Wichman HA et al. Distribution of LINEs and other repetitive elements in the karyotype of the bat Carollia: implications for X-chromosome inactivation. Cytogenet Genome Res 2002; 96(1-4):191-7.
144. Wichman HA, Van den Bussche RA, Hamilton MJ et al. Transposable elements and the evolution of genome organization in mammals. Genetica 1992; 86(1-3):287-93.
145. Bailey JA, Carrel L, Chakravarti A et al. Molecular evidence for a relationship between LINE-1 elements and X chromosome inactivation: the Lyon repeat hypothesis. Proc Natl Acad Sci USA 2000; 97(12):6634-9.
146. Carrel L, Willard HF. X-inactivation profile reveals extensive variability in X-linked gene expression in females. Nature 2005; 434(7031):400-4.
147. Wang Z, Willard HF, Mukherjee S et al. Evidence of influence of genomic DNA sequence on human X chromosome inactivation. PLoS Comput Biol 2006; 2(9):e113.
148. Hall LL, Clemson CM, Byron M et al. Unbalanced X; autosome translocations provide evidence for sequence specificity in the association of XIST RNA with chromatin. Hum Mol Genet 2002; 11(25):3157-65.
149. Sharp AJ, Spotswood HT, Robinson DO et al. Molecular and cytogenetic analysis of the spreading of X inactivation in X; autosome translocations. Hum Mol Genet 2002; 11(25):3145-56.
150. Sharp A, Robinson DO, Jacobs P. Absence of correlation between late-replication and spreading of X inactivation in an X; autosome translocation. Hum Genet 2001; 109(3):295-302.
151. White WM, Willard HF, Van Dyke DL et al. The spreading of X inactivation into autosomal material of an x; autosome translocation: evidence for a difference between autosomal and X-chromosomal DNA. Am J Hum Genet 1998; 63(1):20-8.
152. Solari AJ, Rahn IM, Ferreyra ME et al. The behavior of sex chromosomes in two human X-autosome translocations: failure of extensive X-inactivation spreading. Biocell 2001; 25(2):155-66.
153. Dobigny G, Ozouf-Costaz C, Bonillo C et al. Viability of X-autosome translocations in mammals: an epigenomic hypothesis from a rodent case-study. Chromosoma 2004; 113(1):34-41.
154. Chureau C, Prissette M, Bourdet A et al. Comparative sequence analysis of the X-inactivation center region in mouse, human and bovine. Genome Res 2002; 12(6):894-908.
155. Cantrell MA, Carstens BC, Wichman HA. X chromosome inactivation and Xist evolution in a rodent lacking LINE-1 activity. PLoS One 2009; 4(7):e6252.
156. Ke X, Collins A. CpG islands in human X-inactivation. Ann Hum Genet 2003; 67(Pt 3):242-9.
157. Cao R, Zhang Y. SUZ12 is required for both the histone methyltransferase activity and the silencing function of the EED-EZH2 complex. Mol Cell 2004; 15(1):57-67.
158. de la Cruz CC, Fang J, Plath K et al. Developmental regulation of Suz 12 localization. Chromosoma 2005; 114(3):183-92.
159. Wang J, Mager J, Chen Y et al. Imprinted X inactivation maintained by a mouse Polycomb group gene. Nat Genet 2001; 28(4):371-5.
160. Kalantry S, Magnuson T. The Polycomb group protein EED is dispensable for the initiation of random X-chromosome inactivation. PLoS Genet 2006; 2(5):e66.
161. Schoeftner S, Sengupta AK, Kubicek S et al. Recruitment of PRC1 function at the initiation of X inactivation independent of PRC2 and silencing. EMBO J 2006; 25(13):3110-22.
162. de Napoles M, Mermoud JE, Wakao R et al. Polycomb group proteins Ring1A/B link ubiquitylation of histone H2A to heritable gene silencing and X inactivation. Dev Cell 2004; 7(5):663-76.
163. Fang J, Chen T, Chadwick B et al. Ring1b-mediated H2A ubiquitination associates with inactive X chromosomes and is involved in initiation of X inactivation. J Biol Chem 2004; 279(51):52812-5.
164. Hernández-Muñoz I, Lund AH, van der Stoop P et al. Stable X chromosome inactivation involves the PRC1 Polycomb complex and requires histone MACROH2A1 and the CULLIN3/SPOP ubiquitin E3 ligase. Proc Natl Acad Sci USA 2005; 102(21):7635-40.
165. Nusinow DA, Hernández-Muñoz I, Fazzio TG et al. Poly(ADP-ribose) polymerase 1 is inhibited by a histone H2A variant, MacroH2A and contributes to silencing of the inactive X chromosome. J Biol Chem 2007; 282(17):12851-9.
166. Kim MY, Zhang T, Kraus WL. Poly(ADP-ribosyl)ation by PARP-1: 'PAR-laying' NAD+ into a nuclear signal. Genes Dev 2005; 19(17):1951-67.
167. Rouleau M, Aubin RA, Poirier GG. Poly(ADP-ribosyl)ated chromatin domains: access granted. J Cell Sci 2004; 117(Pt 6):815-25.
168. Hassa PO, Haenni SS, Elser M et al. Nuclear ADP-ribosylation reactions in mammalian cells: where are we today and where are we going? Microbiol Mol Biol Rev 2006; 70(3):789-829.
169. Kim MY, Mauro S, Gévry N et al. NAD+-dependent modulation of chromatin structure and transcription by nucleosome binding properties of PARP-1. Cell 2004; 119(6):803-14.
170. Helbig R, Fackelmayer FO. Scaffold attachment factor A (SAF-A) is concentrated in inactive X chromosome territories through its RGG domain. Chromosoma 2003; 112(4):173-82.
171. Ganesan S, Silver DP, Greenberg RA et al. BRCA1 supports XIST RNA concentration on the inactive X chromosome. Cell 2002; 111(3):393-405.

172. Silver DP, Dimitrov SD, Feunteun J et al. Further evidence for BRCA1 communication with the inactive X chromosome. Cell 2007; 128(5):991-1002.
173. Vincent-Salomon A, Ganem-Elbaz C, Manié E et al. X inactive-specific transcript RNA coating and genetic instability of the X chromosome in BRCA1 breast tumors. Cancer Res 2007; 67(11):5134-40.
174. Xiao C, Sharp JA, Kawahara M, Davalos AR et al. The XIST noncoding RNA functions independently of BRCA1 in X inactivation. Cell 2007; 128(5):977-89.
175. Blewitt ME, Gendrel AV, Pang Z et al. SmcHD1, containing a structural-maintenance-of-chromosomes hinge domain, has a critical role in X inactivation. Nat Genet 2008; 40(5):663-9.
176. Garrick D, Sharpe JA, Arkell R et al. Loss of Atrx affects trophoblast development and the pattern of X-inactivation in extraembryonic tissues. PLoS Genet 2006; 2(4):e58.
177. Baumann C, De La Fuente R. ATRX marks the inactive X chromosome (Xi) in somatic cells and during imprinted X chromosome inactivation in trophoblast stem cells. Chromosoma 2009; 118(2):209-22.
178. Han HJ, Russo J, Kohwi Y et al. SATB1 reprogrammes gene expression to promote breast tumour growth and metastasis. Nature 2008; 452(7184):187-93.
179. Agrelo R, Souabni A, Novatchkova M et al. SATB1 defines the developmental context for gene silencing by Xist in lymphoma and embryonic cells. Dev Cell 2009; 16(4):507-16.
180. Disteche CM. Escape from X inactivation in human and mouse. Trends Genet 1995; 11(1):17-22.
181. Disteche CM. Escapees on the X chromosome. Proc Natl Acad Sci USA 1999; 96(25):14180-2.
182. Disteche CM, Filippova GN, Tsuchiya KD. Escape from X inactivation. Cytogenet Genome Res 2002; 99(1-4):36-43.
183. Brown CJ, Greally JM. A stain upon the silence: genes escaping X inactivation. Trends Genet 2003; 19(8):432-8.
184. Looijenga LH, Gillis AJ, van Gurp RJ et al. X inactivation in human testicular tumors. XIST expression and androgen receptor methylation status. Am J Pathol 1997; 151(2):581-90.
185. Chow JC, Hall LL, Clemson CM et al. Characterization of expression at the human XIST locus in somatic, embryonal carcinoma and transgenic cell lines. Genomics 2003; 82(3):309-22.
186. Migeon BR, Chowdhury AK, Dunston JA et al. Identification of TSIX, encoding an RNA antisense to human XIST, reveals differences from its murine counterpart: implications for X inactivation. Am J Hum Genet 2001; 69(5):951-60.
187. Chow JC, Hall LL, Lawrence JB et al. Ectopic XIST transcripts in human somatic cells show variable expression and localization. Cytogenet Genome Res 2002; 99(1-4):92-8.
188. Hall LL, Byron M, Sakai K et al. An ectopic human XIST gene can induce chromosome inactivation in postdifferentiation human HT-1080 cells. Proc Natl Acad Sci USA 2002; 99(13):8677-82.
189. Hall LL, Byron M, Butler J et al. X-inactivation reveals epigenetic anomalies in most hESC but identifies sublines that initiate as expected. J Cell Physiol 2008; 216(2):445-52.
190. Shen Y, Matsuno Y, Fouse SD et al. X-inactivation in female human embryonic stem cells is in a nonrandom pattern and prone to epigenetic alterations. Proc Natl Acad Sci USA 2008; 105(12):4709-14.
191. Silva SS, Rowntree RK, Mekhoubad S et al. X-chromosome inactivation and epigenetic fluidity in human embryonic stem cells. Proc Natl Acad Sci USA 2008; 105(12):4820-5.
192. Enver T, Soneji S, Joshi C et al. Cellular differentiation hierarchies in normal and culture-adapted human embryonic stem cells. Hum Mol Genet 2005; 14(21):3129-40.
193. Hoffman LM, Hall L, Batten JL et al. X-inactivation status varies in human embryonic stem cell lines. Stem Cells 2005; 23(10):1468-78.
194. Dhara SK, Benvenisty N. Gene trap as a tool for genome annotation and analysis of X chromosome inactivation in human embryonic stem cells. Nucleic Acids Res 2004; 32(13):3995-4002.
195. Takagi N. Imprinted X-chromosome inactivation: enlightenment from embryos in vivo. Semin Cell Dev Biol 2003; 14(6):319-29.
196. West JD, Papaioannou VE, Frels WI et al. Preferential expression of the maternally derived X chromosome in the mouse yolk sac. Cell 1977; 12(4):873-82.
197. van den Berg IM, Laven JS, Stevens M et al. X chromosome inactivation is initiated in human preimplantation embryos. Am J Hum Genet 2009; 84(6):771-9.
198. Looijenga LH, Gillis AJ, Verkerk AJ et al. Heterogeneous X inactivation in trophoblastic cells of human full-term female placentas. Am J Hum Genet 1999; 64(5):1445-52.
199. Ropers HH, Wolff G, Hitzeroth HW. Preferential X inactivation in human placenta membranes: is the paternal X inactive in early embryonic development of female mammals? Hum Genet 1978; 43(3):265-73.
200. Daniels R, Zuccotti M, Kinis T et al. XIST expression in human oocytes and preimplantation embryos. Am J Hum Genet 1997; 61(1):33-9.
201. Ray PF, Winston RM, Handyside AH. XIST expression from the maternal X chromosome in human male preimplantation embryos at the blastocyst stage. Hum Mol Genet 1997; 6(8):1323-7.
202. Bamforth F, Machin G, Innes M. X-chromosome inactivation is mostly random in placental tissues of female monozygotic twins and triplets. Am J Med Genet 1996; 61(3):209-15.

203. Migeon BR, Wolf SF, Axelman J et al. Incomplete X chromosome dosage compensation in chorionic villi of human placenta. Proc Natl Acad Sci USA 1985; 82(10):3390-4.
204. Migeon BR, Axelman J, Jeppesen P. Differential X reactivation in human placental cells: implications for reversal of X inactivation. Am J Hum Genet 2005; 77(3):355-64.
205. Goto T, Wright E, Monk M. Paternal X-chromosome inactivation in human trophoblastic cells. Mol Hum Reprod 1997; 3(1):77-80.
206. Harrison KB. X-chromosome inactivation in the human cytotrophoblast. Cytogenet Cell Genet 1989; 52(1-2):37-41.
207. Harrison KB, Warburton D. Preferential X-chromosome activity in human female placental tissues. Cytogenet Cell Genet 1986; 41(3):163-8.
208. Liu W, Sun X. Skewed X chromosome inactivation in diploid and triploid female human embryonic stem cells. Hum Reprod 2009; 24(8):1834-43.
209. Sato N, Sanjuan IM, Heke M et al. Molecular signature of human embryonic stem cells and its comparison with the mouse. Dev Biol 2003; 260(2):404-13.
210. Ginis I, Luo Y, Miura T et al. Differences between human and mouse embryonic stem cells. Dev Biol 2004; 269(2):360-80.
211. Xu RH, Peck RM, Li DS et al. Basic FGF and suppression of BMP signaling sustain undifferentiated proliferation of human ES cells. Nat Methods 2005; 2(3):185-90.
212. Ying QL, Nichols J, Chambers I et al. BMP induction of Id proteins suppresses differentiation and sustains embryonic stem cell self-renewal in collaboration with STAT3. Cell 2003; 115(3):281-92.
213. Boyer LA, Lee TI, Cole MF et al. Core transcriptional regulatory circuitry in human embryonic stem cells. Cell 2005; 122(6):947-56.
214. Brons IG, Smithers LE, Trotter MW et al. Derivation of pluripotent epiblast stem cells from mammalian embryos. Nature 2007; 448(7150):191-5.
215. Tesar PJ, Chenoweth JG, Brook FA et al. New cell lines from mouse epiblast share defining features with human embryonic stem cells. Nature 2007; 448(7150):196-9.
216. Rossant J. Stem cells and early lineage development. Cell 2008; 132(4):527-31.
217. Lovell-Badge R. Many ways to pluripotency. Nat Biotechnol 2007; 25(10):1114-6.
218. Xu RH, Chen X, Li DS et al. BMP4 initiates human embryonic stem cell differentiation to trophoblast. Nat Biotechnol 2002; 20(12):1261-4.
219. Guo G, Yang J, Nichols J et al. Klf4 reverts developmentally programmed restriction of ground state pluripotency. Development 2009; 136(7):1063-9.
220. Bao S, Tang F, Li X et al. Epigenetic reversion of post-implantation epiblast to pluripotent embryonic stem cells. Nature 2009; 461(7268):1292-5.
221. Hanna J, Markoulaki S, Mitalipova M et al. Metastable pluripotent states in NOD-mouse-derived ESCs. Cell Stem Cell 2009; 4(6):513-24.
222. Nichols J, Smith A. Naive and primed pluripotent states. Cell Stem Cell 2009; 4(6):487-92.
223. Zvetkova I, Apedaile A, Ramsahoye B et al. Global hypomethylation of the genome in XX embryonic stem cells. Nat Genet 2005; 37(11):1274-9.
224. Lagarkova MA, Volchkov PY, Lyakisheva AV et al. Diverse epigenetic profile of novel human embryonic stem cell lines. Cell Cycle 2006; 5(4):416-20.
225. Rugg-Gunn PJ, Ferguson-Smith AC, Pedersen RA. Epigenetic status of human embryonic stem cells. Nat Genet 2005; 37(6):585-7.
226. Allegrucci C, Wu YZ, Thurston A et al. Restriction landmark genome scanning identifies culture-induced DNA methylation instability in the human embryonic stem cell epigenome. Hum Mol Genet 2007; 16(10):1253-68.

CHAPTER 11

ADULT STEM CELLS AND THEIR NICHES

Francesca Ferraro, Cristina Lo Celso and David Scadden*

Abstract: Stem cells participate in dynamic physiologic systems that dictate the outcome of developmental events and organismal stress, Since these cells are fundamental to tissue maintenance and repair, the signals they receive play a critical role in the integrity of the organism. Much work has focused on stem cell identification and the molecular pathways involved in their regulation. Yet, we understand little about how these pathways achieve physiologically responsive stem cell functions. This chapter will review the state of our understanding of stem cells in the context of their microenvironment regarding the relation between stem cell niche dysfunction, carcinogenesis and aging.

THE NICHE CONCEPT, DEFINITION AND HISTORICAL BACKGROUND

The stem cell niche is the in vivo microenvironment where stem cells both reside and receive stimuli that determine their fate. Therefore, the niche should not be considered simply a physical location for stem cells, rather as the place where extrinsic signals interact and integrate to influence stem cell behavior. These stimuli include cell-to-cell and cell-matrix interactions and signals (molecules) that activate and/or repress genes and transcription programs. As a direct consequence of this interaction, stem cells are maintained in a dormant state, induced to self-renewal or commit to a more differentiated state.

Schoefield first postulated the hypothesis of a specialized stem cell microenvironment in 1978.[1] He proposed that niches have a defined anatomical location and also that removal of stem cells from their niche results in differentiation. The first demonstration and characterization of niche components was conducted in the invertebrate model of *Drosophila melanogaster* and *Caenorhabditis elegans* gonads.[2,3] Examination of

*David Scadden— Harvard Stem Cell Institute, Massachusetts General Hospital, Department of Stem Cell and Regenerative Biology, Harvard University, 185 Cambridge Street, Boston, Massachusettes, USA. Email: dscadden@mgh.harvard.edu

The Cell Biology of Stem Cells, edited by Eran Meshorer and Kathrin Plath.
©2010 Landes Bioscience and Springer Science+Business Media.

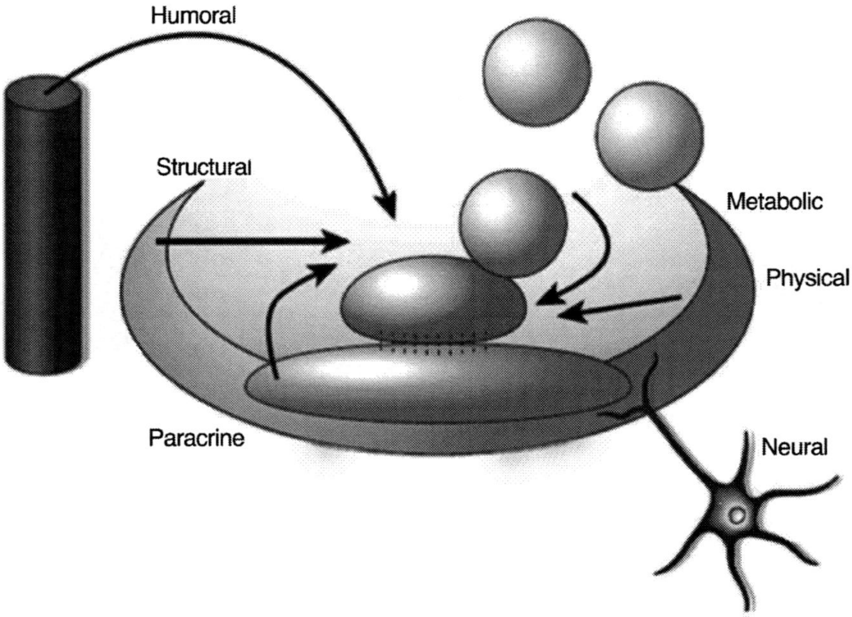

Figure 1. Representation of a stem cell niche: the stem cell niche is the place where humoral, neuronal, local (paracrine), positional (physical) and metabolic cues interact with each other to regulate stem cell fate. (Adapted from Scadden DT. Nature 441:1075-1079).

these systems, characterized in less complex animals, has led to pivotal insights into understanding the more complex mammalian niche architecture. It appears that the fundamental anatomical components and molecular pathways of the niche environment are highly conserved among species, although their respective roles within the niche may show distinct variations. Therefore, it has been proposed that it is possible to identify common niche components that are associated with similar functions (Fig. 1).

The general niche model involves the association between resident stem cells and heterologous cell types—the niche cells. However, the existence of a heterologous cell type is not essential and components of the extracellular matrix (or other noncellular components) may determine the niche for stem cells. Notably, a niche environment may retain its key functions and properties, even in the temporary absence of stem cells (such as following stem cell depletion through radiation treatment) allowing recruitment and homing of exogenous stem cells to the pre-existing stem cell niche.

Conserved components of the niche are:

1. Stromal support cells, including cell-cell adhesion molecules and secreted soluble factors, which are found in close proximity to stem cells.
2. Extracellular matrix (ECM) proteins that act as a stem cell "anchor" and constitute a mechanical scaffolding unit to transmit stem cell signaling.
3. Blood vessels that carry nutritional support and systemic signals to the niche from other organs and also participate in the recruitment of circulating stem cells from and to the niche.

4. Neural inputs that favor the mobilization of stem cells out of their niches and integrate signals from different organ systems. Neuronal cues appear to be particularly important in hematopoietic stem cells trafficking.[4]

Given the profound effect of the niche environment on stem cell behavior, newer work is exploring how niche perturbations may cause stem cell dysfunctions, as it is seen in aging or neoplastic transformation.[5-9]

STEM CELL NICHE COMPONENTS

In **the invertebrate model** of *Drosophila* ovary, germinal stem cells (GSCs) located in the germarium are in physical contact with cap cells and terminal filaments cells. During the process of asymmetric division, GSCs that physically contact cap cells through E-cadherin junctions retain their stem cell properties, whereas those cells that lose contact with cap cells differentiate into mature follicle cells. A similar system, driven by polarity cues, applies also for *Drosophila* testis, where two sets of stem cells, germinal stem cells (GSCs) and somatic stem cells (SSCs) are associated at the apical tip of the testis with hub cells. Daughter cells that detach from the hub initiate a differentiation program to become, respectively, spermatogonia and somatic cyst cells. In *C. elegans* 225 germ cells are associated to distal tip cells (DTC) and they are maintained stem cells through signals from these cells.

Several niches have been identified also in many **mammalian tissues**: hematopoietic system, skin, intestine, brain and muscle (Fig. 2).

In **trabecular bone marrow,** hematopoietic stem and progenitor cells (HSPCs) reside along the endosteal surface close to osteoblasic cells[10,11] and in proximity to the blood vessels.[12,13] Since a niche is defined by its functional regulation of stem cells, the mere physical proximity of stem cells to other components is insufficient to determine if those components are niche elements. To date, data indicating a regulatory role for osteoblastic cells has several lines of support but a role for endothelium is less clear.[10,11,14-16]

In the **skin,** epithelial stem cells (ESCs) are found in the bulge area of the hair follicles.[17] While the exact components of skin niche have not been fully identified yet, although critical regulatory cues derive from the dermal papilla. These stem cells are important in regeneration of hair follicles while scattered stem cells attached to the basal membrane that separates epidermis from dermis (basal keratinocytes) are involved in replacement of interfollicular epidermidis.[18] Sebaceous glands are maintained by cells at the base of each gland,[19] but their niche is still largely unknown.

In adult **central nervous system,** neural stem cells (NSCs) have been identified in the lateral subventricular zone (SVZ) and in the subgranular zone (SGZ) of the dentate gyrus within the hippocampus.[20-22] Within these areas neural stem cells have been shown to express the astorcyte marker glial fibrillary acidic protein (GFAP). GFAP-positive astrocytes in SVZ and SGZ are able to give rise to neuroblasts and subsequently mature neurons. Located in close proximity to NSCs, endothelial cells are considered niche cells in the central nervous system.[22]

In the **gut,** intestinal stem cells (ISCs) reside in the bottom part of the intestinal crypt interdigitated between Paneth cells.[23] The area surrounding the crypts is particularly rich of enteric neurons and blood vessels. Specialized mesenchymal cells also known

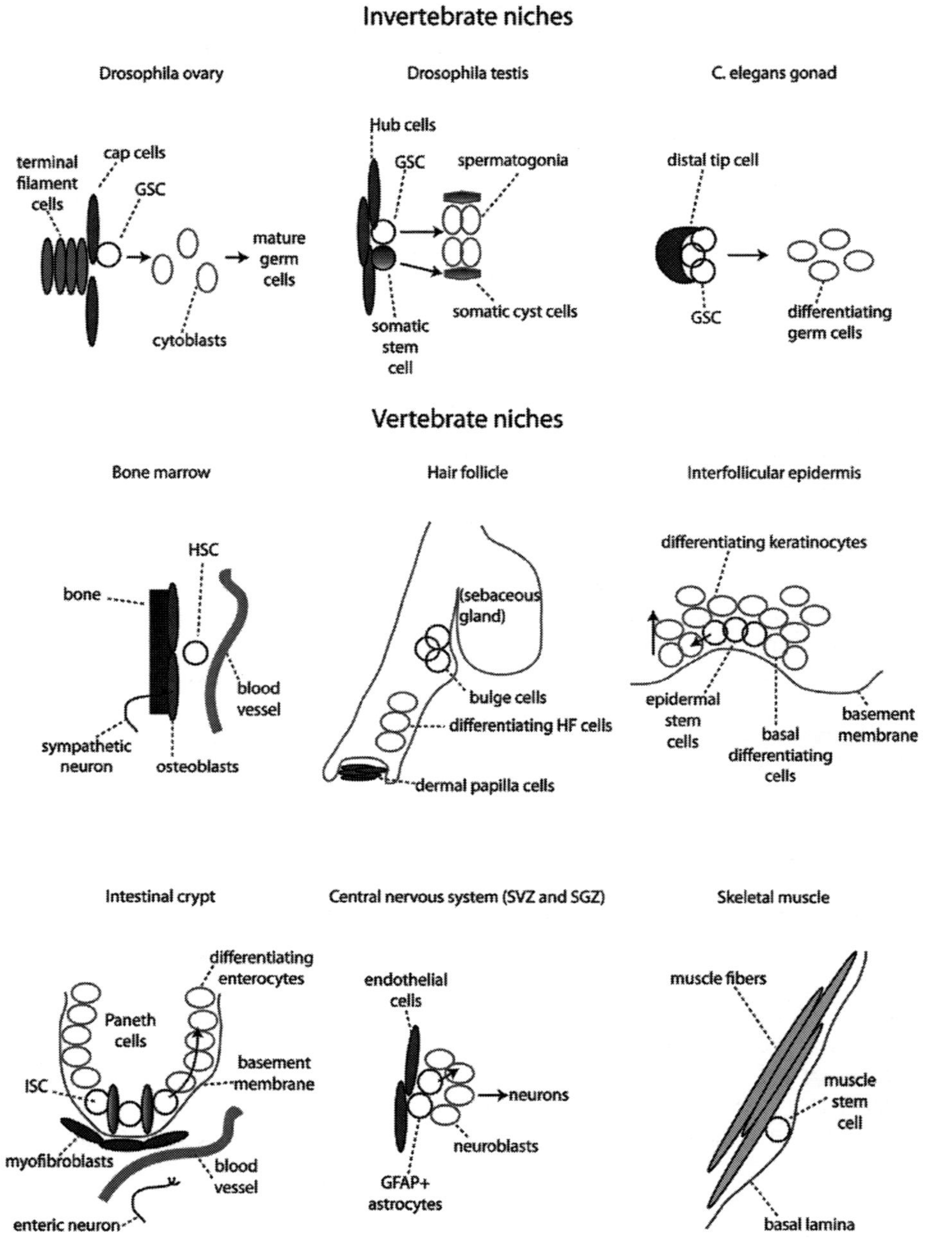

Figure 2. Stem cell niches. In this figure are shown vertebrate and invertebrate stem cell niches along with their identified components. GSC: germinal stem cell; HSC: hematopoietic stem cell; ISC: intestinal stem cell; SVZ: sub-ventricular zone; SGZ: sub-granular zone; GFAP: glial fibrillary acidic protein.

as myofibroblasts in close proximity to crypt cells and have been suggested to be niche cells for ISCs.[24]

In the **muscle**, stem cells, known as satellite cells, are located along muscle fiber tracts attached to the plasma membrane that surrounds each muscle fiber bundle. In this case the basal lamina may represent the niche for satellite cells.[25,26]

MOLECULAR PATHWAYS ASSOCIATED WITH NICHE FUNCTION

Molecules that influence fate determination of stem cells are secreted from niche cells and may exert their function either through paracrin effects or through neural output (neuro-endocrine effect).

Molecular pathways recognized to be important modulators of stem cell maintenance and function are redundant in different niches but have different roles according to the specific niche. These pathways include Wnt/beta-catenin, bone morphogenetic protein (BMP), Notch, Angiopoietin-1 (Ang-1) and several growth factors, such as fibroblast growth factor (FGF), insulin growth factor (IGF), vascular endothelial growth factor (VEGF), transforming growth factor-alpha (TGF-alpha) and platelet derived growth factor (PDGF). Among these, BMP and Wnt signaling appear to be highly conserved controlling self-renewal and lineage commitment in both invertebrates and mammals.

Wnt/beta-catenin signaling may exert differential effects depending on the tissue. For instance, in the hematopoietic system[27] and intestine[28,29] Wnt/beta-catenin is an important mediator of self-renewal and proliferation of stem cells, while in the skin it promotes differentiation of hair follicole precursors.[30-32] In mammalian brain, over-expression of beta-catenin through Wnt signaling leads to the expansion of neuronal stem cells populations.[33]

Bone morphogenic protein (BMP) signaling plays an important role in the control of *D. melanogaster* GCS expansion by repressing expression of bam,[34] the mediator of cystoblast differentiation. BMP signals are also generated in *Drosophila* testis' hub cells and are required for the control of GCS self-renewal. In the hematopoietic system, BMP plays an important role in control of HSC number[11] while in the skin, BMP signals act opposite to Wnt signaling, inhibiting the activation of follicle stem cells and favoring epidermal cell fate.[35,36] In central nervous system, BMP signals favor the differentiation of NSC towards astrocytes while the BMP inhibitor, Noggin, promotes a neurogenic fate.[37] Notch signaling pathways are required to maintain stem cells in undifferentiated states in most of these systems, however it triggers differentiation of epidermal progenitor cells.

In addition to the above-mentioned secreted proteins, other molecules such as ions, oxygen and reactive oxygen species (ROS) act on stem cells to affect their behavior. In bone marrow for instance, high calcium concentrations are found in proximity of the endosteal surface, the site of active bone remodeling where both osteoblasts and osteoclasts are found in close proximity. HSCs normally express the calcium sensing receptor (CaR) and its deletion results in HSC abnormal function leading to an impaired marrow engraftment.[38] Further research indicates that HSC preferential localization follows a hypoxic gradient and also that ROS can lead to a premature HSC senescence.[39] These components are summarized in Table 1.

Table 1. Summary of cellular and molecular components of known stem cell niches

Tissue	Stem Cell	Support Cell	Signalling Pathways	Adhesion
C. elegans gonad	GSC	Distal tip cell	Notch	NI
D. melanogaster testis	GSC	Hub cells	JAK-STAT	DE-cadherin, β-catenin
D. melanogaster ovary	GSC	Cap cells, ESCs	DPP-BMP	DE-cadherin, β-catenin
D. melanogaster testis	CPC	Hub cells	JAK-STAT	DE-cadherin, β-catenin
D. melanogaster ovary	ESC	NI	JAK-STAT	NI
D. melanogaster ovary	FSC	NI	Hedgehog	DE-cadherin, β-catenin
Mouse skeletal muscle	Satellite cells	NI	Notch	β-1 integrin
Mouse bone marrow	HSC	Osteoblastic cells, endothelial cells	Wnt, Notch, ANG1, OPN	β-1 integrin
Mouse small intestine	CBC	Crypt fibroblasts, Paneth cells	Wnt, BMP	β-catenin
Mouse skin	Inteollicular kerinocyte	NI	Wnt, Shh, Notch	E-cadherin, β-catenin, β-1 integrin
Mouse skin	Follicular bulge stem cells	Dermal fi-broblasts	Wnt, BMP	β-catenin, β-1 integrin
Mouse brain (latral ventricle)	SVZ stem cells	Vascular cells, astrocytes	Shh, BMP	N-cadherin, β-catenin
Rat brain (hippocampus)	SVZ stem cells	Vascular cells, astrocytes	Shh, Wnt	N-cadherin, β-catenin

ANG1: angiopoietin-1; BNP: bone morphogenetic protein; CBC: crypt base columnar cell; *C. elegans*: *Caenorabditis elegans*; CPC: cyst progenitor cell (somatic stem cell); DPP: Decapentaplegic; *D. melanogaster*: *Drosophila melanogaster*; ECM: extracellular matrix; ESC: escort stem cell; FSC: follicle stem cell; GSC: germinal stem cell; HSC: hematopoietic stem cell; ISC: intestinal stem cell; JAK: Janus kinase; NI: none identified; OPN: osteopontin; SGZ: subgranular zone; Shh: sonic hedgehog; STAT: Signal transducer and activator of transcriptions; SVZ: subventricular zone.

EXTRACELLULAR MATRIX AND CELL-CELL INTERACTIONS

The ECM acts as a scaffolding system, in which stem cells, stromal cells and molecular cues are embedded. Its role is to retain the stem cells in place, to localize signals and to create gradients that guide stem cells in their processes of self-renewal and differentiation. Examples of these key properties are represented by beta-1 integrins that are expressed in various stem cells types, such as HSCs,[40] skin[41-43] and muscle stem cells[44] and that mediates stem cells adhesion to matrix components, regulating stem cell maintenance. Deletion of beta integrins, however, do not lead to loss of marrow stem cells,[45] suggesting that additional factors are involved in HSC localization.[46,47] Tenascin-C is another ECM component expressed in the stromal compartment of the brain[48] and bone marrow.[49] In the brain, it increases sensitivity of NSCs to fibroblast growth factor 2 (FGF2) and bone morphogenetic protein- 4 (BMP4).[48] In bone marrow, osteopontin (OPN) is another matrix glycoprotein that interacts with cell adhesion molecules expressed on HSCs, like CD44 and other integrins[50,51] to facilitate HSC retention in their niche.

It has recently been reported that several mechanical characteristic of ECM, such as the grade of stiffness and elasticity, are involved in stem cell differentiation by means of affecting lineage commitment.[52] This might be particularly important after tissue injury, where subsequent scar formation might negatively affect the ability of stem cells in repair activity.

As previously indicated in invertebrate models, cell-cell interactions via cadherins proteins and spindle cell orientation represent a well-established system through which cells can undergo symmetric cell division (self-renewal) or asymmetric cell division (commitment/differentiation).[53,54] It has been suggested that the same system might also be active in the mammalian niche even though substantial evidence for this is still missing. For instance, preliminary studies have shown that N-cadherin facilitates the association between hematopoietic stem cells and osteoblasts,[11] and that M-cadherin is involved in the association between muscle stem cells and muscle fibers.[55] However, the loss of N-cadherin and M-cadherin in hematopoietic system and in muscles, respectively, does not translate into overt stem cell dysfunction in either system.[56-58]

STEM CELL NICHE DYNAMISM

Stem cell niches receive and mediate messages from the periphery about the necessity for tissue repair following stress of injury, so that stem cell function is flexible and adaptable.

In Drosophila, ovary niche cells can induce somatic stem cells to enter the niche and replace lost germline stem cells[59] and testis niche cells can induce de-differentiation of spermatogonia if the niche needs to be replenished.[60]

In mammalian epidermis, bulge stem cells can migrate upwards and regenerate all epidermal compartments following injury, even though this involvement is only transient and during homeostasis they are only responsible for hair follicle maintenance.[61] Artificially induced Wnt signaling activation in the basal layer of the epidermis leads to the induction of new dermal papillae and ectopic development of hair follicles, leading to the generation of new stem cell niches as well.[62]

In the *Drosophila* intestine and mammalian muscle it has been shown that differentiated cells feedback to stem cells inducing them to proliferate in response to injury.[63,64]

The mammalian HSC niche provides perhaps some of the most striking examples of niche dynamism in response to multiple stimuli. For example, it has been shown that the HSC mobilizing agent G-CSF acts in the first place on osteoblasts, inducing them to proliferate. A wave of HSC proliferation follows the osteoblast one and the return of osteoblast numbers to normal accompanies HSC egression from the niche and mobilization to the peripheral blood.[65] Moreover, the HSC niche is able to survive lethal irradiation, attract freshly transplanted HSC and regenerate itself and the whole hematopoietic tree thanks to complex molecular interactions interactions between osteoblasts, megakaryocytes, endothelial and perivascular cells.[15,66] The chemokine CXCL12 (also called SDF1) has a crucial role in the HSC niche regeneration process as it is responsible for recruitment and retention of transplanted HSC and regulates neo-vascularization and survival of megakaryocytes and osteoblast progenitors.[67-69]

The development of wound repair assays and of an enormous number of transgenic and knock out model organisms has allowed revealing the plasticity of numerous stem cell niches and studying the molecular mechanisms of tissue regeneration. More recently, the advancement of in vivo imaging technologies lead to direct visualization of transplanted hematopoietic stem cells reaching the bone marrow and initiating the engraftment process.[70-73] The combination of in vivo cell tracking and precise stem cell and lineage marking is currently the most promising strategy to evaluate the dynamic response on adult stem cell niches to tissue injury.[74] Further, setting of physiologic stress can see the development of new niches. In hematopoiesis, the infiltration of the bone marrow by abnormal cells (e.g., myelofibrosis) or high demand of blood cell production (e.g., hemoglobinopathies) can result in hematopoiesis occurring in ectopic sites like limph node, spleen and liver.

STEM CELL NICHE AGING

The aging process affects not only the stem cells but also their microenvironment. In several aged tissues stem cells are more proliferative but less efficient in terms of self renewal and progeny production (see for example *Drosophila* intestinal stem cells[75] and mammalian hematopoietic stem cells).[76] Works on *Drosophila* and mouse gonads demonstrated that stem cells do not age by themselves, but rather aged niches have reduced supportive properties.[77-80] In fact, when mouse germline stem cells are serially transplanted into the testis of young mice they are able to maintain their function for years.[80] Young hematopoietic stem cells transplanted into old recipient mice present an aged phenotype at least transiently.[81,82] Moreover, young HSC transiently exposed in vitro to osteoblasts from old mice perform more similarly to older HSC in transplantation assays.[9]

The deregulation of local signaling pathways and accumulation of stress-induced damage, including reactive oxygen species (ROS), have been indicated as major causes of stem cell and niche loss of function during aging. Old niches appear to constitutively maintain damage response pathways and mechanisms, which younger niches activate only temporarily in response to injury. In this sense, niche aging might be perceived as a progressive switch from the ability to flexibly respond to damage to a constant, yet inefficient, repair mode, which eventually exhaust the stem cell regenerative potential.[75,83-85] The same molecular pathways regulating young stem cell-niche interactions are still present in older niches, however signaling the latter is deregulated. Examples the Notch pathway in the *Drosophila* intestine[83] and the Wnt pathway in mammalian muscle.[6]

Not only local, but also systemic factors play a role in stem cell and niche aging. While harder to identify, these factors have allowed the first successful experiments leading to stem cell and niche rejuvenation. Older mice exposed to younger systemic factors through parabiosis experiments (in which two animals are surgically joined and develop communal circulation) present rejuvenated muscle and hematopoietic stem cells and niches.[6,9]

MALIGNANT STEM CELL NICHES

In the same way the microenvironment has a critical role in regulating stem cell function, the niche and niche alterations can play a role in the development of cancer. The idea that the microenvironment might be one of the factors involved in cancer development was proposed over one century ago.[86] The role of local vasculature in the support of tumor growth has long been recognized[87,88] and development of anti-angiogenic molecules has been a very fertile field of basic and translational research aiming at the generation of efficient anticancer therapies.[89] Transcriptional profiling of basal cell carcinoma cells and stroma indicated that tumor fibroblasts express high levels of Gremlin1, antagonizing BMP 2 and 4, suppressing epidermal cells differentiation and promoting local proliferation.[90] Also in breast cancer, cancer associated fibroblasts have been demonstrated to be able to promote tumor progression.[91] Immune cells play a dual role in the regulation of cancer development. On one side they provide immunosurveillance and can actively remove transformed cells from tissues; on the other side chronic inflammation contributes to skewing local molecular signals towards a chronic stress-response, conducive to cell damage and premature aging, all the way to the generation of a modified microenvironment now supportive of tumor growth.[92,93]

Moreover, as stem cells have been described to be the cell of origin of certain malignancies, it is likely that defect in their microenvironment could have contributed to their malignant phenotype. There are some examples of niche/microenvironment induced malignancies from the hematopoietic stem cell field, where some transgenic or knock out mice have been described to develop myeloproliferative or myelodisplastic syndromes which are not transplantable, but rather present again when the mice receive wild type bone marrow transplantation.[94-96] For example deletion of Dicer1 in a select subset of mesenchymal cells, osetoprogenitors, results in the development of myelodisplasia and remarkably, acute leukemia. The leukemia had secondary genetic changes but had normal Dicer1. Further transplant of the leukemia was only successful if the recipient had deletion of Dicer1 in the marrow osteoprogenitors. Therefore, this niche can be the source of an initiating oncogenic event in cancer and be required for its maintenance.[97]

Recently, the application of the stem cell model to cancer naturally raises the question whether cancer stem cells reside in and are regulated by specific microenvironments and whether they do so in competition with normal stem cells. The answer is complex and variable depending on the cancer analyzed. Human B cell leukemia cells transplanted into mice localize in the bone marrow following the expression pattern of SDF1 and therefore appear similar to normal HSC.[13] However, it has been shown in a mouse model of chronic myeloid leukemia that malignant, but not normal HSC depend on CD44 function in order to localize to the bone marrow and give rise to leukemia,[98] indicating the presence of leukemia-specific mechanisms of interaction with the microenvironment. Moreover, it is likely that more aggressive and advanced stages of disease become independent from niche support. Even though leukemia could develop independent of the influence of the microenvironment, in certain cases cytokines and other secreted factors produced by stroma cells have been shown to confer resistance to chemotherapeutic interventions.[99,100] The differential support of normal and malignant cells by niche alterations represents new promising area for research. If distinct sensitivity to niche signals exists then distinct sensitivity to niche signals can be exploited in therapy.

Tumor progression can alter existing niches at the expenses of normal stem cells. For example, at advanced stages of disease leukemia cells in the bone marrow prevent normal transplanted HSC from correctly localizing in their niches and actively secrete SCF to form a new, inhibitory niche.[101] Myeloma cells disrupt the endosteal HSC niche by secreting the Wnt inhibitor Dkk1.[102]

Invasive primary tumors secrete soluble factors that can act at long distances to induce premetastatic niches, which in turns will recruit tumor cells and support metastatic growth.[103] For example, melanoma cells produce factors that activate hepatic stellate cells to become similar to myofibroblasts and supportive of metastatic melanoma cells.[104,105] Finally, normal niches can attract malignant cells with similar characteristics to the stem cells normally residing in the niche. This is the case of the bone marrow, where osteotropic cancers such as breast, ovarian, prostate and neuroblastoma metastatize to based on the SDF1-CXCR4 signaling axis.[103]

CONCLUSION

The stem cell niche concept has gained experimental support and conceptual complexity since proposed by Schofield. The crosstalk between different cell types intrinsic to the stem cell niche offers the opportunity to target these cell communication networks and tailor the dynamics of normal stem cells to boost their ability to respond to injury as well as to manage the competitive advantage of malignant cells. The niche is a point of intervention still under-explored, that offers a uniquely drug-gable opportunity to affect regenerative medicine and anticancer treatments.

REFERENCES

1. Schofield R. The relationship between the spleen colony-forming cell and the haemopoietic stem cell. Blood Cells 1978; 4(1-2):7-25.
2. Xie T, Spradling AC. decapentaplegic is essential for the maintenance and division of germline stem cells in the Drosophila ovary. Cell 1998; 94(2):251-260.
3. Kimble JE, White JG. On the control of germ cell development in Caenorhabditis elegans. Dev Biol 1981; 81(2):208-219.
4. Katayama Y, Battista M, Kao WM et al. Signals from the sympathetic nervous system regulate hematopoietic stem cell egress from bone marrow. Cell 2006; 124(2):407-421.
5. Conboy IM, Conboy MJ, Wagers AJ et al. Rejuvenation of aged progenitor cells by exposure to a young systemic environment. Nature 2005; 433(7027):760-764.
6. Brack AS, Conboy MJ, Roy S et al. Increased Wnt signaling during aging alters muscle stem cell fate and increases fibrosis. Science 2007; 317(5839):807-810.
7. Zhu Y, Ghosh P, Charnay P et al. Neurofibromas in NF1: Schwann cell origin and role of tumor environment. Science 2002; 296(5569):920-922.
8. Trimboli AJ, Cantemir-Stone CZ, Li F et al. Pten in stromal fibroblasts suppresses mammary epithelial tumours. Nature 2009; 461(7267):1084-1091.
9. Mayack SR, Shadrach JL, Kim FS et al. Systemic signals regulate ageing and rejuvenation of blood stem cell niches. Nature 2010; 463(7280):495-500.
10. Calvi LM, Adams GB, Weibrecht KW et al. Osteoblastic cells regulate the haematopoietic stem cell niche. Nature 2003; 425(6960):841-846.
11. Zhang J, Niu C, Ye L et al. Identification of the haematopoietic stem cell niche and control of the niche size. Nature 2003; 425(6960):836-841.
12. Kiel MJ, Yilmaz OH, Iwashita T et al. SLAM family receptors distinguish hematopoietic stem and progenitor cells and reveal endothelial niches for stem cells. Cell 2005; 121(7):1109-1121.

13. Sipkins DA, Wei X, Wu JW et al. In vivo imaging of specialized bone marrow endothelial microdomains for tumour engraftment. Nature 2005; 435(7044):969-973.
14. Visnjic D, Kalajzic Z, Rowe DW et al. Hematopoiesis is severely altered in mice with an induced osteoblast deficiency. Blood 2004; 103(9):3258-3264.
15. Hooper AT, Butler JM, Nolan DJ et al. Engraftment and reconstitution of hematopoiesis is dependent on VEGFR2-mediated regeneration of sinusoidal endothelial cells. Cell Stem Cell 2009; 4(3):263-274.
16. Jung Y, Wang J, Song J et al. Annexin II expressed by osteoblasts and endothelial cells regulates stem cell adhesion, homing and engraftment following transplantation. Blood 2007; 110(1):82-90.
17. Cotsarelis G, Sun TT, Lavker RM. Label-retaining cells reside in the bulge area of pilosebaceous unit: implications for follicular stem cells, hair cycle and skin carcinogenesis. Cell 1990; 61(7):1329-1337.
18. Levy V, Lindon C, Harfe BD, et al. Distinct stem cell populations regenerate the follicle and interfollicular epidermis. Dev Cell 2005; 9(6):855-861.
19. Blanpain C, Fuchs E. Epidermal homeostasis: a balancing act of stem cells in the skin. Nat Rev Mol Cell Biol 2009; 10(3):207-217.
20. Doetsch F, Caille I, Lim DA, et al. Subventricular zone astrocytes are neural stem cells in the adult mammalian brain. Cell 1999; 97(6):703-716.
21. Palmer TD, Takahashi J, Gage FH. The adult rat hippocampus contains primordial neural stem cells. Mol Cell Neurosci 1997; 8(6):389-404.
22. Shen Q, Goderie SK, Jin L et al. Endothelial cells stimulate self-renewal and expand neurogenesis of neural stem cells. Science 2004; 304(5675):1338-1340.
23. Barker N, van Es JH, Kuipers J et al. Identification of stem cells in small intestine and colon by marker gene Lgr5. Nature 2007; 449(7165):1003-1007.
24. Mills JC, Gordon JI. The intestinal stem cell niche: there grows the neighborhood. Proc Natl Acad Sci USA 2001; 98(22):12334-12336.
25. Mauro A. Satellite cell of skeletal muscle fibers. J Biophys Biochem Cytol 1961; 9:493-495.
26. Kuang S, Kuroda K, Le Grand F et al. Asymmetric self-renewal and commitment of satellite stem cells in muscle. Cell 2007; 129(5):999-1010.
27. Reya T, Duncan AW, Ailles L et al. A role for Wnt signalling in self-renewal of haematopoietic stem cells. Nature 2003; 423(6938):409-414.
28. Brittan M, Wright NA. Gastrointestinal stem cells. J Pathol 2002; 197(4):492-509.
29. Brittingham J, Phiel C, Trzyna WC et al. Identification of distinct molecular phenotypes in cultured gastrointestinal smooth muscle cells. Gastroenterology 1998; 115(3):605-617.
30. Huelsken J, Vogel R, Erdmann B et al. beta-Catenin controls hair follicle morphogenesis and stem cell differentiation in the skin. Cell 2001; 105(4):533-545.
31. Merrill BJ, Gat U, DasGupta R et al. Tcf3 and Lef1 regulate lineage differentiation of multipotent stem cells in skin. Genes Dev 2001; 15(13):1688-1705.
32. Niemann C, Owens DM, Hulsken J et al. Expression of DeltaNLef1 in mouse epidermis results in differentiation of hair follicles into squamous epidermal cysts and formation of skin tumours. Development 2002; 129(1):95-109.
33. Chenn A, Walsh CA. Regulation of cerebral cortical size by control of cell cycle exit in neural precursors. Science 2002; 297(5580):365-369.
34. Chen D, McKearin D. Dpp signaling silences bam transcription directly to establish asymmetric divisions of germline stem cells. Curr Biol 2003; 13(20):1786-1791.
35. Botchkarev VA, Botchkareva NV, Nakamura M et al. Noggin is required for induction of the hair follicle growth phase in postnatal skin. FASEB J 2001; 15(12):2205-2214.
36. Kulessa H, Turk G, Hogan BL. Inhibition of Bmp signaling affects growth and differentiation in the anagen hair follicle. EMBO J 2000; 19(24):6664-6674.
37. Temple S. The development of neural stem cells. Nature 2001; 414(6859):112-117.
38. Adams GB, Chabner KT, Alley IR et al. Stem cell engraftment at the endosteal niche is specified by the calcium-sensing receptor. Nature 2006; 439(7076):599-603.
39. Ito K, Hirao A, Arai F et al. Regulation of oxidative stress by ATM is required for self-renewal of haematopoietic stem cells. Nature 2004; 431(7011):997-1002.
40. Wagers AJ, Allsopp RC, Weissman IL. Changes in integrin expression are associated with altered homing properties of Lin(-/lo)Thy1.1(lo)Sca-1(+)c-kit(+) hematopoietic stem cells following mobilization by cyclophosphamide/granulocyte colony-stimulating factor. Exp Hematol 2002; 30(2):176-185.
41. Brakebusch C, Grose R, Quondamatteo F et al. Skin and hair follicle integrity is crucially dependent on beta 1 integrin expression on keratinocytes. EMBO J 2000; 19(15):3990-4003.
42. Jones PH, Watt FM. Separation of human epidermal stem cells from transit amplifying cells on the basis of differences in integrin function and expression. Cell 1993; 73(4):713-724.

43. Jensen UB, Lowell S, Watt FM. The spatial relationship between stem cells and their progeny in the basal layer of human epidermis: a new view based on whole-mount labelling and lineage analysis. Development 1999; 126(11):2409-2418.
44. Sherwood RI, Christensen JL, Conboy IM et al. Isolation of adult mouse myogenic progenitors: functional heterogeneity of cells within and engrafting skeletal muscle. Cell 2004; 119(4):543-554.
45. Bungartz G, Stiller S, Bauer M et al. Adult murine hematopoiesis can proceed without beta1 and beta7 integrins. Blood 2006; 108(6):1857-1864.
46. Fleming WH, Alpern EJ, Uchida N et al. Steel factor influences the distribution and activity of murine hematopoietic stem cells in vivo. Proc Natl Acad Sci USA 1993; 90(8):3760-3764.
47. Sugiyama T, Kohara H, Noda M et al. Maintenance of the hematopoietic stem cell pool by CXCL12-CXCR4 chemokine signaling in bone marrow stromal cell niches. Immunity 2006; 25(6):977-988.
48. Garcion E, Halilagic A, Faissner A et al. Generation of an environmental niche for neural stem cell development by the extracellular matrix molecule tenascin C. Development 2004; 131(14):3423-3432.
49. Ohta M, Sakai T, Saga Y et al. Suppression of hematopoietic activity in tenascin-C-deficient mice. Blood 1998; 91(11):4074-4083.
50. Stier S, Ko Y, Forkert R et al. Osteopontin is a hematopoietic stem cell niche component that negatively regulates stem cell pool size. J Exp Med 2005; 201(11):1781-1791.
51. Nilsson SK, Johnston HM, Whitty GA et al. Osteopontin, a key component of the hematopoietic stem cell niche and regulator of primitive hematopoietic progenitor cells. Blood 2005; 106(4):1232-1239.
52. Engler AJ, Sen S, Sweeney HL et al. Matrix elasticity directs stem cell lineage specification. Cell 2006; 126(4):677-689.
53. Deng W, Lin H. Spectrosomes and fusomes anchor mitotic spindles during asymmetric germ cell divisions and facilitate the formation of a polarized microtubule array for oocyte specification in Drosophila. Dev Biol 1997; 189(1):79-94.
54. Yamashita YM, Jones DL, Fuller MT. Orientation of asymmetric stem cell division by the APC tumor suppressor and centrosome. Science 2003; 301(5639):1547-1550.
55. Irintchev A, Zeschnigk M, Starzinski-Powitz A et al. Expression pattern of M-cadherin in normal, denervated and regenerating mouse muscles. Dev Dyn 1994; 199(4):326-337.
56. Hollnagel A, Grund C, Franke WW et al. The cell adhesion molecule M-cadherin is not essential for muscle development and regeneration. Mol Cell Biol 2002; 22(13):4760-4770.
57. Hooper AT, Butler J, Petit I et al. Does N-cadherin regulate interaction of hematopoietic stem cells with their niches? Cell Stem Cell 2007; 1(2):127-129.
58. Kiel MJ, Radice GL, Morrison SJ. Lack of evidence that hematopoietic stem cells depend on N-cadherin-mediated adhesion to osteoblasts for their maintenance. Cell Stem Cell 2007; 1(2):204-217.
59. Kai T, Spradling A. An empty Drosophila stem cell niche reactivates the proliferation of ectopic cells. Proc Natl Acad Sci USA 2003; 100(8):4633-4638.
60. Brawley C, Matunis E. Regeneration of male germline stem cells by spermatogonial dedifferentiation in vivo. Science 2004; 304(5675):1331-1334.
61. Ito M, Liu Y, Yang Z et al. Stem cells in the hair follicle bulge contribute to wound repair but not to homeostasis of the epidermis. Nat Med 2005; 11(12):1351-1354.
62. Silva-Vargas V, Lo Celso C, Giangreco A et al. Beta-catenin and Hedgehog signal strength can specify number and location of hair follicles in adult epidermis without recruitment of bulge stem cells. Dev Cell 2005; 9(1):121-131.
63. Jiang H, Patel PH, Kohlmaier A et al. Cytokine/Jak/Stat signaling mediates regeneration and homeostasis in the Drosophila midgut. Cell 2009; 137(7):1343-1355.
64. Bischoff R. Interaction between satellite cells and skeletal muscle fibers. Development 1990; 109(4):943-952.
65. Mayack SR, Wagers AJ. Osteolineage niche cells initiate hematopoietic stem cell mobilization. Blood 2008; 112(3):519-531.
66. Dominici M, Rasini V, Bussolari R et al. Restoration and reversible expansion of the osteoblastic hematopoietic stem cell niche after marrow radioablation. Blood 2009; 114(11):2333-2343.
67. Peled A, Petit I, Kollet O et al. Dependence of human stem cell engraftment and repopulation of NOD/SCID mice on CXCR4. Science 1999; 283(5403):845-848.
68. Kortesidis A, Zannettino A, Isenmann S et al. Stromal-derived factor-1 promotes the growth, survival and development of human bone marrow stromal stem cells. Blood 2005; 105(10):3793-3801.
69. Jin DK, Shido K, Kopp HG et al. Cytokine-mediated deployment of SDF-1 induces revascularization through recruitment of CXCR4+ hemangiocytes. Nat Med 2006; 12(5):557-567.
70. Kohler A, Schmithorst V, Filippi MD et al. Altered cellular dynamics and endosteal location of aged early hematopoietic progenitor cells revealed by time-lapse intravital imaging in long bones. Blood 2009; 114(2):290-298.

71. Lewandowski D, Barroca V, Duconge F et al. In vivo cellular imaging pinpoints the role of reactive oxygen species in the early steps of adult hematopoietic reconstitution. Blood 2009.
72. Lo Celso C, Fleming HE, Wu JW et al. Live-animal tracking of individual haematopoietic stem/progenitor cells in their niche. Nature 2009; 457(7225):92-96.
73. Xie Y, Yin T, Wiegraebe W et al. Detection of functional haematopoietic stem cell niche using real-time imaging. Nature 2009; 457(7225):97-101.
74. Voog J, Jones DL. Stem Cells and the Niche: a dynamic duo. Cell Stem Cell, in press 2010; 6.
75. Biteau B, Hochmuth CE, Jasper H. JNK activity in somatic stem cells causes loss of tissue homeostasis in the aging Drosophila gut. Cell Stem Cell 2008; 3(4):442-455.
76. Janzen V, Forkert R, Fleming HE et al. Stem-cell ageing modified by the cyclin-dependent kinase inhibitor p16INK4a. Nature 2006; 443(7110):421-426.
77. Boyle M, Wong C, Rocha M et al. Decline in self-renewal factors contributes to aging of the stem cell niche in the Drosophila testis. Cell Stem Cell 2007; 1(4):470-478.
78. Hsu HJ, Drummond-Barbosa D. Insulin levels control female germline stem cell maintenance via the niche in Drosophila. Proc Natl Acad Sci USA. 2009; 106(4):1117-1121.
79. Pan L, Chen S, Weng C et al. Stem cell aging is controlled both intrinsically and extrinsically in the Drosophila ovary. Cell Stem Cell 2007; 1(4):458-469.
80. Ryu BY, Orwig KE, Oatley JM et al. Effects of aging and niche microenvironment on spermatogonial stem cell self-renewal. Stem Cells 2006; 24(6):1505-1511.
81. Liang Y, Van Zant G, Szilvassy SJ. Effects of aging on the homing and engraftment of murine hematopoietic stem and progenitor cells. Blood 2005; 106(4):1479-1487.
82. Rossi DJ, Bryder D, Zahn JM et al. Cell intrinsic alterations underlie hematopoietic stem cell aging. Proc Natl Acad Sci USA 2005; 102(26):9194-9199.
83. Ohlstein B, Spradling A. Multipotent Drosophila intestinal stem cells specify daughter cell fates by differential notch signaling. Science 2007; 315(5814):988-992.
84. Brack AS, Rando TA. Intrinsic changes and extrinsic influences of myogenic stem cell function during aging. Stem Cell Rev Fall 2007; 3(3):226-237.
85. Wagner W, Horn P, Bork S et al. Aging of hematopoietic stem cells is regulated by the stem cell niche. Exp Gerontol 2008; 43(11):974-980.
86. Paget S. The distribution of secondary growths in cancer of the breast. Lancet 1889; 1:571-573.
87. Folkman J. Tumor angiogenesis: a possible control point in tumor growth. Ann Intern Med 1975; 82(1):96-100.
88. Bergers G, Javaherian K, Lo KM et al. Effects of angiogenesis inhibitors on multistage carcinogenesis in mice. Science 1999; 284(5415):808-812.
89. Abdelrahim M, Konduri S, Basha R et al. Angiogenesis: an update and potential drug approaches (review). Int J Oncol 2010; 36(1):5-18.
90. Sneddon JB, Zhen HH, Montgomery K et al. Bone morphogenetic protein antagonist gremlin 1 is widely expressed by cancer-associated stromal cells and can promote tumor cell proliferation. Proc Natl Acad Sci USA 2006; 103(40):14842-14847.
91. Hu M, Yao J, Carroll DK et al. Regulation of in situ to invasive breast carcinoma transition. Cancer Cell 2008; 13(5):394-406.
92. Mantovani A, Romero P, Palucka AK et al. Tumour immunity: effector response to tumour and role of the microenvironment. Lancet 2008; 371(9614):771-783.
93. Raulet DH, Guerra N. Oncogenic stress sensed by the immune system: role of natural killer cell receptors. Nat Rev Immunol 2009; 9(8):568-580.
94. Walkley CR, Olsen GH, Dworkin S et al. A microenvironment-induced myeloproliferative syndrome caused by retinoic acid receptor gamma deficiency. Cell 2007; 129(6):1097-1110.
95. Walkley CR, Shea JM, Sims NA et al. Rb regulates interactions between hematopoietic stem cells and their bone marrow microenvironment. Cell 2007; 129(6):1081-1095.
96. Lane SW, Sykes SM, Shahrour F et al. The APCmin mouse has altered hematopoietic stem cell function and provides a model for cell extrinsic MPD/MDS. Blood, in press. 2010.
97. Raaijmakers MHGP, Mukherjee S, Guo S et al. Bone progenitor dysfunction induces myelodysplasia and secondary leukemia Nature, 2010.
98. Krause DS, Lazarides K, von Andrian UH et al. Requirement for CD44 in homing and engraftment of BCR-ABL-expressing leukemic stem cells. Nat Med 2006; 12(10):1175-1180.
99. Iwamoto S, Mihara K, Downing JR et al. Mesenchymal cells regulate the response of acute lymphoblastic leukemia cells to asparaginase. J Clin Invest 2007; 117(4):1049-1057.
100. Williams RT, den Besten W, Sherr CJ. Cytokine-dependent imatinib resistance in mouse BCR-ABL+, Arf-null lymphoblastic leukemia. Genes Dev 2007; 21(18):2283-2287.
101. Colmone A, Amorim M, Pontier AL et al. Leukemic cells create bone marrow niches that disrupt the behavior of normal hematopoietic progenitor cells. Science 2008; 322(5909):1861-1865.

102. Qiang YW, Chen Y, Stephens O et al. Myeloma-derived Dickkopf-1 disrupts Wnt-regulated osteoprotegerin and RANKL production by osteoblasts: a potential mechanism underlying osteolytic bone lesions in multiple myeloma. Blood 2008; 112(1):196-207.
103. Wels J, Kaplan RN, Rafii S et al. Migratory neighbors and distant invaders: tumor-associated niche cells. Genes Dev 2008; 22(5):559-574.
104. Olaso E, Salado C, Egilegor E et al. Proangiogenic role of tumor-activated hepatic stellate cells in experimental melanoma metastasis. Hepatology 2003; 37(3):674-685.
105. Olaso E, Santisteban A, Bidaurrazaga J et al. Tumor-dependent activation of rodent hepatic stellate cells during experimental melanoma metastasis. Hepatology 1997; 26(3):634-642.

CHAPTER 12

ADULT STEM CELL DIFFERENTIATION AND TRAFFICKING AND THEIR IMPLICATIONS IN DISEASE

Ying Zhuge, Zhao-Jun Liu and Omaida C. Velazquez*

Abstract: Stem cells are unspecialized precursor cells that mainly reside in the bone marrow and have important roles in the establishment of embryonic tissue. They also have critical functions during adulthood, where they replenish short-lived mature effector cells and regeneration of injured tissue. They have three main characteristics: self-renewal, differentiation and homeostatic control. In order to maintain a pool of stem cells that support the production of blood cells, stromal elements and connective tissue, stem cells must be able to constantly replenish their own number. They must also possess the ability to differentiate and give rise to a heterogeneous group of functional cells. Finally, stem cells must possess the ability to modulate and balance differentiation and self-renewal according to environmental stimuli and whole-organ needs to prevent the production of excessive number of effector cells.[1] In addition to formation of these cells, regulated movement of stem cells is critical for organogenesis, homeostasis and repair in adulthood. Stem cells require specific inputs from particular environments in order to perform their various functions. Some similar trafficking mechanisms are shared by leukocytes, adult and fetal stem cells, as well as cancer stem cells.[1,2] Achieving proper trafficking of stem cells will allow increased efficiency of targeted cell therapy and drug delivery.[2] In addition, understanding similarities and differences in homing and migration of malignant cancer stem cells will also clarify molecular events of tumor progression and metastasis.[2] This chapter focuses on the differentiation, trafficking and homing of the major types of adult bone marrow stem cells: hematopoietic stem cells (HSCs), mesenchymal stem cells (MSCs) and endothelial progenitor cells (EPCs) and the term "stem cell" will refer to "adult stem cells" unless otherwise specified.

*Corresponding Author: Omaida C. Velazquez—Division of Vascular and Endovascular Surgery, Room 3016 Holtz Center—JMH East Tower, 1611 N.W. 12th Avenue, Miami, Florida 33136, USA. Email: ovelazquez@med.miami.edu

The Cell Biology of Stem Cells, edited by Eran Meshorer and Kathrin Plath.
©2010 Landes Bioscience and Springer Science+Business Media.

DIFFERENTIATION

Hematopoietic Stem Cells

The process of stem cell differentiation is most-studied and well-established for HSCs that reside in the bone marrow and become progressively more specific to constitute mature circulating blood cells, which consist of red blood cells, megakaryocytes, myeloid cells and lymphocytes (Fig. 1). Since functional blood cells are short-lived, HSCs are required to replenish the circulating blood elements throughout life. When HSCs are transplanted, they are able to reconstitute the entire blood system of the appropriately-conditioned recipient indefinitely.

The process of hematopoiesis is well-conserved throughout vertebrate evolution and animal models such as the mouse and zebrafish have complemented and greatly extended studies of human hematopoiesis.[3] Formation of HSCs as well as subsequent lineage-restricted differentiation is mainly regulated by transcription factors, which encompass virtually all classes of DNA-binding proteins.[4] Transcription factors that are necessary for the formation of HSCs include mixed lineage-leukemia gene (MLL), Runt-related transcript factor-1 (Runx1), translocation ets leukemia/ets variant gene 6 (Tel/ETV6), stem cell leukemia/T-cell acute leukemia 1 (SCL/tal1) and LIM domain only 2 (LMO2).[3] The genes encoding the SET-domain containing histone methyltransferase

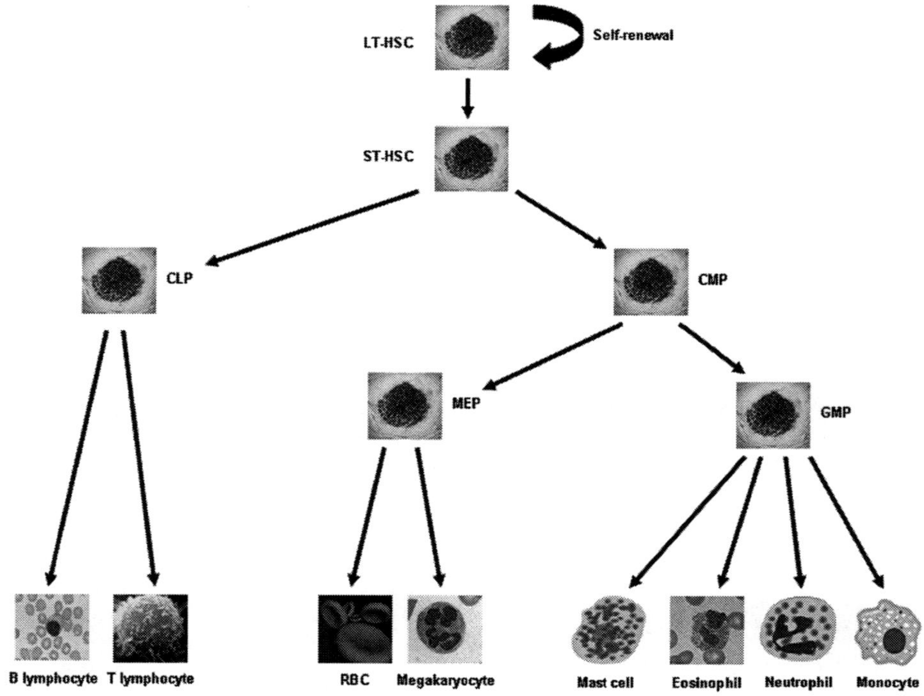

Figure 1. Pluripotent capacity of HSCs to differentiate into various short-lived mature effectors blood cells (solid arrows). HSCs are also capable of self-renewal (block arrow). Adapted from Orkin SH, Zon LI. Hematopoiesis: an evolving paradigm for stem cell biology.

MLL and runt-domain Runx1 proteins are necessary for HSC generation in the various sites of hematopoiesis.[4] The basic-helix-loop-helix factor SCL/tal1 and its associated protein partner LMO2 are both essential for the development of the hematopoietic systems.[5] In the absence of Runx1, SCL/tal1, or LMO2, no blood cells are formed. On the contrary, enforced expression of SCL/tal1 and LMO2 converts mesoderm into hematopoietic lineages. While critical for HSC specification, these transcription factors may not be continuously required to for their survival and proliferation. For example, *SCL/tal1*, which is needed early on during developing, does not affect the maintenance of self-renewing progenitors if inactivated in adult HSCs.[6] Similarly, inaction of *Runx1* in adult HSCs merely causes disturbances in differentiation of some lineages, but does not abolish HSC properties all together.[7] Therefore, a large repertoire of factors is required in temporal-dependent ways to achieve regulation of HSC speciation.

The above-mentioned, as well as additional transcription factors are involved in differentiation of individual blood lineages from multipotent progenitors (Table 1). The zinc finger factor GATA-binding factor 1 (GATA-1) is highly-produced in megakaryocytic/erythroid progenitors, which give rise to megakaryocyte and red blood cell precursors.[8] In contrast, CCAAT/enhancer binding protein-α (C/EBPα) is present in granulocyte/myeloid precursors and PU.1 induces myeloid development.[9] Finally, paired box protein 5 (Pax5) is required for B-cell commitment and differentiation.[10] In addition the presence of factors, their concentration at a given time in development is also critical to differentiation.[4] For example, eosinophils form at low levels of GATA-1, whereas at higher levels, erythroid and megakaryocytic development occurs.[8] Conversely, the maturation of erythroid precursors is retarded by a threefold decrease in the expression of GATA-1.[11] In addition, a high level of PU.1 favors macrophage development while low levels favor B cell generation.[4]

While it is simpler to identify a one-to-one correspondence of lineage-restricted transcription factor and progenitor in later generations of differentiation, this is challenging in earlier generations, as one cell expresses different lineage markers.[12] For example, GATA-1, FOG-1, Ikaros and PU.1 are all expressed in progenitors with multilineage potential, leaving multiple development options for these cells. Once a tentative decision is reached, mutual reinforcement of a stable pathway for subsequent differentiation

Table 1. Regulatory signals for differentiaion of hematopoietic stem cells

HSC Speciation	
Mixed lineage leukemia gene (MLL)	
Runx1	
Tel/ETV6	
SCL/tal1	
LMO2	
HSC Differentiation Transcription Factor	**Lineage**
GATA-1	Erythroid, megakaryocytic
C/EBPα	Myeloid, granulocyte
PU.1	Myeloid
Pax5	B lymphocyte

may be provided by auto-upregulation of one lineage factor and by cross-antagonism of others.[4] For example, both GATA-1 and PU.1 both positive feedback their own production to promote erythroid and myeloid cells, respectively.[13,14] In addition to allowing for crosstalk at the molecular level between different cellular fates, the presence of multiple lineage markers on common progenitor cells also demonstrates the principle of lineage priming, which refers to the idea that lineage selection is a process of extinguishing alternative possibilities, rather than imposing one dominant pathway on an otherwise blank slate.[3] This is accomplished by maintaining chromatin in early HSCs in an open confirmation, followed by transient repression and more permanent silencing, all the while maintaining the plasticity of progenitor cells.[3] Although cellular differentiation was once considered unidirectional, evidence now favor the notion of cellular reprogramming.[3] For example, enforced GATA-1 expression in early myeloid progenitor cells drives them to redifferentiate to erythroid, eosinophilic, or megakaryocytic precursors.[8,15] Similarly, committed B- and T-lymphoid cells can be reprogrammed to macrophages via enforced C/EBPα expression.[16,17] Cells that are reprogrammed transit through an intermediate state where different lineage markers are expressed, indicating this process occurs in a stepwise fashion.[3]

Another characteristic of key lineage-restricted transcription factors involved in HSC differentiation is their simultaneous ability to promote one pathway while antagonizing another, providing an efficient mechanism of resolving and reinforcing lineage choices. For example, upregulation of eosinophil markers upon expression of GATA-1 is associated with a concomitant downregulation of myeloid lineage markers.[8] Also, in the absence of absence Pax5, progenitors originally destined to become B cells fail to restrict their lineage choice and develop into a variety of hematopoietic cells such as macrophages, osteoclasts and granulocytes. Pax5 both drives B-cell development and suppresses alternative lineage choices.[18] This cross-regulation by regulators of hematopoiesis also occurs at the protein level. For example, GATA-1 and PU.1 physically interact via association of the amino terminus of PU.1 with the carboxy-finger of GATA-1, which blocks GATA-1's ability to recognize DNA.[19] At the same time, PU.1 and GATA-1 impairs PU.1-dependent transactivation due to displacement of a cofactor.[4,20] In addition, Pax5 is critical for the development of B cells as it represses other growth factors that are responsible for allowing cells without Pax5 to differentiate to T-, NK- or dendritic cells, macrophages, neutrophils, or erythroid precursors.[3] Finally, GATA-3 is necessary for the production of T_H2 from $CD4^+$ T-cells and can switch T_H1 cells to the T_H2 phenotype.[21] Conversely, the transcription factor T-bet converts T_H2 cells to T_H1 phenotype.[22] Differentiation is an ongoing process that necessitates continuous and active participation of key regulators.[4]

Almost all hematopoietic transcription factors are directly associated with hematopoietic malignancies, or leukemias. Disturbance of the homeostatic balance of the critical transcription factors is the defining feature of leukemias.[3] Most of these altered mutations consist of chromosomal translocations or somatic mutations of key transcription factors in differentiation. In chromosomal translocations, chimeric transcription factors inappropriately activate or repress genes, which then cause improper downstream effects. For example, *Scl/tal-1*, *Lmo2*, *Tel*, *E2A* and *runx1* are all involved in chromosomal translocations, which create fusion proteins that function in a dominant-negative fashion to block the action of lineage-determining factors.[4] Somatic mutations in *GATA-1, PU.1* and *Ikaros* cause misexpressed or dysregulated transcription factors that alter control of differentiation. *GATA-1* mutation is involved in Down syndrome-associated megakaryocytic leukemias,[23] *PU.1* and *C-EBPα* is involved in myeloid leukemias,[24,25] and *Pax5* mutations are involved in B-lymphoid leukemias.[26]

Mesenchymal Stem Cells

MSCs are a heterogeneous group of progenitor cells with pluripotent capacity to differentiate into mesodermal and nonmesodermal cell lineages, including osteocytes, adipocytes, chondrocytes, myocytes, cardiomyocytes, fibroblasts, myofibroblasts, epithelial cells and neurons (Fig. 2).[27] The International Society for Cytotherapy has identified three criteria for identifying these cells: plastic adherence of cells in culture; expression of CD105, CD73 and CD90 in greater than 95% of culture and lack of expression of CD34, CD45, CD14 or CD11b, CD79α or CD19 and HLD-DR in greater than 95% of the culture; and differentiation into bone, fat and cartilage.[28] MSCs reside primarily in the bone marrow, but also exist in other sites such as adipose tissue, peripheral blood, cord blood, liver, periosteum, synovial membrane, synovial fluid, skeletal muscle, dermis, deciduous teeth, pericytes, trabecular bone, infrapatellar fat pad, articular cartilage, placenta, spleen, thymus and fetal tissues.[27] When stimulated by specific signals, they can be released from their niche in the bone marrow into circulation and recruited to specific tissues to undergo in situ differentiation and contribute to tissue regeneration.[29-31]

Although MSCs isolated from different tissues show similar phenotypic characteristics, they show different propensities differentiate in response to stimulation by various growth

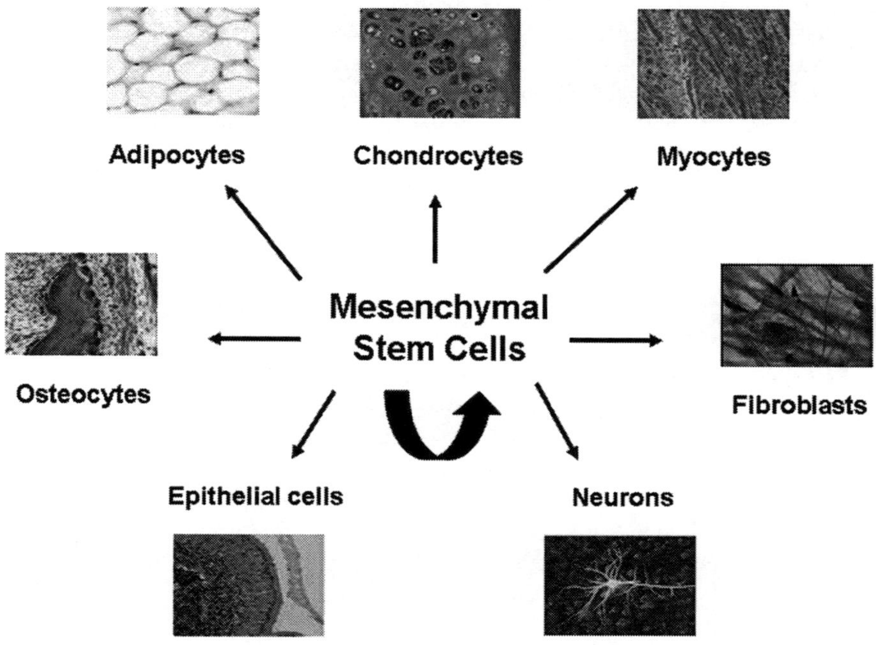

Figure 2. Pluripotent capacity of MSCs to differentiate into mesodermal and nonmesodermal cell lineages (solid arrows). MSCs are also capable of self-renewal (block arrow). Reprinted with permission from Liu Z, Zhuge Y, Velazquez OC. Trafficking and differentiation of mesenchymal stem cells. J Cell Biochem 2009; 106:984-91.

Table 2. Regulatory signals for differentiation of mesenchymal stem cells

Biological Signals	Differentiation Potential
TGF-β	Chondrogenic
IGF-1	Chondrogenic
bFGF	Chondrogenic, osteogenic, neural
EGF	Chondrogenic
PDGF	Condrogenic, myofibroblastic
VEGF	Endothelial
Wnt	Chondrogenic, osteogenic, neural

Adapted from Liu Z, Zhuge Y, Velazquez OC. Trafficking and differentiation of mesenchymal stem cells. J Cell Biochem 2009; 106:984-91.

factors. Growth factors that have regulatory effects on MSCs include members of the transforming growth factor-beta (TGF-β) superfamily, the insulin-like growth factor (IGF), the fibroblast growth factors (FGF), the epidermal growth factor (EGF), the platelet-derived growth factor (PDGF), the vascular endothelial growth factor (VEGF) and family of growth factors known as Wnt (Table 2).[32-37] The most potent inducers of chondrogenesis are the TGF-β family, including TGF-β1, TGF-β2 and TGF-β3, as well as bone morphogenic protein (BMPs). Adipose tissue-derived MSCs lack expression of TGF-β Type I receptor and have reduced expression of BMP-2, BMP-4 and BMP-6 when compared to bone marrow MSCs.[27] In addition, canonical and noncanonical Wnts have been shown to crosstalk with each other in regulating stem cell proliferation and osteogenic differentiation.[32] Additional regulatory signals of growth factors are delivered through specific signal transduction pathways that control downstream transcription factors. For example, mitogen-activated protein kinase (MAPK) and Smads are activated, which induces transcription factors such as Sox9, Sox5 and Sox6, leading to the production of extracellular matrix proteins such as collage Type II, aggrecan and cartilage oligomeric matrix protein, leading ultimately to chondrogenesis.[27] In fact, Sox9 is one of the most important molecules for expression of cartilaginous phenotype and is considered the "master switch" in chondrogenesis.[38]

TRAFFICKING

Stem cell trafficking is defined as oriented and directed movement of cells toward a specific stimuli. There are two kinds of stem cell trafficking: homing and interstitial migration. The former is a process whereby stem/progenitor cells are disseminated throughout the body by the flowing blood until they recognize and interact with microvascular endothelial cells in a target organ.[2] The homing process is guided by specific "sensor(s)" on the cell surface, which senses chemotactic gradient in the target tissue. It is preceded and followed by a more active migratory phase during which cells access and exit the bloodstream.[2]

Mesenchymal Stem Cells

Recruitment of MSCs to repair damaged tissue is complex and involves sensing the signal from the remote injured tissue, release of MSCs from their bone marrow niche, homing of circulating MSCs to the target tissue and in situ differentiation of MSCs into mature, functional cells (Fig. 3).

One hypothesis regarding mobilization involves cytokines and/or chemokines that are upregulated under injury and are released into the circulation from remote tissues, stimulating MSCs to down-regulate their adhesion molecules that hold them at their niche.[27] This is contrast to quiescent conditions when progenitor cells are maintained inactive by contacting the bone marrow. The process of mobilization depends on many different molecules, such as matrix metalloproteinase protein (MMP)-9 and stromal cell-derived factor (SDF)-1α/CXCL12 and its receptor CXCR4.[39] It has been found that over-expression of CXCR4 on MSCs augments myoangiogenesis in the infarcted myocardium.[40] Similarly, over-expression of IGF-1 in MSCs induces massive stem cell mobilization via SDF-1α signaling and culminates in extensive myoangiogenesis in the infarcted heart.[41] In addition to SDF-1α, high mobility group box-1 (HMGB-1) is a nuclear protein that is released extracellularly upon activation of cells by inflammatory cytokines and during cell necrosis and is known to act as a chemoattractant for inflammatory cells, stem cells and EPC in vitro and in vivo.[42,43]

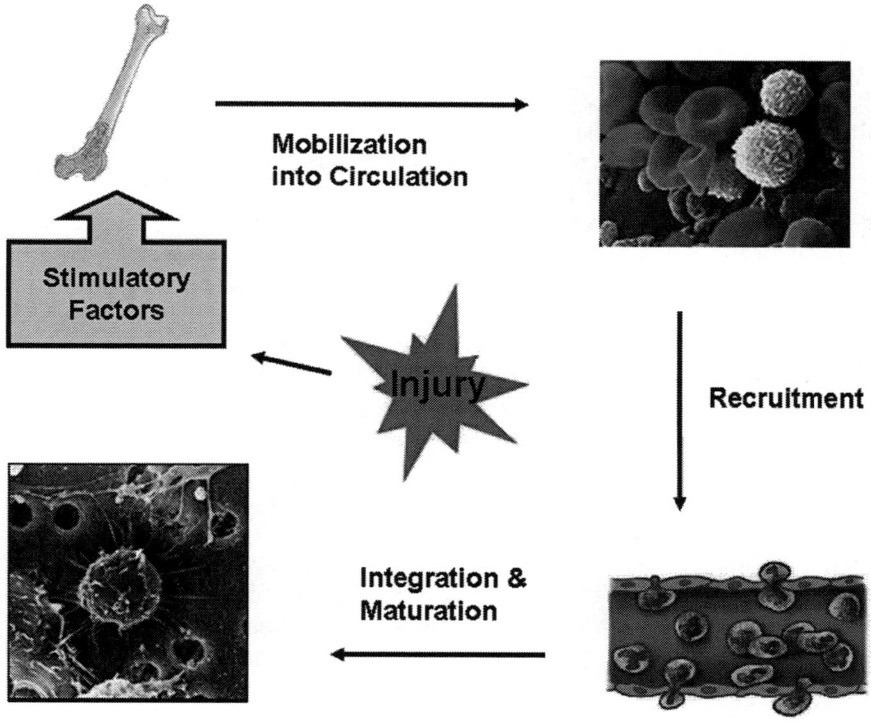

Figure 3. Injury in the periphery releases stimulatory factors that cause mobilization of MSCs from the bone marrow into the circulation. At the site of injury, certain molecules on the endothelium causes recruitment of MSCs, whereby they transmigrate from blood vessels and undergo in situ maturation and integrate into the injured tissue to bring about regeneration and healing. Reprinted with permission from Liu Z, Zhuge Y, Velazquez OC. Trafficking and differentiation of mesenchymal stem cells. J Cell Biochem 2009; 106:984-91.

After injury or ischemic insult, local resting resident fibroblasts are activated, which play a critical role in recruiting MSCs to the injured site. Platelet-derived growth factor B (PDGF-B)-activated fibroblasts were recently studied in regulating recruitment, migration and differentiation of murine bone marrow MSCs in an in vitro wound healing assay and a novel three-dimensional model. PDGF-B-activated fibroblasts caused significant increases in MSC migration velocity compared to control and invasion of MSCs into 3-dimensional (3D) collagen gels was enhanced in the presence of PDGF-B-activated fibroblasts.[27] In addition, PDGF-B-activated fibroblasts induced differentiation of MSCs into myofibroblasts. These effects are likely mediated by basic fibroblast growth factor (bFGF) and epithelial neutrophils activating peptide-78 (ENA-78 or CXCL5) as protein array analysis indicated elevated levels of these two soluble factors. Blocking antibodies were able to inhibit trafficking and differentiation of MSCs into 3D collagen gels while supplement of exogenous bFGF and/or CXCL5 promoted invasion of MSCs into 3D collagen gels.[37]

Homing mechanism of MSCs to the sites of target tissues involve a cascade of events, including rolling of MSCs in the blood vessels, adhesion onto the endothelial cell surface lining the capillaries, transendothelial migration, extravasation from the blood vessels and migration through the extracellular matrix to the target injured area. Studies show that the adhesion molecules are similar to those engaged by leukocytes for recruitment to sites of inflammation.[44,45] These molecules include integrins, selectins, CAMs and chemokine receptors. Rolling slows down subsets of MSCs, allowing their subsequent endothelial adhesion. P-selectin has been suggested to be involved in the rolling of MSCs and adhesion to endothelial cells, as neutralizing this molecule significantly reduces the rolling of MSCs under shear flow conditions.[46] Other adhesion molecules mediating MSC-endothelial cell interaction include various integrins and CAMs, such as α1, α2, α3, α4, α5, αv, β1, β3, β4, VCAM-1, ICAM-1, ICAM-3, ALCAM and endoglin/CD105.[47,48] These adhesion molecules and their counter-ligands are expressed on either MSCs or endothelial cells. Approximately half of human MSCs express the integrin very late antigen (VLA)-4 (α4β1, CD49d) and it has been shown via in vitro studies that under conditions of shear flow, binding of human MSCs to endothelial cells is mediated by VLA-4.[46] VLA-4 and its counterpart adhesion molecular VCAM-1 are responsible for mediating firm adhesion of human MSCs to endothelial cells.

Recently, CD44 has been demonstrated as a homing molecule for bone-marrow derived MSCs, as a glycoform CD44 mediates trafficking of human MSCs to bone through interactions with E-selecting, which is present on marrow vasculature.[49] CD44 adhesion molecules represent a large family of transmembrane glycoproteins and a stereospecific fucosylation of CD44 was observed to confer robust rolling of MSCs on vascular E-selectin in vitro and induces significantly increased homing of intravenously injected human MSCs to bone in NOD/SCID mice.[49] The receptor/ligand specificity allows for direction of migration of stem cells to one organ versus another and manipulation of these homing signatures may allow for precise targeting of stem cells to damaged areas.

It is unclear whether bone marrow-derived circulating MSCs and tissue-derived MSCs utilize identical trafficking mechanisms. The Rho family of GTPases performs key function in regulating actin cytoskeletal dynamics and may affect cell migration and adhesion. However, while it has been generally believed that signaling through the Rho family would influence the migration response of MSCs, experimental data has not been consistent.

In the clinical setting, MSCs can be therapeutically utilized via several approaches, most notably by either direct site delivery or systemic intravascular administration. In the case of the former, MSCs are delivered to the local tissue directly via local or intralesional implantation. In this situation, mobilization and homing are not required. Alternatively, when a systemic intravascular administration approach is utilized, this bypasses the mobilization process but requires homing/recruitment. Understanding the molecular mechanisms underlying homing of both bone marrow-derived and circulating MSCs will help to expand the clinical application of MSCs.

Endothelial Progenitor Cells

Similar to MSCs, the process of mobilization and homing of EPCs are regulated by chemokines, adhesion molecules and growth factors that guide them to the vessel wall after injury and during ischemia.[39] By using cell sorting techniques, it has been recently revealed that functional endothelial outgrowth was entirely derived from circulating $CD34^+$ $CD45^-$ mononuclear cells that were positive for VEGFR2, but negative for CD133.[50]

The SDF-1α/CXCR4 axis has also been shown to be critical for mobilization of EPCs to participate in vasculogenesis and angiogenesis.[51,52] The gene regulation of SDF-1α/CXCL12 is regulated by the transcription factor hypoxia-inducible factor (HIF)-1, which is up-regulated by injured tissues , thereby recruiting $CXCR4^+$ progenitor cells by the hypoxic gradients via HIF-1-induced expression of SDF-1α/CXCL12.[35,53] Inhibition of the interaction partially blocks the homing of these progenitor cells to the ischemic myocardium.[51,54] Suppression of CXCR4 by anti-CXCR4 neutralizing antibodies significantly reduces SDF-1α/CXCL12-induced adhesion of EPCs to mature endothelial cell monolayers, the migration of EPC in vitro and in vivo homing of myeloid EPC to the ischemic limb in the hind limb ischemia model.[53,55] Vascular trauma also mediates transient increase in circulating EPCs.[56] SDF-1α/CXCL12 produced at the injured site induces release of nitric oxide (NO) in endothelial cells as well in the bone marrow, which mobilizes EPCs.[51,52]

EPCs function in endothelial regeneration and has been shown to be helpful in many situations, such as attenuating neointimal hyperplasia after carotid injury in mice, postnatal neovascularization in ischemic tissue and the treatment of acute and chronic myocardial ischemia.[27,51,57,58] It has also been shown that EPCs decreases plaque stability in apolipoprotein E knockout ($ApoE^{-/-}$) mice.[59] Therefore, therapeutic use of EPCs must be context-dependent, as wide-spread mobilization of EPCs may destabilize plaques while local stimulation of EPCs could improve arterial injury or during ischemia by promoting regeneration.[39]

Hematopoietic Stem Cells

The process of mobilization of HSCs from the bone marrow is controlled by specific signals such as chemo-attractant cytokines, growth factors and hormones both in the bone marrow and the periphery. For example, SDF-1α/CXCR4 axis plays a pivotal role for hematopoietic precursor cell positioning in the bone marrow niche, and disrupting this relationship results in rapid mobilization of these stem cells from their niches.[60,61] Also, ischemic peripheral tissue leads to upregulation of SDF-1α/CXCL12 expression on endothelial cells, which mediates recruitment of HSCs to sites of vascular injury.[53,62] It has been shown that different progenitor cells respond differently to chemokines in terms of mobilization. For example, nitric oxide (NO) has been implicated in the mobilization of hematopoietic progenitor cells from the bone marrow, but does not alter mobilization of less-differentiated stem cells.[63,64]

The regulation of adhesion between cells is critical for the transition of stem cells between different tissues and the breaking of existing junctions between cells represents the earliest step in migration.[2] For HSCs, deadhesion from the niche requires the function of proteolytic enzymes such as matrix metalloproteinase (MMP)-9 and cysteine protease cathepsin K.[65,66] In addition, osteoclasts as well as protease activity by HSCs themselves contribute to the silencing of HSC retention signals in the marrow by cleaving and inactivating SDF-1α/CXCL12.[2,66,67]

After deadhesion, hematopoietic stem cells exit the bone marrow and become disseminated in the circulation. Once blood-borne, they may home to target sites, where they become firmly attached to the microvascular endothelial cells. This process requires that HSCs recognize tissue-specific microvascular features in target organs and must adhere to endothelial cells with sufficient strength to overcome the shear force exerted by the flowing blood.[2] Adhesion molecules located on the surface of HSCs and endothelial cells are critical to this process, which resembles the homing of mature blood leukocytes in peripheral tissues.[68] The first step in this multi-step cascade involves tethering and rolling, mediated by primary adhesion molecules (selectins or α4-integrin) with fast binding kinetics and high tensile strength but short bond lifetime.[2] Subsequently, chemotactic/activating stimuli are provided by soluble or surface-bound chemoattractants. Finally, more steadfast sticking is mediated by secondary adhesion molecules, mostly integrins (β2 or α4) that interact with endothelial ligands of the immunoglobulin superfamily.[2]

After performing their function in the periphery, progenitor cells must home back to the bone marrow with high efficiency to replenish the bone marrow pool for consecutive proliferation. The α4β1 integrin (VLA-4) is expressed on most HSCs and it binds to vascular cell adhesion molecule (VCAM)-1 on stromal cells and endothelium of the bone marrow.[69] Interrupting of this axis results in increased HSC mobilization and inability of bone marrow cells to engraft within their niche.[70-73] Other adhesive pathways such as selectins (E-selectin and P-selectin) and β2-integrin also contribute to bone marrow homing of precursor cells.[74,75]

HSCs arrive in peripheral tissues via the blood, but exit via draining lymphatics. Sphingosine-1-phosphage (S1P) has been shown to be critical to controlling the egress of tissue hematopoietic precursor cells into the draining lymphatics.[64] S1P lyase maintains low levels of S1P in tissues and high levels in the blood and lymph,[76,77] thereby regulating the movement of such cells as mature lymphocytes from thymus, spleen and lymph nodes.[78,79] Comparable mechanism also controls the exit of HSCs from peripheral tissues into the draining lymph.[80] For example, murine hematopoietic stem cells migrate towards steep gradients of S1P in a predominantly S1P1 receptor-dependent manner and blocking these receptors blunts the egress of tissue-resident HSCs into lymphatics and inhibits their recirculation.[80]

There is a constitutive recirculation of HSCs between the bone marrow, blood, extramedullary tissues and lymph compartment.[80] Hematopoietic cells can switch between dormancy and self-renewal as needed to preserve homeostasis and replenish the pool of circulating cells.[3] Therefore, circulating HSCs may help to constitutively replenish tissue-resident myeloid and other specialized cells.[64] In addition to the constitutive replenishment of tissue-resident myeloid cells under steady-state conditions, circulating hematopoietic stem cells may also participate in immune responses during tissue injury and/or infection.[64] HSCs express functional Toll-like receptors (TLR) that recognize foreign molecules on bacterial outer membranes, whose binding promotes stem cells to entire into cell cycle and trigger myeloid differentiation.[81,82] Therefore, migratory progenitor cells can survey peripheral tissues and respond to situations rapidly to promptly produce large numbers of immune cells.[64]

Recent data has shown that some cancers have a population of cancer stem cells (CSCs) that maintain production of different malignant cells and that these stem cells are capable of causing cancer in normal recipients.[1] It has been found that these CSCs rely on their dysregulated adhesion and migration to disseminate disease throughout the body. These malignant cells use similar mechanisms of migration as those normally used by normal tissue,[2] and specific adhesion/deadhesion pathways determine the efficiency of metastasis. For example, CXCR4 expression is associated with trafficking and enhanced metastasis of some blood as well as solid tumors.[83,84] In addition, marrow-derived nonneoplastic cells may be necessary to form a "premetastatic niche" that directs the organ-specific homing patterns of malignant carcinoma cells.[85] This niche secrets chemoattractants such as SDF-1α/CXCL12 to recruit metastatic cancer stem cells to establish secondary tumors at other sites.[85] Therefore, interventions to prevent the establishment of this niche can block in vivo metastasis, pointing out the importance of targeting the homing mechanism of malignant cells in an effort to limit cancer spread.[2]

CONCLUSION

Complex interactions between adhesive, chemotactic and signaling pathways act cooperatively to bring about the proper differentiation, trafficking and homing of stem cells. These systems are often conserved through embryogenesis, across different organisms and can be extended to cancer stem cells as well as gene vector formation and drug delivery.

MSCs are uniquely suited for various therapeutic possibilities, such as tissue regeneration, correcting inherited disorders, dampening chronic inflammation and delivering biological agents. Understanding differentiation and trafficking of MSCs may allow the development of therapeutic strategies to enhance the recruitment of bone marrow-derived and/or tissue-derived MSCs. The efficiency of cell-based therapy requires not only sufficient amount of MSCs, but also efficient delivery of these cells to the desired target site. Fulfilling the promise of regenerative medicine critically depends on identifying the mechanisms and the molecules that control and mediate MSC lineage-specific differentiation as well tissue-specific stem cell homing.

Similarly, the importance of HSC homing is clear as clinical use of bone-marrow transplant for patients with various types of cancers relies on the ability of transplanted HSCs to traffic efficiently to the bone marrow.[2] Donor stem cells are injected intravenously into the recipient and in order to properly expand repopulate the bloodstream of the recipient, donor HSCs must properly engraft into the recipient bone marrow niche. It appears that this medical therapy uses the same pre-existing pathways that normally support recirculation of HSCs during homeostasis.

Finally, EPCs is a well-established source of endothelial cells that participate in both reparative and pathologic postnatal neovascularization in the setting of injury, ischemia and tumor formation. They play a pivotal role not only in wound healing, but also in limb ischemia,[86-88] postmyocardial infarction,[89-91] endothelialization of vascular grafts,[92,93] atherosclerosis,[94] retinal and lymphoid organ neovascularization,[95,96] vascularization during neonatal growth,[97] and tumor growth.[98,99] Improved understanding of the differentiation and trafficking of these cells can revolutionize the treatment of these various processes.

SDF-1α/CXCR4 is the one axis that is broadly conserved in multiple tissues in both embryo and adult. During development, SDF-1α/CXCR4 signals homing of fetal mouse HSCs to the fetal liver and bone marrow and in the adult, it regulates mobilization of mouse and human HSCs as well as their re-entry into the marrow.[2,60,100,101] This axis is also involved in the dissemination of tumor-forming cells in metastatic cancers.[83] It is therefore likely that further study of this axis will yield novel treatments for synergistic treatment of various types of hematopoietic and solid cancers.

REFERENCES

1. Dalerba P, Cho RW, Clarke MF. Cancer stem cells: models and concepts. Annu Rev Med 2007; 58:267-84.
2. Laird DJ, von Andrian UH, Wagers AJ. Stem cell trafficking in tissue development, growth and disease. Cell 2008; 132(4):612-30.
3. Orkin SH, Zon LI. Hematopoiesis: an evolving paradigm for stem cell biology. Cell 2008; 132(4):631-44.
4. Orkin SH. Diversification of haematopoietic stem cells to specific lineages. Nat Rev Genet 2000; 1(1):57-64.
5. Kim SI, Bresnick EH. Transcriptional control of erythropoiesis: emerging mechanisms and principles. Oncogene 2007; 26(47):6777-94.
6. Mikkola HK, Klintman J, Yang H et al. Haematopoietic stem cells retain long-term repopulating activity and multipotency in the absence of stem-cell leukaemia SCL/tal-1 gene. Nature 2003; 421(6922):547-51.
7. Ichikawa M, Asai T, Saito T et al. AML-1 is required for megakaryocytic maturation and lymphocytic differentiation, but not for maintenance of hematopoietic stem cells in adult hematopoiesis. Nat Med 2004; 10(3):299-304.
8. Kulessa H, Frampton J, Graf T. GATA-1 reprograms avian myelomonocytic cell lines into eosinophils, thromboblasts and erythroblasts. Genes Dev 1995; 9(10):1250-62.
9. Nerlov C, Graf T. PU.1 induces myeloid lineage commitment in multipotent hematopoietic progenitors. Genes Dev 1998; 12(15):2403-12.
10. Nutt SL, Heavey B, Rolink AG et al. Commitment to the B-lymphoid lineage depends on the transcription factor Pax5. Nature 1999; 401(6753):556-62.
11. McDevitt MA, Shivdasani RA, Fujiwara Y et al. A "knockdown" mutation created by cis-element gene targeting reveals the dependence of erythroid cell maturation on the level of transcription factor GATA-1. Proc Natl Acad Sci USA 1997; 94(13):6781-5.
12. Orkin SH. Priming the hematopoietic pump. Immunity 2003; 19(5):633-4.
13. Chen H, Ray-Gallet D, Zhang P et al. PU.1 (Spi-1) autoregulates its expression in myeloid cells. Oncogene 1995; 11(8):1549-60.
14. Tsai SF, Strauss E, Orkin SH. Functional analysis and in vivo footprinting implicate the erythroid transcription factor GATA-1 as a positive regulator of its own promoter. Genes Dev 1991; 5(6):919-31.
15. Iwasaki H, Akashi K. Myeloid lineage commitment from the hematopoietic stem cell. Immunity 2007; 26(6):726-40.
16. Laiosa CV, Stadtfeld M, Xie H et al. Reprogramming of committed T-cell progenitors to macrophages and dendritic cells by C/EBP alpha and PU.1 transcription factors. Immunity 2006; 25(5):731-44.
17. Xie H, Ye M, Feng R et al. Stepwise reprogramming of B cells into macrophages. Cell 2004; 117(5):663-76.
18. Busslinger M, Nutt SL, Rolink AG. Lineage commitment in lymphopoiesis. Curr Opin Immunol 2000; 12(2):151-8.
19. Zhang P, Zhang X, Iwama A et al. PU.1 inhibits GATA-1 function and erythroid differentiation by blocking GATA-1 DNA binding. Blood 2000; 96(8):2641-8.
20. Nerlov C, Querfurth E, Kulessa H et al. GATA-1 interacts with the myeloid PU.1 transcription factor and represses PU.1-dependent transcription. Blood 2000; 95(8):2543-51.
21. Zheng W, Flavell RA. The transcription factor GATA-3 is necessary and sufficient for Th2 cytokine gene expression in CD4 T-cells. Cell 1997; 89(4):587-96.
22. Szabo SJ, Kim ST, Costa GL et al. A novel transcription factor, T-bet, directs Th1 lineage commitment. Cell 2000; 100(6):655-69.

23. Wechsler J, Greene M, McDevitt MA et al. Acquired mutations in GATA1 in the megakaryoblastic leukemia of Down syndrome. Nat Genet 2002; 32(1):148-52.
24. Mueller BU, Pabst T, Osato M et al. Heterozygous PU.1 mutations are associated with acute myeloid leukemia. Blood 2002;100(3):998-1007.
25. Pabst T, Mueller BU, Zhang P et al. Dominant-negative mutations of CEBPA, encoding CCAAT/enhancer binding protein-alpha (C/EBPalpha), in acute myeloid leukemia. Nat Genet 2001; 27(3):263-70.
26. Mullighan CG, Goorha S, Radtke I et al. Genome-wide analysis of genetic alterations in acute lymphoblastic leukaemia. Nature 2007; 446(7137):758-64.
27. Liu ZJ, Zhuge Y, Velazquez OC. Trafficking and differentiation of mesenchymal stem cells. J Cell Biochem 2009;106(6):984-91.
28. Dominici M, Le Blanc K, Mueller I et al. Minimal criteria for defining multipotent mesenchymal stromal cells. The International Society for Cellular Therapy position statement. Cytotherapy 2006; 8(4):315-7.
29. Bianco P, Robey PG, Simmons PJ. Mesenchymal stem cells: revisiting history, concepts and assays. Cell Stem Cell 2008; 2(4):313-9.
30. Chen FH, Rousche KT, Tuan RS. Technology Insight: adult stem cells in cartilage regeneration and tissue engineering. Nat Clin Pract Rheumatol 2006; 2(7):373-82.
31. Mimeault M, Batra SK. Recent progress on normal and malignant pancreatic stem/progenitor cell research: therapeutic implications for the treatment of type 1 or 2 diabetes mellitus and aggressive pancreatic cancer. Gut 2008; 57(10):1456-68.
32. Baksh D, Tuan RS. Canonical and noncanonical Wnts differentially affect the development potential of primary isolate of human bone marrow mesenchymal stem cells. J Cell Physiol 2007; 212(3):817-26.
33. Barry F, Boynton RE, Liu B et al. Chondrogenic differentiation of mesenchymal stem cells from bone marrow: differentiation-dependent gene expression of matrix components. Exp Cell Res 2001; 268(2):189-200.
34. Ito T, Sawada R, Fujiwara Y et al. FGF-2 increases osteogenic and chondrogenic differentiation potentials of human mesenchymal stem cells by inactivation of TGF-beta signaling. Cytotechnology 2008; 56(1):1-7.
35. Kratchmarova I, Blagoev B, Haack-Sorensen M et al. Mechanism of divergent growth factor effects in mesenchymal stem cell differentiation. Science 2005; 308(5727):1472-7.
36. Matsuda C, Takagi M, Hattori T et al. Differentiation of Human Bone Marrow Mesenchymal Stem Cells to Chondrocytes for Construction of Three-dimensional Cartilage Tissue. Cytotechnology 2005; 47(1-3):11-7.
37. Nedeau AE, Bauer RJ, Gallagher K et al. A CXCL5- and bFGF-dependent effect of PDGF-B-activated fibroblasts in promoting trafficking and differentiation of bone marrow-derived mesenchymal stem cells. Exp Cell Res 2008; 314(11-12):2176-86.
38. Tsuchiya H, Kitoh H, Sugiura F et al. Chondrogenesis enhanced by overexpression of sox9 gene in mouse bone marrow-derived mesenchymal stem cells. Biochem Biophys Res Commun 2003; 301(2):338-43.
39. Hristov M, Weber C. Progenitor cell trafficking in the vascular wall. J Thromb Haemost 2009; 7 Suppl 1:31-4.
40. Zhang D, Fan GC, Zhou X et al. Over-expression of CXCR4 on mesenchymal stem cells augments myoangiogenesis in the infarcted myocardium. J Mol Cell Cardiol 2008; 44(2):281-92.
41. Haider H, Jiang S, Idris NM et al. IGF-1-overexpressing mesenchymal stem cells accelerate bone marrow stem cell mobilization via paracrine activation of SDF-1alpha/CXCR4 signaling to promote myocardial repair. Circ Res 2008; 103(11):1300-8.
42. Chavakis E, Hain A, Vinci M et al. High-mobility group box 1 activates integrin-dependent homing of endothelial progenitor cells. Circ Res 2007; 100(2):204-12.
43. Palumbo R, Bianchi ME. High mobility group box 1 protein, a cue for stem cell recruitment. Biochem Pharmacol 2004; 68(6):1165-70.
44. Ley K, Laudanna C, Cybulsky MI et al. Getting to the site of inflammation: the leukocyte adhesion cascade updated. Nat Rev Immunol 2007; 7(9):678-89.
45. Luster AD, Alon R, von Andrian UH. Immune cell migration in inflammation: present and future therapeutic targets. Nat Immunol 2005; 6(12):1182-90.
46. Ruster B, Gottig S, Ludwig RJ et al. Mesenchymal stem cells display coordinated rolling and adhesion behavior on endothelial cells. Blood 2006; 108(12):3938-44.
47. Krampera M, Pizzolo G, Aprili G et al. Mesenchymal stem cells for bone, cartilage, tendon and skeletal muscle repair. Bone 2006; 39(4):678-83.
48. Minguell JJ, Erices A, Conget P. Mesenchymal stem cells. Exp Biol Med (Maywood) 2001; 226(6):507-20.
49. Sackstein R, Merzaban JS, Cain DW et al. Ex vivo glycan engineering of CD44 programs human multipotent mesenchymal stromal cell trafficking to bone. Nat Med 2008; 14(2):181-7.

50. Timmermans F, Van Hauwermeiren F, De Smedt M et al. Endothelial outgrowth cells are not derived from CD133+ cells or CD45+ hematopoietic precursors. Arterioscler Thromb Vasc Biol 2007; 27(7):1572-9.
51. Gallagher KA, Liu ZJ, Xiao M et al. Diabetic impairments in NO-mediated endothelial progenitor cell mobilization and homing are reversed by hyperoxia and SDF-1 alpha. J Clin Invest 2007; 117(5):1249-59.
52. Petit I, Jin D, Rafii S. The SDF-1-CXCR4 signaling pathway: a molecular hub modulating neo-angiogenesis. Trends Immunol 2007; 28(7):299-307.
53. Ceradini DJ, Kulkarni AR, Callaghan MJ et al. Progenitor cell trafficking is regulated by hypoxic gradients through HIF-1 induction of SDF-1. Nat Med 2004; 10(8):858-64.
54. Abbott JD, Huang Y, Liu D et al. Stromal cell-derived factor-1alpha plays a critical role in stem cell recruitment to the heart after myocardial infarction but is not sufficient to induce homing in the absence of injury. Circulation 2004; 110(21):3300-5.
55. Walter DH, Haendeler J, Reinhold J et al. Impaired CXCR4 signaling contributes to the reduced neovascularization capacity of endothelial progenitor cells from patients with coronary artery disease. Circ Res 2005; 97(11):1142-51.
56. Hristov M, Weber C. Ambivalence of progenitor cells in vascular repair and plaque stability. Curr Opin Lipidol 2008; 19(5):491-7.
57. Hristov M, Zernecke A, Bidzhekov K et al. Importance of CXC chemokine receptor 2 in the homing of human peripheral blood endothelial progenitor cells to sites of arterial injury. Circ Res 2007; 100(4):590-7.
58. Hristov M, Zernecke A, Liehn EA et al. Regulation of endothelial progenitor cell homing after arterial injury. Thromb Haemost 2007; 98(2):274-7.
59. George J, Afek A, Abashidze A et al. Transfer of endothelial progenitor and bone marrow cells influences atherosclerotic plaque size and composition in apolipoprotein E knockout mice. Arterioscler Thromb Vasc Biol 2005; 25(12):2636-41.
60. Broxmeyer HE, Orschell CM, Clapp DW et al. Rapid mobilization of murine and human hematopoietic stem and progenitor cells with AMD3100, a CXCR4 antagonist. J Exp Med 2005; 201(8):1307-18.
61. Lapidot T, Kollet O. The essential roles of the chemokine SDF-1 and its receptor CXCR4 in human stem cell homing and repopulation of transplanted immune-deficient NOD/SCID and NOD/SCID/B2m(null) mice. Leukemia 2002; 16(10):1992-2003.
62. Ceradini DJ, Gurtner GC. Homing to hypoxia: HIF-1 as a mediator of progenitor cell recruitment to injured tissue. Trends Cardiovasc Med 2005; 15(2):57-63.
63. Aicher A, Heeschen C, Mildner-Rihm C et al. Essential role of endothelial nitric oxide synthase for mobilization of stem and progenitor cells. Nat Med 2003; 9(11):1370-6.
64. Schulz C, von Andrian UH, Massberg S. Hematopoietic stem and progenitor cells: their mobilization and homing to bone marrow and peripheral tissue. Immunol Res 2009; 44(1-3):160-8.
65. Heissig B, Hattori K, Dias S et al. Recruitment of stem and progenitor cells from the bone marrow niche requires MMP-9 mediated release of kit-ligand. Cell 2002; 109(5):625-37.
66. Kollet O, Dar A, Shivtiel S et al. Osteoclasts degrade endosteal components and promote mobilization of hematopoietic progenitor cells. Nat Med 2006; 12(6):657-64.
67. Christopherson KW, 2nd, Cooper S, Broxmeyer HE. Cell surface peptidase CD26/DPPIV mediates G-CSF mobilization of mouse progenitor cells. Blood 2003; 101(12):4680-6.
68. Springer TA. Traffic signals for lymphocyte recirculation and leukocyte emigration: the multistep paradigm. Cell 1994; 76(2):301-14.
69. Potocnik AJ, Brakebusch C, Fassler R. Fetal and adult hematopoietic stem cells require beta1 integrin function for colonizing fetal liver, spleen and bone marrow. Immunity 2000; 12(6):653-63.
70. Levesque JP, Takamatsu Y, Nilsson SK et al. Vascular cell adhesion molecule-1 (CD106) is cleaved by neutrophil proteases in the bone marrow following hematopoietic progenitor cell mobilization by granulocyte colony-stimulating factor. Blood 2001; 98(5):1289-97.
71. Papayannopoulou T, Craddock C, Nakamoto B et al. The VLA4/VCAM-1 adhesion pathway defines contrasting mechanisms of lodgement of transplanted murine hemopoietic progenitors between bone marrow and spleen. Proc Natl Acad Sci USA 1995; 92(21):9647-51.
72. Papayannopoulou T, Nakamoto B. Peripheralization of hemopoietic progenitors in primates treated with anti-VLA4 integrin. Proc Natl Acad Sci USA 1993; 90(20):9374-8.
73. Wagers AJ, Allsopp RC, Weissman IL. Changes in integrin expression are associated with altered homing properties of Lin(-/lo)Thy1.1(lo)Sca-1(+)c-kit(+) hematopoietic stem cells following mobilization by cyclophosphamide/granulocyte colony-stimulating factor. Exp Hematol 2002; 30(2):176-85.
74. Katayama Y, Hidalgo A, Furie BC et al. PSGL-1 participates in E-selectin-mediated progenitor homing to bone marrow: evidence for cooperation between E-selectin ligands and alpha4 integrin. Blood 2003; 102(6):2060-7.

75. Mazo IB, Gutierrez-Ramos JC, Frenette PS et al von Andrian UH. Hematopoietic progenitor cell rolling in bone marrow microvessels: parallel contributions by endothelial selectins and vascular cell adhesion molecule 1. J Exp Med 1998; 188(3):465-74.
76. Hla T, Venkataraman K, Michaud J. The vascular S1P gradient-cellular sources and biological significance. Biochim Biophys Acta 2008; 1781(9):477-82.
77. Yatomi Y, Igarashi Y, Yang L et al. Sphingosine 1-phosphate, a bioactive sphingolipid abundantly stored in platelets, is a normal constituent of human plasma and serum. J Biochem 1997; 121(5):969-73.
78. Pappu R, Schwab SR, Cornelissen I et al. Promotion of lymphocyte egress into blood and lymph by distinct sources of sphingosine-1-phosphate. Science 2007; 316(5822):295-8.
79. Rosen H, Gonzalez-Cabrera P, Marsolais D et al. Modulating tone: the overture of S1P receptor immunotherapeutics. Immunol Rev 2008; 223:221-35.
80. Massberg S, Schaerli P, Knezevic-Maramica I et al. Immunosurveillance by hematopoietic progenitor cells trafficking through blood, lymph and peripheral tissues. Cell 2007; 131(5):994-1008.
81. Hoshino K, Takeuchi O, Kawai T et al. Cutting edge: Toll-like receptor 4 (TLR4)-deficient mice are hyporesponsive to lipopolysaccharide: evidence for TLR4 as the Lps gene product. J Immunol 1999; 162(7):3749-52.
82. Nagai Y, Garrett KP, Ohta S et al. Toll-like receptors on hematopoietic progenitor cells stimulate innate immune system replenishment. Immunity 2006; 24(6):801-12.
83. Balkwill F. Cancer and the chemokine network. Nat Rev Cancer 2004; 4(7):540-50.
84. Burger JA, Kipps TJ. CXCR4: a key receptor in the crosstalk between tumor cells and their microenvironment. Blood 2006; 107(5):1761-7.
85. Kaplan RN, Riba RD, Zacharoulis S et al. VEGFR1-positive haematopoietic bone marrow progenitors initiate the premetastatic niche. Nature 2005; 438(7069):820-7.
86. Kalka C, Masuda H, Takahashi T et al. Transplantation of ex vivo expanded endothelial progenitor cells for therapeutic neovascularization. Proc Natl Acad Sci USA 2000; 97(7):3422-7.
87. Majka SM, Jackson KA, Kienstra KA et al. Distinct progenitor populations in skeletal muscle are bone marrow derived and exhibit different cell fates during vascular regeneration. J Clin Invest 2003; 111(1):71-9.
88. Takahashi T, Kalka C, Masuda H et al. Ischemia- and cytokine-induced mobilization ofbone marrow-derived endothelial progenitor cells for neovascularization. Nat Med 1999; 5(4):434-8.
89. Kocher AA, Schuster MD, Szabolcs MJ et al. Neovascularization of ischemic myocardium by human bone-marrow-derived angioblasts prevents cardiomyocyte apoptosis, reduces remodeling and improves cardiac function. Nat Med 2001; 7(4):430-6.
90. Orlic D, Kajstura J, Chimenti S et al. Transplanted adult bone marrow cells repair myocardial infarcts in mice. Ann N Y Acad Sci 2001; 938:221-9; discussion 29-30.
91. Orlic D, Kajstura J, Chimenti S et al. Bone marrow cells regenerate infarcted myocardium. Nature 2001; 410(6829):701-5.
92. Kaushal S, Amiel GE, Guleserian KJ et al. Functional small-diameter neovessels created using endothelial progenitor cells expanded ex vivo. Nat Med 2001; 7(9):1035-40.
93. Shi Q, Rafii S, Wu MH et al. Evidence for circulating bone marrow-derived endothelial cells. Blood 1998; 92(2):362-7.
94. Sata M, Saiura A, Kunisato A et al. Hematopoietic stem cells differentiate into vascular cells that participate in the pathogenesis of atherosclerosis. Nat Med 2002; 8(4):403-9.
95. Grant MB, May WS, Caballero S et al. Adult hematopoietic stem cells provide functional hemangioblast activity during retinal neovascularization. Nat Med 2002; 8(6):607-12.
96. Otani A, Kinder K, Ewalt K et al. Bone marrow-derived stem cells target retinal astrocytes and can promote or inhibit retinal angiogenesis. Nat Med 2002; 8(9):1004-10.
97. Young PP, Hofling AA, Sands MS. VEGF increases engraftment of bone marrow-derived endothelial progenitor cells (EPCs) into vasculature of newborn murine recipients. Proc Natl Acad Sci USA 2002; 99(18):11951-6.
98. Lyden D, Hattori K, Dias S et al. Impaired recruitment ofbone-marrow-derived endothelial and hematopoietic precursor cells blocks tumor angiogenesis and growth. Nat Med 2001; 7(11):1194-201.
99. Reyes M, Dudek A, Jahagirdar B et al. Origin of endothelial progenitors in human postnatal bone marrow. J Clin Invest 2002; 109(3):337-46.
100. Peled A, Petit I, Kollet O et al. Dependence of human stem cell engraftment and repopulation of NOD/SCID mice on CXCR4. Science 1999; 283(5403):845-8.
101. Sugiyama T, Kohara H, Noda M et al. Maintenance of the hematopoietic stem cell pool by CXCL12-CXCR4 chemokine signaling in bone marrow stromal cell niches. Immunity 2006; 25(6):977-88.

CHAPTER 13

VERTEBRATES THAT REGENERATE AS MODELS FOR GUIDING STEM CELLS

Christopher L. Antos* and Elly M. Tanaka*

Abstract: There are several animal model organisms that have the ability to regenerate severe injuries by stimulating local cells to restore damaged and lost organs and appendages. In this chapter, we will describe how various vertebrate animals regenerate different structures (central nervous system, heart and appendages) as well as detail specific cellular and molecular features concerning the regeneration of these structures.

VERTEBRATE MODELS OF REGENERATION: THEIR ATTRIBUTES

There have been significant advances in our understanding of how differentiated cells can be programmed to behave as stem cells and much effort is focused on how cell reprogramming can be used for regeneration therapies.[1-3] However, only a few therapeutic strategies exist that allow successful restructuring of the compound architecture such as an organ or an appendage, and it is not clear what long-term consequences (such as carcinogenic predisposition) exist after the in vivo transplantation of iPS-derived cells.[3] Thus, there is a need to understand to what extent tissue cells must be reprogrammed and how these cells can be instructed to create the complexity of an organ or an appendage.

In the animal kingdom, there is a wide variation in the ability to regenerate lost or damaged tissues. While some animals replace the loss with durable scar tissue, others possess the ability to reactivate developmental mechanisms that result in the redevelopment (regeneration) of missing structures. The ability of the residual cells to

*Corresponding Authors: Christopher L. Antos—DFG-Center for Regenerative Therapies Dresden, Technische Universität Dresden, Tatzberg 47/49, 01307 Dresden, Germany. Email: christopher.antos@crt-dresden.de
Elly M. Tanaka—Max-Planck Institute for Molecular Cell Biology and Genetics, Pfotenhauerstrasse 108 and DFG-Center for Regenerative Therapies Dresden, Technische Universität Dresden, Tatzberg 47/49, 01307 Dresden, Germany. Email: elly.tanaka@crt-dresden.de

The Cell Biology of Stem Cells, edited by Eran Meshorer and Kathrin Plath.
©2010 Landes Bioscience and Springer Science+Business Media.

initiate a regeneration response and to reproduce the original architecture rather than a neoplasm requires exacting control of the new growth. Thus, animal models that amend themselves to cell and molecular studies offer the opportunity to observe the fundamental mechanisms underlying organ and appendage regeneration.

Among vertebrates, mammals regenerate only a limited number of tissues while other vertebrates including fish (goldfish and zebrafish) and amphibians (newts, salamanders and frogs—animals that evolved before the appearance of early reptiles) can regenerate a broad spectrum of complex organs such as damaged retina, severed spinal cords, injured heart and amputated limbs. Based on histological and molecular observations, this capability involves the activation of quiescent or differentiated cells, the induction morphological changes and transcriptional changes to produce precursor cells. Consequently, these animals provide an opportunity to understand the cellular and molecular mechanisms of reprogramming differentiated cells into regeneration-competent precursors and to comprehend how these precursors are directed to reconstruct the appropriate architecture of a lost structure.

The regeneration of all tissues involves the formation of one or more pools of proliferating cell populations and these cells come either from sequestered stem (progenitor) cells or from mature tissue cells that revert to a progenitor state. A mass of progenitor cells that are competent to regenerate a complex structure is produced and their production must involve changes in epigenetic and gene transcription programs. The overlying questions in regards to the formation of these progenitors are: (i) What is the extent of the reprogramming necessary and (ii) what are the mechanisms involved? In this chapter, we first survey regeneration in particular organ systems focusing on what type of cell mechanisms underlie the regenerative event with some reference to molecular mechanism. In the second part of the chapter, we focus on appendage regeneration due to the detail of our understanding of this complex regenerative system and discuss what is known about the cellular and molecular mechanisms that are involved in generating from adult tissues a set of proliferative precursors that are competent to reconstruct multi-tissue structures.

Defining Reprogramming Processes in Regenerating Tissues

The various organ systems use different strategies to produce regenerative progenitor cells ranging from stem cell activation to dedifferentiation and even to transdifferentiation (Fig. 1). Here we set out our working definition of these terms for this chapter. The strictest definition for "dedifferentiation" posits that a classically postmitotic cell type would lose its differentiated features to form an undifferentiated, proliferating cell. More recently, the term "dedifferentiation" has also been used to describe the conversion of any cell type (even a proliferating progenitor) to a more primitive state. In this chapter, we use "dedifferentiation" to mean postmitotic cells losing differentiated character and acquiring proliferative capacity irrespective of what cell types they make later (Fig. 1B). While others have restricted transdifferentiation to mean the direct conversion of one cell type to another without an intermediate progenitor, we use a broader definition of transdifferentiation. We use "transdifferentiation" to describe the conversion from one differentiated cell type to a different differentiated cell type irrespective of whether there is or is not an obvious, less differentiated intermediate. We have chosen this definition, because it is sometimes difficult to determine whether or not an intermediate less-differentiated state exists and also because historically, lens regeneration has been described as a transdifferentiation event even though an undifferentiated intermediate seems to be formed. We find the term appropriate since the nature and potency of this intermediate is not clear. Therefore, in our definition, transdifferentiation can be accomplished via a dedifferentiation step (Fig. 1C).

REGENERATION MECHANISMS OF MATURE TISSUES

Eye—A Model of Transdifferentiation

Regeneration of amphibian eye tissues (lens and retina) provides the best studied examples of transdifferention between two defined cell types in vertebrates. In newts, lens removal (lentectomy) induces the dorsal iris pigment epithelium (dIPE) to lose its pigmentation, to proliferate and to transdifferentiate into lens (Fig. 2).[4] This is noteworthy, since IPE and the lens originate from two different developmental lineages during embryonic gastrulation: the IPE comes from the neuroectoderm while the lens forms from the nonneural surface ectoderm.[5] The transdifferentiation potential of the dIPE appears, however, to be lineage restricted, because when dIPE is implanted into the regenerating limb blastema, lens is formed. However, due to the tracking methods used, these experiments can not exclude that some cells may form other tissue types. Interestingly, transplantation of IPE into the brain or the fin does not result in lens formation, which suggests that regeneration-permissive cues are required to promote an intrinsic lens formation program of the IPE.[6]

Clonal cell culture experiments have delineated factors required for dIPE to lens transdifferentiation. In clonal cultures of IPE from chick embryos, transdifferentiation from pigment epithelium to lens was achieved by applying basic FGF and the depigmenting enzyme PTU.[7] In the newt system, FGF2 has been shown to induce dIPE to form ectopic lenses in vitro and in intact newt eyes while other growth factors (VEGF, IGF and EGF) did not.[8,9]

Defining the intracellular factors underlying this transdifferentiation event—how they respond to extracellular cues and how they operate to switch cellular phenotype—are important issues, and some insight into the molecular requirements for this transdifferentiation event has been provided by comparing the dorsal and ventral iris. Full transdifferentiation in vivo is seen only from the dorsal IPE. While the ventral IPE responds to injury via cell cycle re-entry, it ultimately does not form a lens.[8] However, the combination of retinoic acid, the overexpression of the transcription factor sine oculis homoebox-3 gene (six-3) and the inhibition of BMP signaling was sufficient to induce ectopic lens regeneration by the ventral IPE,[10] indicating that six3 levels and retinoic acid signaling are crucial factors for competence to generate lens by the ventral iris in the absence of BMP signaling. The inhibition of Bmp signaling is a key limiting step, because restoration of competence can also be induced by the inhibition of BMP signaling but at a much lower frequency.[10] Another transcription factor that is important for formation of the lens is the paired-box transcription factor Pax6.[11,12] Pax6 is expressed very early in the IPE cells as they begin to reform the regenerating lens.[13] However, experiments show that Pax6 is not involved in the early induction events of lens regeneration:[10] transfection of Pax6 was unable to induce the initial dedifferentiation process and morpholino knockdown did not prevent the dedifferentiation (pigment lost and cell elongation) of the dIPE.[10,14] Rather, Pax-6 is required for the subsequent proliferation and respecification of the dedifferentiated cells to lens.[14]

An intriguing question is whether any of the factors used to induce pluripotent stem cells in mammalian experiments are used to regenerate the lens in the newt. Early studies using differential hybridization of IPE with dedifferentiated pigment epithelium showed that c-Myc is highly upregulated during the pigment epithelium to lens cell transition.[15] More recently, expression profiling found that Sox2, Myc and Klf4 are upregulated during lens regeneration.[16] In addition to these transcription factors, Maki et al have found that the stem cell-associated nucleolar factor nucleostemin is expressed during lens regeneration and showed its upregulation several days before cell cycle re-entry of the dIPE.[17] While it is intriguing that factors promoting iPS cell formation are expressed during lens regeneration, the lack of Oct4 or Nanog expression suggests that IPE cells do not become pluripotent.[16] Furthermore, functional tests are required to determine the exact roles of Sox2, Klf4, Myc and nucleostemin in the transdifferentiation of the IPE to lens.

In addition to the newt, *Xenopus laevis* larva can regenerate the lens; however, this amphibian reforms a new lens by transdifferentiating cells from the corneal epithelium.[18] The transdifferentiation of corneal epithelium is triggered by factors present in the vitreous chamber that are produced by the neural retina in the larva and the optic vesicle in the embryo.[19-22] Gargioli et al have shown that the competence to generate lens from corneal epithelium or from head epidermis is conferred by Pax6,[23] which may reflect the general role of Pax6 in imparting eye field competence as it does during development.[11,12] Future work needs to examine which signals and which signal transduction pathways lead to the activation of Pax6 and to determine whether there are other competence-promoting factors that confer this regeneration capability to the epithelium surrounding the lens.

Several amphibians and fish also regenerate the retina. In amphibians, retina regeneration occurs through the delamination, dedifferentiation and then transdifferentiation of retinal pigment epithelium (RPE) to reform neuronal cell types.[24,25] Evidence indicates that this transdifferentiation process involves transcriptional programs that are not found during retinal development: dedifferentiated RPE cells express CRALBP (cellular retinal-binding protein), which is not present in the embryonic retinal progenitors,[26] but is found in embryonic pigment epithelium and Müller glial cells,[26,27] which suggests that the dedifferentiation process of the RPE does not go through an embryonic retinal state. Detailed expression profiling of the conversion from RPE to neural retinal tissue will help trace the genetic reprogramming during the transdifferentiation process.

Transdifferentiation has also been induced in avian and mammalian embryonic RPE by addition of bFGF.[28-30] Although this phenomenon appears similar to amphibian RPE transdifferentiation, differences have been observed: the transdifferentiation of avian and mammalian RPE is a complete conversion to retinal cells without producing a second, delaminated cell layer. Consequently, the RPE is depleted to form a retinal layer.[29,31] Another difference involves a limited duration in the responsiveness by avian and mammalian RPE to FGF-induced retina formation during eye development.[28,29,32] This restriction was delayed when activin signaling, a signal transduction pathway responsible for RPE differentiation, is inhibited.[33] While embryonic chick RPE can be induced to regenerate retinal neurons, the generation of new retina neurons in postnatal chicks can only be generated by the Müller glia within the retina and not cells in the RPE.[34]

In contrast to newts, retina regeneration in fish comes from the Müller glia residing in the inner nuclear layer of the uninjured retina. Damage to the photoreceptor cells induces the Müller glia to proliferate and to migrate into the site of injury.[35,36] This process also appears to be involved in the reactivation of the retinal progenitor markers in the mature Müller glia that are normally expressed in immature Müller glia and retinal progenitor cells of the ciliary marginal zone, a region where new retinal cells are normally born.[37] These results suggest that Müller glia behave as progenitor cells that regenerate the retina.

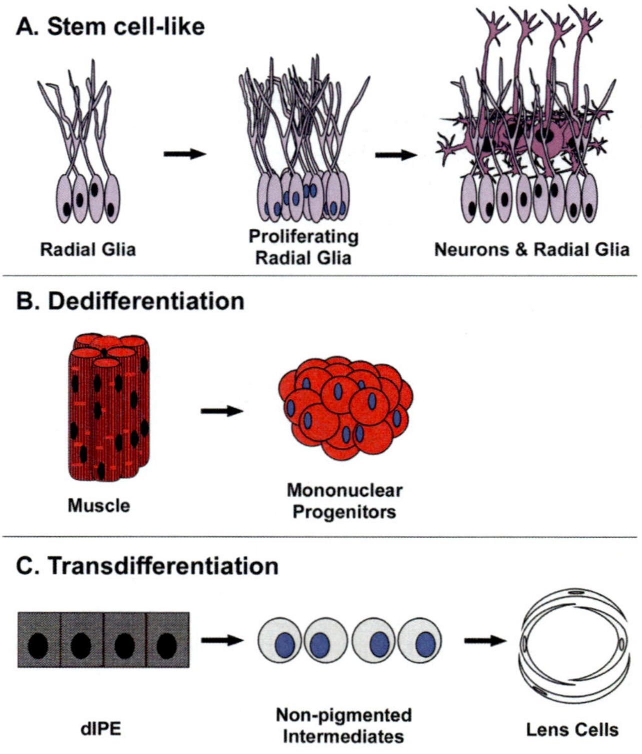

Figure 1. Please see legend on following page.

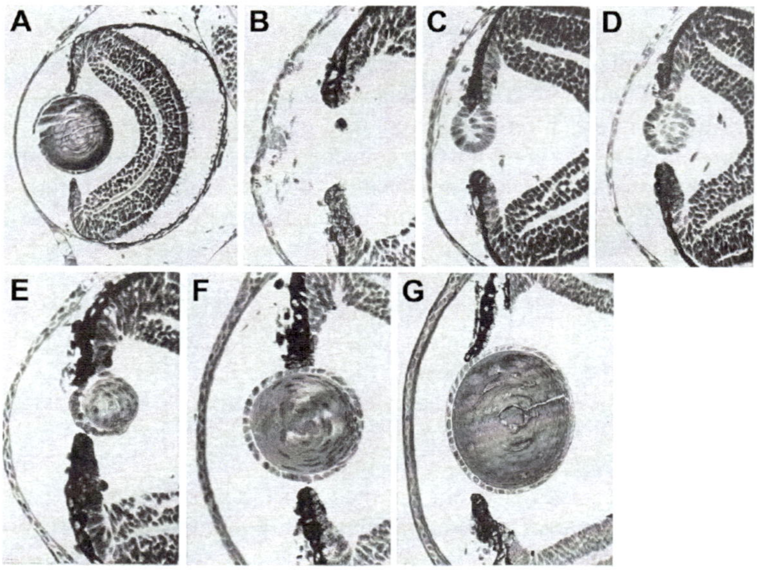

Figure 2. Please see legend on following page.

Figure 1, viewed on previous page. Cell reprogramming process observed in regenerating tissues. A) One mode of regenerating new tissue is through resident stem cell populations. After injury, radial glia of in the central nervous system proliferate and some cells become neurons while others remain radial glia. B) Some tissue cells undergo dedifferentiation to regenerate new tissues. Skeletal muscle multinucleated cells that contain the characteristic contraction apparatus of striated filaments. These cells dedifferentiate by dismantling the contraction apparatus, fragmenting into mononucleated cells and proliferating to generate more cells. C) Lentectomy induces differentiated dorsal iris pigment epithelial cells (dIPE) to lose their epithelial characteristics and pigment. These nonpigmented cells proliferate and subsequently form into lens cells. Current evidence indicates that these cells may be unipotent.

Damage to fish rod cells, to photoreceptor cells and to dopaminergic neurons within the retina initiates a regeneration response that leads to their replacement.[38] Surprisingly, while selective damage either to rod cells or cone cells results in their restoration, selective loss of the dopaminergic neurons does not. Instead, the neurons are only replaced when the injury encompasses other cell types.[39] Other resident precursor cells such as committed rod precursor cells are also present in the fish retina;[40,41] however, because rod cells are undergoing continual replacement throughout life and because rod cells are replaced only after the regenerating retinal neurons become postmitotic, the formation of new rod cells is more like homeostatic replacement. (For review, see ref. 42)

The Nervous System—A Model for Sequestered Progenitor Cells

Maintenance of the nervous system is key to an animal's survival, because it is the system for recognition, interpretation and reaction to the environment. Surprisingly, some vertebrates (reptiles, amphibians and fish) have maintained the ability to regenerate central and peripheral elements of the nervous system. Advancement of genetic and molecular technologies in amphibians and fish are providing an understanding of how these animals regenerate brain and spinal cord structures after injury.

Brain

Amphibians have an extensive capacity to generate new brain tissue after injury.[42-45] Surgical removal of part of the optic tectum in newts or of the telencephalon in Axolotl induces a regeneration response that results in a significant restoration of the removed structure.[45,46] Histological identification of mitotic bodies, BrdU-incorporation and ^{3}H-thymidine-radiolabeling experiments show that the proliferation of cells in distinct zones of the brain is associated with regeneration.[45,47] Similar resection experiments in *Xenopus* larva also result in the regeneration of the optic tectum and telencephalon.[44] In

Figure 2, viewed on previous page. Regeneration of the lens from the dorsal pigment epithelium in the newt. A) Cross sectionthrough the eye of the newt shows the cornea, lens, dorsal and ventral regions of the iris, the retina and retinal pigment epithelium. B) Lentectomy surgically removes the lens from the rest of the eye. C) Upon loss of the lens, cells at the tip of the dorsal iris begin to elongate to form a columnar epithlium and lose their pigmentation. D) Cells from the inner wall of the newly forming vesicle enter the lumen. E) Cells of the vesicle start to differentiate to lens fibers. F) Primary lens fibers form internally to the secondary fibers that appear as a monolayer on the periphery of the regenerating lens. G) The nuclei of the lens fibers disappear throughout most of the lens (a few remaining nuclei are present in the younger lens cells. Thus, a new lens has replaced the lost lens. [The panels were reproduced and adapted from Reyer RW. Quart Rev Biol 29:1-46. ©1954 with permission from The University of Chicago Press.]

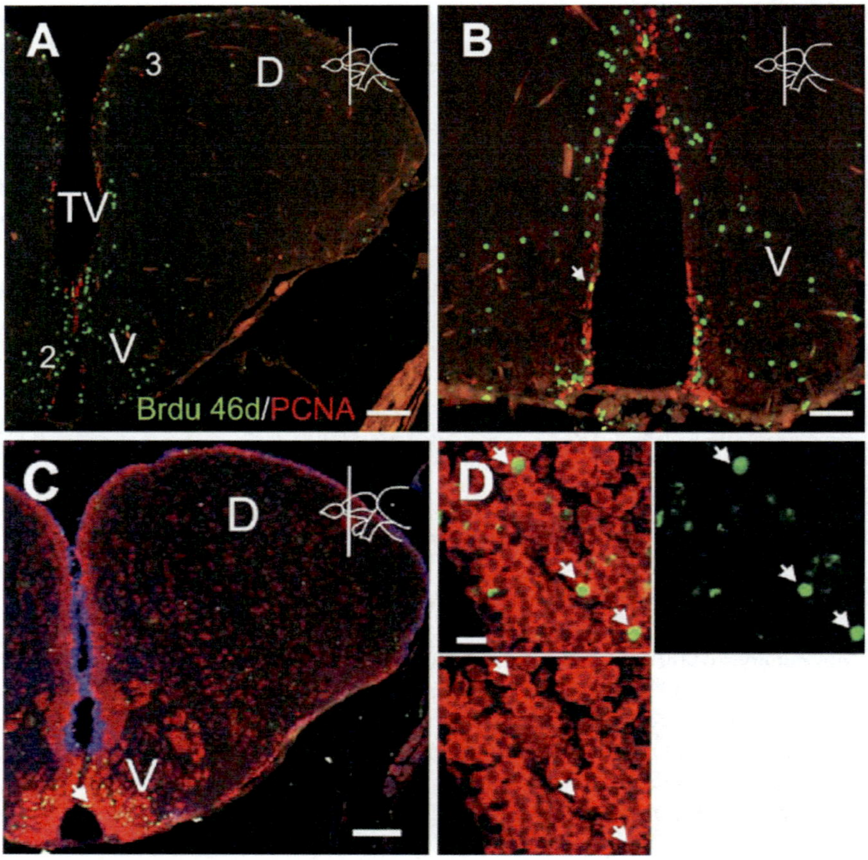

Figure 3. Please see legend on following page.

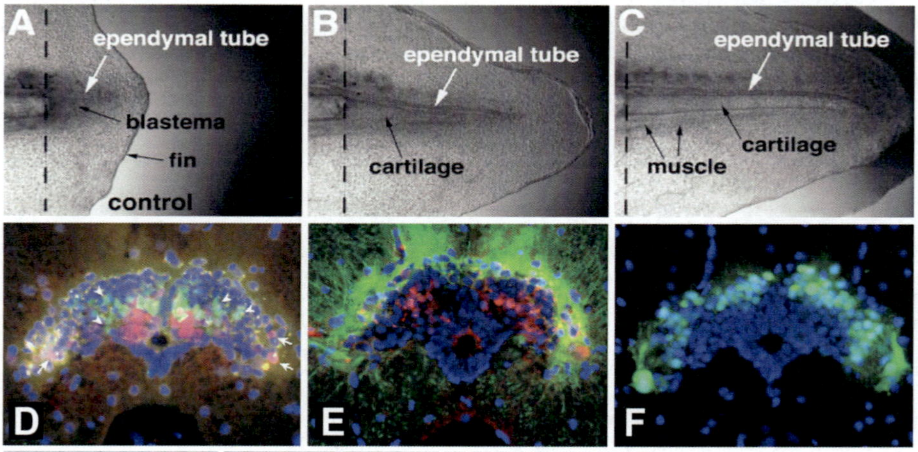

Figure 4. Please see legend on following page.

Figure 3, viewed on previous page. The zebrafish has endogenous proliferation zones within the adult brain. A) Cross section through the telencephelon of the adult zebrafish brain as shown by the immunohistological staining for proliferating cell nuclear antigen (PCNA: red). BrdU-positive cells (green) migrate from the proliferation zones in the ventral telencephalon (V) along the telecephalic ventricle (TV). B) Higher magnification of the ventricular proliferating zone. C) Near the proliferation zone in the ventral telencephelon (V), BrdU-positive cells are positive for the nerve marker Hu (red) (D) New nerves are produced by proliferation as shown by BrdU incorporation (green). [The panels were reproduced and adapted from Grandel H et al. Dev Biol 295:263-277;[52] ©2006 with permission from Elsevier.]

order to test the regenerative ability of specific nerve cell types, chemicals are used to destroy specific cell types, such as dopaminergic neurons in the brain. Immunohistochemical analysis for Tryosine Hydrolase (the enzyme for dopamineric biosynthesis) show that the regenerated architecture includes dopamineric neurons.[43] Even with a 75% loss of the dopaminergic neurons, a majority of the neurons are replaced after 30 days.[43] As immunohistochemistry for BrdU-incorporation, GFAP (glial marker) and Neu (differentiated nerve marker) illustrate, the new nerves are generated by the proliferation of glial cells of these zones,[43] indicating that these glial cells act as a stem cell population that regenerated new nerves. These data reveal that the regeneration of new neurons in amphibian brains comes from proliferation zones within the adult brain and glial cells act as a stem cell population for neurons.

Like amphibians, fish respond to injury to the brain by a gradual restoration of the layered architecture,[48] and associated with the restoration of the architecture is the functional recovery of movement.[48-50] Cell proliferation assays and cell lineage tracing experiments in uninjured adult fish brains highlight several regions throughout the brain that behave like neural stem cells (Fig. 3).[51-55] The continued proliferation of the cells in these regions suggests that the relative regenerative capacity of these vertebrates is linked to the amount of constitutive neurogenesis. However, injury to the brain results in a significant increase in cell proliferation in cells near the damaged site.[53] Therefore, it is not clear how much of the recuperative capacity that these animals display comes from neurogenesis through sequestered precursor centers or a broader response from the surrounding neural tissue. Current experiments are trying to determine how and where the generation of new nerves occurs after injury, as well as what the guidance cues are that target new born nerves to the correct connections. Tracing experiments for the integration of regenerating optic nerves show an increase in the number of erroneous connections after regeneration.[56] In this regard, data from goldfish indicate that incorrect connections in the brain by the optic nerves eventually disappear.[57] Future experiments need to identify the guidance cues that target new born nerves to make the correct connections.

Figure 4, viewed on previous page. Tail and spinal cord regeneration in the axolotl (*Ambystoma mexicanum*) involves progenitor cells that are distinct from differentiated nerves. A) Four days after tail amputation, the regeneration of a new tail is apparent. The regenerating tail is reforming the ependymal tube (which regenerates the spinal cord), a blastema (which will regenerate the muscle and cartilage of the tail) and the epidermal fin. B) By 8 days post amputation, extensive tail growth has occurred with the appearance of the skeletal cartilage and the ependymal tube. C) By 14 days post amputation, muscle has started to differentiate in the regenerating tail along with the cartilage and ependymal tube. D) Immunohistochemistry staining for Pax6 (red) and Pax 7 (green) in the adult axolotl spinal cord shows that two genes involved in spinal cord development continue to be expressed in cell populations of the adult. E,F) Pax 7 (red) is not expressed by differentiated nerves as marked by III-Tubulin (green) and nerves identified by Neu (green) do not express Pax7 (lack of red). [Panels A-C were reproduced and adapted from Schnapp E et al. Development 132:3242-3253,[68] ©2005; panels D-F from Mchedlishvili L et al. Development 134:2083-2093,[63] ©2007; with permission from The Company of Biologists Ltd.]

Spinal Cord

Radial glia are a non-neuronal cell type that are found in the developing nervous system of all vertebrates, and they contribute to the formation of the central nervous system by guiding the radial migration of new neurons and by acting as neuronal precursors themselves.[58-60] Adult salamanders and larval frogs can successfully regenerate their spinal cords as part of the regeneration of the tail because of the regenerative behavior of radial glia (Fig. 4).[61,62] Injury to the amphibian spinal cord induces the radial glia to transcribe embryonic genes, to undergo epithelial-to-mesenchymal transition, to proliferate and to migrate.[62,63] These cells make a tube of neuroepithelial cells (the ependymal tube) that undergo proliferation and some of them form new neurons.[64-66] Clonal analysis of these cells indicates a degree of plasticity by the cells within the spinal cord in that they produce progeny that do not retain parental gene expression and regional tissue specificity.[63,67] This is an attribute that appears to involve the retention of progenitor cell domains along the dorsal-ventral axis of the spinal cord due to the continued expression of embryonic genes (Shh, Pax7, Pax6 and Msx1) in mature animals (Fig. 4).[63,68]

Shh, Wnt, Bmp and Notch have all been shown to be important for tail regeneration. Wnt and Bmp are sufficient to induce spinal cord and tail regeneration (Fig. 7A-D),[68-70] and Shh is required for regeneration, implicating a role for these factors in spinal regrowth although the direct and indirect effects working through the surrounding tissues have not been disentangled.[68] Notably, Notch activation can induce spinal cord and notochord outgrowth but can not induce surrounding muscle, suggesting a direct affect on spinal regeneration.[69]

Fish also have a significant capacity to regenerate axons in the spinal cord after lesions, which allows them to regain locomoter aptitude within weeks after lesion.[71-74] It is not clear whether neurogenesis of the fish is as extensive as the salamander, but a number of experiments indicate that replacement of lost neurons does occur. Histological analyses of regenerated spinal cords in the fish *Sternarchus albifrons* and *Apteronotus albifrons* show that it is mediated through the outgrowth of the ependymal tube.[75,76] Ependymal tube cultures suggest that these cells are the source of new neurons.[77] Recently, Reimer et al used transgenic markers in zebrafish to track the formation of new neurons in response to Shh.[78] Fish appear not to regenerate all nerves of the central nervous system,[79] because spinal cord lesions in zebrafish larva result in the lack of Mauthner cells and Müller neurons.[80] However, Mauthner cells were stimulated to regenerate when dibutyryl cyclic adenosine monophate (cAMP) was applied to the cell bodies or transected nerves,[80] suggesting the missing nerve types can regenerate once they receive specific signals.

Heart—A Model of Regeneration by Differentiated Cells

Amphibians and fish can regenerate resected ventricles by the replacement of missing cardiomyocytes.[81,82] Despite the formation of scar tissue, these animals form new cardiac muscle that replaces the scar and restores the ventricle (Fig. 5).[82,83] Although definitive genetic cell tracking is still lacking, the incorporation of the DNA synthesis markers, such as tritiated thymidine and BrdU, into cardiomyocyte nuclei in culture and in vivo as well as the division of cardiomyocytes in vitro suggest that these cells retain at least some of their myocyte character during division.[81,82,84,85] Some intriguing studies tracking cultured newt cardiomyocytes labelled with a lipophilic dye suggest the possibility that cardiomyocytes may be capable of further dedifferentiation when placed in the environment of a regenerating limb. When fluorescently marked newt cardiomyocytes were transplanted from the heart to the regenerating limb blastema, the marked cells apparently lost cardiac-specific gene expression and eventually contributed to skeletal muscle and cartilage.[86] However, since these tracking studies were performed using membrane dyes, one can not exclude dye transfer to host cells by cell-cell fusion and contamination of the transplanted cells with other cell types such as connective tissue. Re-examination of this phenomenon with genetic tracking tools will be critical to confirm or refute these findings.

Recently, zebrafish transgenic reporter lines have been used to show that pre-existing cardiomyocytes are the major source of regenerating heart muscle.[87,87a] Some molecular evidence suggests that cardiomyocytes re-express genes that were involved in heart development, such as *gata4*, *hand2* and *nkx2.5*.[88] Upregulation of other genes known to affect progenitor cells such as *notch 1b*, *deltaC* and *msxb* have also been described to be expressed in regenerating hearts.[89] Microarray analyses have shown that several growth factors are expressed in regenerating heart tissues and current focus is on which induce cardiomyocyte to proliferate. In particular, Platelet-derived Growth Factor (PDGF) has been implicated in cardiomyocyte cell cycle progression.[90] More work needs to be done to determine how injury signals influence cardiomyocyte gene expression and expansion.

In addition to the myocardium, the heart consists of other tissue layers that also need to be reconstructed: the endocardium and the epicardium. It is not yet clear how these tissues participate in the regeneration process of amphibian and fish hearts. However, one can surmise that they are involved in processes similar to their roles during heart development. The epicardium has been shown to provide growth factor signals to the myocardium and to contribute to the cardiac vascularization.[91] This is also true for the endocardium.[91] More work needs to be done to determine whether the epicardium, the endocardium and other tissues influence the regeneration process of the heart.

Appendages—Making Progenitor Cells from Mature Tissues

Appendages have been a particular focus of regeneration research due to the complexity of the process and their experimental accessibility. The mature limb consists of multiple tissues and proper morphogenesis of the regenerating limb requires patterning events along the anterior-posterior, dorsal-ventral and proximal-distal axes. Understanding how the complex three-dimensional structure of multiple tissues is reconstituted after amputation is a current endeavor. Histology of regenerating appendages show that immediately after amputation, the wound is sealed through migration of wound epidermis. Subsequently, an underlying mesenchymal cell mass (the blastema) forms, which consists of proliferating progenitor cell populations.[92-94] Tissue grafting and cell lineage tracing experiments in salamanders have shown that blastema cells primarily come from tissue sources within the first few millimeters of the amputation plane,[95-97] and that the cells of these tissues lose their characteristic differentiated morphology.[98]

A major question has been whether blastema formation involves dedifferentiation of classically postmitotic cells or the activation of a stem cell pool. The current evidence suggests that both mechanisms occur. Based on ultrastructural histological analyses, newt muscle tissue was long thought to lack satellite cells,[99] the stem cells that normally repair the striated muscle in other vertebrates,[100-102] but recently molecular analysis has revealed that newt satellite cells do indeed exist.[103] However, the contribution by satellite cells is likely not the dominant mode of muscle cell contribution to the formation of the blastema:

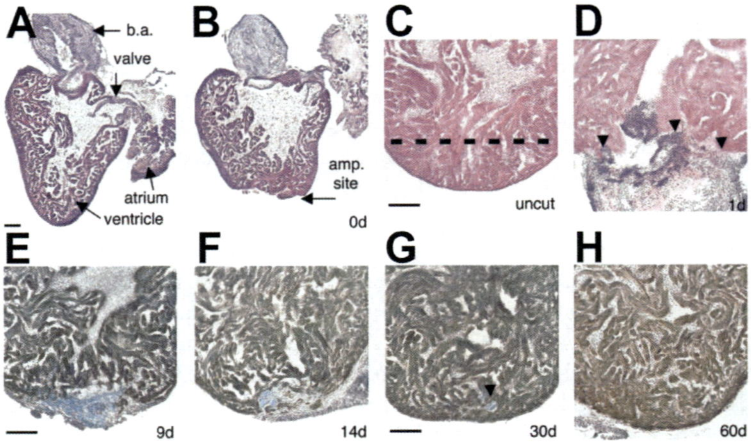

Figure 5. Please see legend on following page.

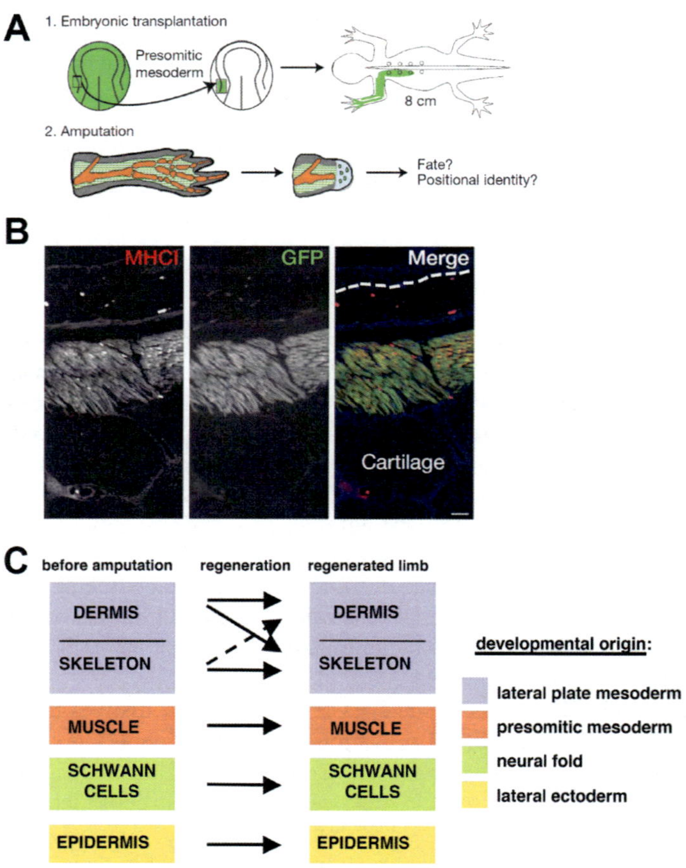

Figure 6. Please see legend on following page.

Figure 5, viewed on previous page. The adult zebrafish heart regenerates the ventricular wall. A) The adult heart of the zebrafish consists of one ventricular chamber, one atrial chamber and the bulbous arteriosus (b.a.), the primative outflow tract. B) Surgical resection of the apex of the ventrical removes the epicardium, myocardium and endocardium of the ventricular wall. C) A higher magnification of the apex of the ventricle with an approximation of the amputation plane (dashed line) through the wall. D) One day post amputation (1d), the loss of ventricular tissue is apparent and a blood clot is pronouced at the amputation plane (arrowheads). E) At 9 days post amputation (9d), fibrin scar tissue is present (light blue) at the tip of the apex. F) As time progresses, the scar tissue is replaced by myocardial tissue, (G) and as this replacement continues, new muscle tissue and cardiac wall are created. H) After approximately 60 days (60d), the ventrical shows no overt sign of injury, despite the original size of the injury. [This figure was reproduced and adapted from Poss et al. Science 298:2188-2190;[82] ©2002 with permission from AAAS.]

a number of in vitro culture and cell tracking studies have documented the fragmenting of multinucleated myotubes and muscle fibers into mononucleated, proliferative cells,[104-108] indicating that muscle dedifferentiation is a mechanism for generating at least one population of progenitor cells in the blastema.[98,106,109] Furthermore, early transplantation experiments grafting different tissues (trackable by radiolabelling or by tripoid nucleoli of donor tissues in host diploid animals) showed that dermal fibroblasts, cartilage, Schwann and connective tissue cells, in addition to muscle, all contribute to the blastema,[96,97,110,111] and subsequently contribute to the tissues of the regenerated appendage.[112-114]

The loss of differentiated characteristics by the cells of various tissues raises important questions about the potency of blastema cells: do all blastema cells dedifferentiate to multipotency so that each cell becomes capable of giving rise to all limb tissues, or does each tissue give rise to its own tissue-restricted blastema cells? It had been difficult to fully answer these questions, because previous experimental techniques (cell lineage tracing by tritiated-thymidine labeled tissues or by triploid cells in diploid animals) produced data that could not definitively rule out the contribution of cells from alternative sources. For example, in Steen's early experiments, tritiated-thymidine-labeled cartilage grafts gave rise to a small number of labeled cells in muscle tissue,[115] suggesting that the cartilage-derived cells may not be highly multipotent. Similarly, when limb blastemas lacking differentiated skeletal muscle were grafted to the epithelium of the dorsal fin, they did not produce muscle but did produce cartilage.[116,117] Analogous results were obtained when the blastemas were grafted to the orbit of the eye.[118] Muscle was only regenerated when differentiated muscle from the residual stump was attached to the grafted blastema.[116,117] These results were interpreted as indicating that muscle is required to regenerate muscle and implied that cartilage-forming cells are not competent to produce muscle, although other interpretations are possible. In contrast, muscle grafts suggested that muscle derived blastema cells may contribute to other tissues.[119] These conclusions had been promoted by recent experiments implanting cultured satellite cells which apparently contributed to other tissue types.[103,120,121] However, such experiments have been hampered by lack of

Figure 6, viewed on previous page. Limited plasticity of regenerating limb cell. A) The method of tissue labeling by embryonic grafting. Embryonic transplantation of cells constitutively expressing GFP results in the contribution of GFP-positive cells to tissues of the adult animal. Tissue-specific expression of the GFP transgene allows cell lineage tracing of these tissues through the regeneration. B) GFP-labeled muscle (before amputation) contributes to muscle but not to skin or cartilage after regeneration as shown by co-immunohistochemical labeling of cells with the muscle-specific myosin heavy chain I (MHCI) and GFP in the merged overlay. C) Summary of cell fates during axolotl limb regeneration. [The panels were reproduced and adapted from Kragl et al. Nature 460:60-65;[113] ©2009 with permission from Nature Publishing Group.]

comprehensive and long-lasting lineage labels and by the lack of precise tissue-specific molecular markers to track cell identity.

The development of transgenic axolotl lines expressing GFP, combined with the increased availability of molecular markers has allowed this issue to be re-explored with more confidence and results indicate that blastema cells are highly restricted in their differentiation potential. Kragl et al applied embryonic grafting in order to label the major tissue types of the limb with high purity. This technique has an advantage over direct limb grafting, because it was possible to identify embryonic stages where the Anlage of each major tissue such as muscle, dermis/connective tissues, etc. can be grafted without contamination of other cell layers (Fig. 6). Previous experiments used adult limb tissue grafts that usually consisted of several different cell populations. Also, the use of Pax7 as a molecular marker of muscle progenitor cells was necessary to distinguish muscle progenitors from interspersed connective tissue fibroblasts. The results of this tissue-specific cell labeling was that each tissue, including muscle, dermis, cartilage and Schwann cells, produced restricted progenitor cells that contributed to a very limited spectrum of tissues.[113] Therefore, blastema cells are not multipotent but rather highly restricted in differentiation potential. Because these experiments were performed only in the axolotl, an Ambyostomid that represents one major branch of the salamander family, it will be important to confirm the results in the other branches of the salamander family—i.e., the newt—where many important regeneration experiments have also been performed.

Guiding of Progenitor Cells through Regeneration

The regeneration of three-dimensional structures involves tissues that provide extracellular information as well as intracellular factors that transduce this information to create and guide progenitor cells. Thus, another important aspect to regenerating a compound structure requires the maintenance of an environment that promotes regeneration. While cells that contribute to the blastema lose their differentiated morphology, become migratory and proliferate dramatically, these cells still correctly contribute to the formation of a highly organized structure of appropriate size and do not create a tumor or become cancerous. Work in limb regeneration has demonstrated that the epidermis and peripheral nerves regulate blastema growth and has addressed how blastema cells behave along specific body axes, which includes how cells at the amputation stump reprogram to reform missing distal structures.

Epidermis

After wound healing, amputated limbs form a specialized thickened epidermis.[122-125] This specialized epidermis is required, because its substitution with grafts of uninjured skin halts regeneration.[126] Cell lineage tracing experiments show that cells within the epidermis only participate in the regeneration of new epidermis,[113,114,127,128] indicating that the wound epidermis regulates the blastema but does not contribute to it. The most distal region of the wound epidermis forms an apical ectodermal cap (AEC),[122,123] and it is likened to the apical ectodermal ridge (AER) of the developing chick limb during embryogenesis,[129] a required thickened epithelial structure that guides limb development. However, there is an important distinction between the embryonic chicken AER and the regenerative amphibian AEC: the AER does not regenerate after loss and its removal results in limb truncation,[130] while a regenerating limb will reform the AEC after its removal.[123,131] This indicates that regeneration competency is linked with the reformation of the AEC. Transplantation experiments implicate the maintenance of patterning information in the reformation of the AEC.[131,132] However, it is still not known what the regeneration-specific maintenance program is.

The role of the wound epidermis in regeneration appears to be guidance. Asymmetrical engraftment of the AEC to a posterior position on the blastema results in a corresponding asymmetrical regenerated limb that project at sharp angles from the limb stump,[133] and transplanting the AEC to the base of the blastema results in the regeneration of ectopic limb structures,[134] indicating that this specialized wound epidermis directs regenerative outgrowth. The wound epidermis in amphibians and fish expresses several growth factors that can promote the migration, proliferation and genetic programming of underlying cells.[135-138] One molecule provided by the wound epidermis is retinoic acid,[136] a derivative of vitamin A. Retinoic acid is a potent morphogen, because treatment either of developing or regenerating appendages with distinct levels of retinoic acid adjusts the proximal-distal patterning information by reprogramming blastemas that would form distal structures into blastemas that form proximal and distal structures.[139-141] Other growth factors in the wound epidermis include FGFs (axolotl), Wnt5 (axolotl and zebrafish), *shh* (zebrafish) and *bmp2* (zebrafish).[137,138,142,143] These growth factors are in distinct expression domains within the epidermis of the regenerating appendage, indicating that the epidermis is providing regional signals to the underlying precursor cells of the blastema and consequently regulating patterning.[135]

Peripheral Nerves

The importance of the peripheral nerves for appendage regeneration comes from denervation experiments in amphibians and fish: removal of the nerves that channel into appendages prevents their regenerative outgrowth.[144-147] Furthermore, redirecting the peripheral nerve of an amphibian limb to an ectopic site near the skin results in the formation of a proliferative growth adjacent to the tip of the nerve.[148] This outgrowth shares characteristics of the limb blastema: it initially expresses Msx2, Tbx5 and Hoxa13 and will form differentiated tissues.[149,150] While nerve deviation initiates a blastemal growth, this growth eventually regresses and the cells recruited to the growth will differentiate to form ectopic bone, muscle and connective tissues. However, when skin from the contralateral side of the limb is transplanted next to the site of the wound and deviated nerve, a complete ectopic limb will form.[148] These data indicate that the nerve provides signals that are conducive to early regeneration, but the nerve alone is not sufficient to maintain regenerative outgrowth.

A reciprocal relationship between the wound epidermis and the nerves promotes limb regeneration. Histological observations show that the wound epidermis is innervated,[151] and nerve deviation to the skin induces the formation of a thickened epidermis that expresses Sp9, a transcription factor typically upregulated in the wound epidermis of a regenerating limb.[150] Also, the secreted factor anterior-gradient ligand (nAg), which is expressed first in Schwann cells and subsequently in the epidermis is involved in maintaining the regenerative outgrowth.[152] Transfection of nAg into the cells of a denervated limb stump after amputation overcomes the lost ability to regenerate after removal of the peripheral limb nerve (Fig. 7E,F).[153] This secreted factor interacts with Prod1, a cell surface molecule that is expressed by blastema cells and the peripheral nerves in the regenerating limb.[152,154] These data suggest that nAg interaction with Prod1 regulates distal growth of the regenerating limb. Additional evidence suggests that this pathway is involved in the proximal-distal patterning of the regenerating limb as shown by the inhibition of proximal engulfment of distal blastemas after antibody inhibition of Prod1 in hanging drop cultures[154] and by the proximalization of distally fated blastema cells from ectopic expression of Prod1.[155] Prod1 is up-regulated by retinoic acid,[154] suggesting that it is involved in the patterning activity of retinoic acid signaling. Future experiments need to determine how Prod1 transduces proximal identity in the blastema cells.

Several other factors have been found to be expressed by the nerves and have been shown to induce cell proliferation of blastemal cells: Fibroblast growth factor 2 (Fgf2), Glial growth factor (Ggf) and Transferrin. Both Fgf2 and Ggf have been shown to be sufficient to induce cell proliferation and the expression of genes associated with regeneration such as the blastema marker 22/18 and *Distal-less homolog 3* (*Dlx-3*) in denervated axolotl limbs.[156-158] Furthermore, Fgf2 can rescue appendage regeneration after removal of the nerve.[158,159] The nerve-associated iron transport protein Transferrin is also involved in regeneration: it promotes blastema cell proliferation in vitro[160] and is sufficient to maintain the proliferation of blastema cells in denervated limbs.[161] While the molecular mechanisms of Fgf2 and Ggf are likely to be directly involved in the activation of particular signal transduction pathways, the mechanism through which transferrin is involved remains unknown. In any case, the activities of these nerve-associated factors indicate that the nerve is involved in regeneration through multiple molecular pathways. However, the relationship between these pathways still needs to be elucitated.

Beyond the potential role of Fgf in nerve-dependent regeneration, the importance of Fgf signaling has been further demonstrated by chemical and genetic disruption of this signal transduction pathway in zebrafish (Fig. 7G).[162,163] Treating fish with SU5402, a pharmacological inhibitor of Fgf receptor signaling, blocks the formation of the blastema. Furthermore, a genetic mutation in the *fgf20a* gene that perturbs the ability of the Fgf protein ligand to efficiently bind to its receptors prevents proper blastema formation and consequently regeneration but does not affect the original development of fin,[162] suggesting that *fgf20a* has a regeneration-specific function. It still remains to be resolved whether other Fgf members fail to substitute Fgf20a function.

Other Signaling Factors

Wnt

Wnt growth factor signaling has been found in several developmental and regenerative tissue processes that involve the maintenance of progenitor population and subsequent events during their differentiation. For example, based on overexpression and inhibition studies, Wnt2b has been shown to be involved in maintenance of the progenitor cell population in the progenitor zone (CMZ) of the retina by prohibiting their differentiation into retinal neurons.[164,165] The Wnt signaling is also required for appendage regeneration in salamanders, frogs and fish.[137,166,167] Inhibition of the canonical Wnt signal transduction by overexpression of negative regulators of Wnt signaling (Axin-1 or Dkk1) in *Xenopus* limb and tail stumps reduces or inhibits regenerative outgrowth.[166,168] Conversely, activation of the canonical Wnt pathway by overexpression of an activated form of β-catenin can enhance the regenerative capacity in *Xenopus* limb buds at a stage in which they have lost their natural ability to regenerate.[166] These results were also observed in the regenerating

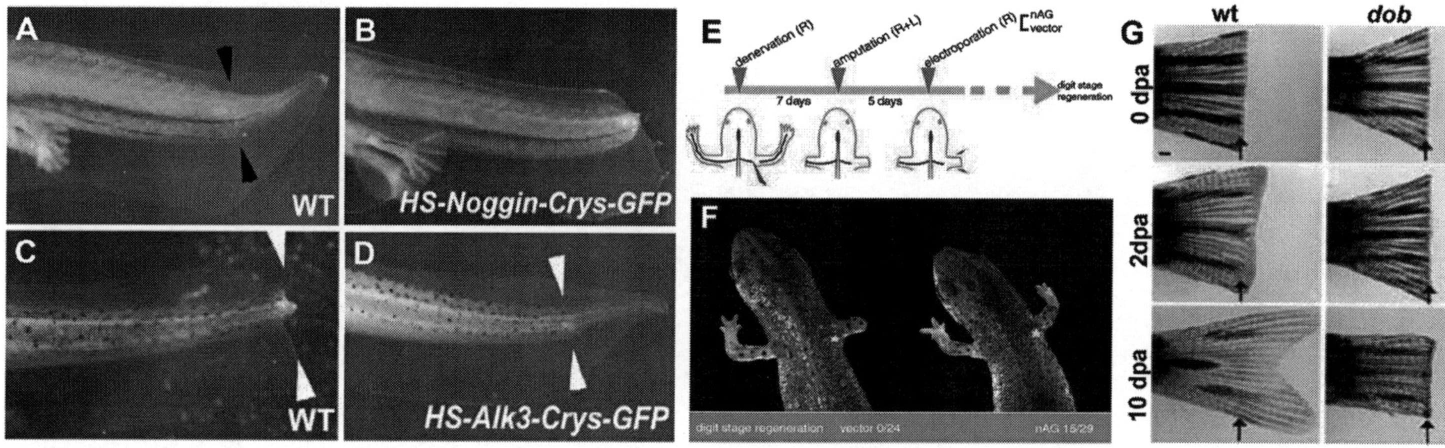

Figure 7. Molecules required for tail, limb and fin regeneration. Requirement for Bmp signaling during *Xenopus* tail regeneration. A) Tail regeneration in the Xenopus tadpole results in the reformation of the spinal cord, neural tube, skeletal muscle and fin tissue. B) Heat-shock-induced overexpression of Noggin, an inhibitor of Bmp signaling, after tail amputation inhibits its regeneration. C) *Xenopus* tadpoles normally show reduced regeneration during a refractory period between deveopmental stages 46-48. D) Conversely, induced expression of the Bmp receptor Alk-3 restores the regenerative capability during the refractory period. The lack of limb regeneration after denervation can be rescued by addition of nAG, a putative ligand for Prod1 transmembrane receptor. E) Experimental design for removing the peripheral nerve supply (denervation) at day 1, amputating the denervated limb 7 days later and then transfecting limb cells with the nAG ligand 5 days after amputation of the limb. F) While limbs not transfected with nAG did not regenerate (vector controls), a majority of animals with nAG transfected limbs did regenerate distal forearm and digit tissues. The yellow star indicates the amputated limb in the vector controls (left animal) and the nAG-transfected animal (right). *dob* (*fgf20a mutant*) in fin regeneration. G) Amputation of the fish caudal fin (0 dpa) results in the progressive regeneration of the fin tissue by wild-type fish (panels in left column). However, the dob mutants (panels in right column) lack the ability to form a blastema and therefore do not regenerate their fins. [Panels A-D were reproduced and adapted from Beck et al. Dev Cell 5:429-439;[69] ©2003 with permission from Elsevier. Panels E and F were reproduced and adapted from Kumar et al. Science 318:722-777;[153] ©2007 with permission from AAAS. Panel G was reproduced and adapted from Whitehead et al. Science 310:1957-1960;[162] ©2005 with permission from AAAS.] A color version of this figure is available at www.landesbioscience.com/curie.

fin of the zebrafish by inhibiting the activation of β-catenin signaling with overexpression of the zebrafish *dkk1*.[137]

Interestingly, the Wnt-5 paralogs display strikingly different effects on regeneration in zebrafish fin and *Xenopus* tail. While *wnt5b* overexpression antagonizes regenerative outgrowth of zebrafish fins,[137] the overexpression of Wnt-5a in *Xenopus* induces the outgrowth of a complete ectopic tail after partial amputation through the tail.[70] More work needs to explore whether the difference observed is due to the divergence in the signal transduction of Wnt-5a and -5b or whether this effect on regenerative growth involves specific differences in the tissues comprising the caudal fin and *Xenopus* tail. In general, it is likely that Wnt signaling has multiple roles in regenerating tissues as it does during embryonic development and some of these roles overlap. It will be important to define the regeneration specific functions of Wnt signaling in the future.

Notch

The notch-delta signal transduction pathway has long been known to be involved in cell fate decisions in several tissues, including stem cell populations of different tissue types in vertebrates. (For review, see ref. 161) It is likely that Notch signaling is involved in the regeneration of the appendages; however, how the Notch signal transduction pathway is involved in appendage regeneration remains to be tested.

Tgf-β

The Transforming Growth Factor family consists of several growth factors that include TGF-βs, activins, inhibins and bone morphogenic proteins (Bmps). Each of these members alters gene transcription by interacting with specific subsets of transmembrane receptors that in turn interact with particular subsets of intracellular signal transduction mediators. The Tgf-β receptor 1 is detectable as early as 6 hours after amputation of the salamander forelimb and its expression continues into early stages of regeneration.[169] Treatment with a chemical inhibitor (SB-43152) that binds to the Tgf-β Type I receptor blocks regeneration at early and later outgrowth stages due to the inhibition of cell proliferation.[169] These early and late requirements for Tgf-β signaling are also observed in the regeneration of *Xenopus* tadpole tails.[170] What is of particular interest is that while inhibiting wound healing with SB-431542 is reversible,[170] the block to regeneration by this inhibitor is irreversible.[169,170] While the chemical inhibitor used for these experiments is selective for Tgf-β signaling, it has affinity for several Tgf-β Type I receptors (Alk4, Alk5 and Alk7).[171] Consequently, this can also affect the signal transduction of other members of the superfamily, such as activin. Activin is up-regulated during early fin regeneration. In situ hybridization experiments show that the transcript of *activin βA* is in the epidermis of the fish fin after amputation injury but before regenerative outgrowth takes place. Inhibition with transcript-specific morpholinos blocks the formation of the blastema, in part due to the induction of cell proliferation within the fin stump.[172] Overall, results from interfering with Activin function are similar to the experimental results with SB-431542 in amphibian appendages: Treatment of amputated zebrafish fins causes a serrated phenotype in which the fins cover the wound with epidermis but the fin tissue at the amputated edge recedes in the interray space.[172] This phenotype mirrors the early expression pattern of *activin βA* after amputation through the fin,[172] and suggests the SB-431342-induced phenotype is caused by the inhibition of activin signaling.

Further genetic tests will be needed to determine whether Tgf-β has a function distinct from activin activity.

The Bmp growth factor members of the Tgf-β family are also important for regeneration. *Xenopus* tadpoles that transgenically overexpresses Noggin, a Bmp antagonist, do not undergo tail regeneration.[69] Similarly, inhibition of Bmp signaling by the overexpression of the Bmp antagonist Chordin reduces the formation of bone.[173] In contrast, overexpression of *bmp2* in the blastemas of regenerating fins results in the increased production of dermal ray bone.[138] These results suggest a complex signaling system that provides information beyond an induction and inhibition of regeneration that based on the presence or absence of the ligand. What needs to be determined is how these signals function in the formation and maintenance of different regenerating tissues.

Shh

In amphibians and fish, sonic hedgehog is found in regenerating tissues and its activity is required. Inhibition of Shh signaling in regenerating fore- and hindlimbs of the salamander results in a severe reduction in the number of distal digits,[174] and chemical inhibition of Shh signaling after amputation of fish fins blocks regenerative outgrowth.[175] Conversely, overexpression of *shh* upon viral infection of regenerating limb tissues results in ectotopic digits, digit duplications and fusion of digit bones,[176] and overexpression of *shh* in regenerating fins causes mispatterning of the dermal fin bones.[175] Amputated *Xenopus* froglet limbs do form blastemas; however, they only regenerate unbranched cartilaginous spike appendages.[177,178] The failure to regenerate all the digits is attributed to a lack of re-expression of Shh in the froglet limb blastema.[177,178] While providing Shh resulted in the regeneration of branched cartilaginous structures within the regenerated spike, this did not create a digitated limb,[179] indicating additional factors are required for correct patterning. Based on gene expression and the current functional data, Shh signaling appears to have the same role in regeneration as it does during development. Thus far, two fundamental questions in regard to Shh (and other members of the hedgehog family) are (i) does Shh have a role in the regeneration of tissue that is different from its roles in embryonic development and (ii) what activates Shh expression in cells during regeneration.

Intracellular Translation of Growth Promoting Signals

The change in the differentiation status of blastema cells is concurrent with the increase in cycling cells, and this increase in successive cell divisions is required for blastema formation. The regenerating structure stops growing once a particular body-to-organ proportional relationship has been reached. As blastema cells proliferate, cells in the proximal region of the blastema differentiate and seamlessly integrate with the residual undamaged differentiated tissues.[180-182] Cell differentiation and proliferation states must be coupled by guidance mechanisms that regulate the re-establishment of tissues in order to reconstruct the proper architecture of the missing structure. Thus, there are two questions in regards to this regulation: what are the mechanisms that activate the cell cycle in quiescent terminally differentiated cells and what inactivates cell proliferation when the correct tissue proportions are reached.

Cell Cycle Control Zebrafish Fins

Experiments measuring the growth rate of regenerating fins show that there is a positional component to the rate of regenerative outgrowth—the rate of growth decreases as the regenerating fin outgrowth reaches distal structures. This positional information is associated with the level of fibroblast growth factor-signaling, because the expression of downstream Fgf transcriptional targets decreases as the rate of growth decreases.[183] Thus, one can infer that a general mechanism for the cessation of growth is due to the down-regulation of Fgfs and other genes that promote the regenerative outgrowth.

One molecule that appears to link the regulation of cell proliferation and tissue patterning is the vertebrate gene *fam53b/simplet* (*smp*). In regenerating fins, *smp* knockdown reduces regenerative outgrowth by inhibiting cell proliferation while it up-regulates the expression of *msx* and of *shh* genes , as well as causes ectopic bone to form.[184] This gene encodes an intracellular protein that contains two conserved domains whose function remains unknown,[185] so it will be interesting to determine how this intracellular factor is conveying the extracellular pro-regenerative cues into the regulation of cell proliferation and the formation of tissue patterning.

Cell Cycle Control in Amphibian Limbs

Naturally, cell cycle regulators are involved in the burst of cell proliferation that governs blastema formation, but there appear to be distinct features in how particular cell cycle regulators function or are regulated during regeneration. p53 is a tumor suppressor gene that has been shown either to suppress cell division or induce apoptosis when cells are stressed or sustain DNA damage.[186] Interestingly, chemical inhibition of p53 impairs limb regeneration.[187] Assuming specificity of the drug used, these inhibition results suggest a requirement for p53 activity in regenerative outgrowth, an event that requires significant cell proliferation. Amphibian (newt) blastema cells continually proliferate in long-term cultures and do not undergo any apparent crisis or senescence.[188] The necessity for p53 and the potential "immortal" property of blastema cells suggest that p53 may have a function distinct from its expected roles as an inhibitor of cell division and promoter of apoptosis. Future studies need to define which cells require p53 activity and determine whether p53 functions as more than a brake on the cell cycle during regeneration. In addition to a potential role of the p53 tumor suppressor, the cell cycle inhibitor retinoblastoma (Rb) protein is regulated differently in the cardiac and striated muscles of adult salamanders compared to their mammalian counterparts. Differentiated newt muscle retains the ability to phosphorylate Rb and this phosphorylation capacity is associated with the re-entry of newt myofibers into the cell cycle after serum stimulation.[187,189] In contrast, differentiated mammalian myotubes can not enter the cell cycle when stimulated with serum[190,191] unless Rb protein has been removed: experiements show that cultured myotubes derived from Rb-null mice can be induced to enter the cell cycle in a serum-dependent manner.[192] These results indicate that the regulatory mechanisms that relieve Rb-mediated cell cycle inhibition are missing in mammalian muscle but are present in newt myofibers. Future work needs to define what these regulator mechanisms are.

Molecular Factors Involved in Progenitor Cell and Blastema Formation

The formation of the progenitor cell populations, such as those in the blastema, likely involves several extracellular and intracellular molecular mechanisms and understanding the molecular factors that convert mature limb tissue cells into blastema cells is a major focus of regeneration research. While we have much left to learn about the epigenetic and transcriptional regulators that induce differentiated tissue cells to become blastema cells, some factors, Msx1, Twist, Pax7, etc., have been implicated in the conversion to or control of progenitor states during appendage regeneration.

Msx1

For a number of years there has been intriguing data pointing to a potential function of Msx1 in muscle dedifferentiation. First defined as a factor that is expressed in the proliferative, undifferentiated tip of the developing limb bud,[193-195] it was subsequently shown to prevent the morphological differentiation of myoblasts to myotubes in vitro.[196,197] The discovery that Msx1 is re-expressed during regeneration prompted the question of whether Msx1 can drive the dedifferentiation of myotubes to myoblasts.[198-200] Odelberg forcibly expressed Msx1 in in vitro-formed mouse myotubes and found that a small fraction appeared to undergo dedifferentiation.[121] One study examining isolated muscle fibers from the salamander also suggests that Msx1 expression may be associated with muscle fragmentation[201] and consequently may be involved in myotube-to-myoblast conversion.[121,202] It is clear that Msx1 is also upregulated in nonmuscle cells during regeneration, since the zebrafish homologs of the Msx genes are also re-activated in regenerating caudal fins,[198] a structure that lacks muscle cells but still requires Msx activity: Morpholino knockdown of a Msx fish homolog (*msxb*) in regenerating fins inhibits the regenerative outgrowth.[203] Future experiments that conditionally activate and repress Msx1 expression during and after blastema formation will be needed to directly test whether is part of a program driving dedifferentiation of cells during blastema formation, or whether it acts once blastema cells are formed to prevent their premature redifferentiation.

Twist

Another molecule implicated in muscle dedifferentiation is Twist. Twist belongs to a gene family whose members are involved in several important tissue specification processes of mesoderm- and ectoderm-derived tissues.[204] The original Twist gene was identified to be required for the formation of mesoderm,[205,206] and its activity is involved in the genesis of myoblasts.[207,208] Overexpression of Twist by adenoviral delivery into cultured myotubes prevents myotube formation by cultured myoblasts and appears to induce myotube fragmentation of previously differentiated myotubes along with transcriptional down regulation of muscle genes.[209] Future experiments need to discern whether this gene is expressed in regenerating appendages, whether it is required for regeneration and whether it is sufficient to promote dedifferentiation of muscle and other tissues at the onset of regeneration.

The salamander *AmTwist* ortholog of a second mammalian Twist gene, Twist-2/Dermo-1, was shown to be expressed in the proximal region of the blastema.[210] Based on its expression in regenerating limb, *AmTwist* may function in the regenerating limb like the mammalianTwist-2 does in limb development. Twist-2 is expressed in the subectodermal mesenchyme during early stages of dermal differentiation and development of skeletal elements.[211,212] In salamander limbs, deviation of the peripheral nerve to an ectopic site near the skin causes dermal fibroblasts to proliferate and produce a structure similar to the early blastema of an amputated limb.[148] While *AmTwist* is expressed along the entire dermis of a newly healed wound, its expression is restricted to the proximal peripheral margin and is not found the central distal region of the nerve-induced blastema.[210] In the regenerating appendage, tissue differentiation is observed in the proximal region of the blastema; thus, the proximal expression of *AmTwist* suggests that it may be involved in the differentiation or maturation of the dermal cells. Future work needs to show what the function of *AmTwist* is in the proximal region of the regenerative outgrowth.

Pax7

Members of the Paired-box transcription factors are expressed in different tissues at different stages of embryogenesis, but a few members continue expression in adult progenitor cells.[120,213-216] For example, skeletal muscle progenitor cell behavior is regulated by Pax3 and Pax7 during development.[217] In adult fish, salamanders, frogs and mammals, Pax7 is expressed in quiescent and newly activated skeletal muscle stem cells, and its presence is required for their maintenance.[217] Pax7 is necessary and sufficient for the myogenic differentiation of the resident muscle stem cells (satellite cells) and CD45/Sca1-positive cells,[218] which are a side population of adult stem cell populations found in muscle and are believed to originate from the hematopoietic lineage.[219-221]

Experiments in *Xenopus* also suggest the importance of Pax7 in the regeneration of striated muscle. Immunohistochemistry and electron microscopy identify Pax7-positive cells in striated muscle, and these cells have the characteristic satellite cell morphology: small cells that are dominated by nuclei on the periphery of muscle myofibers.[222] After amputation, Pax7 expression is significantly upregulated and overexpression of a dominant negative Pax7 construct in all cells of the regenerating tail of the *Xenopus* larvae blocked the regeneration of muscle, indicating that Pax7 is required for muscle regeneration.[222] In mammalian cell culture, Pax7 maintains muscle stem cell populations in part by recruiting the histone methyltransferase MLL2 to myogenic regulatory factor Myf5,[223] a gene involved in commitment to a myoblast fate. Whether this molecular mechanism is involved in muscle regeneration of the amphibian limb remains to be determined.

Meis and Hox Genes

After amputation, proximal limb tissues produce new distal limb tissues; thus, the regeneration ability involves the "reprogramming" of cells in the upper arm to create lower arm structures. Grafting experiments have determined that the dermal fibroblasts contain the necessary information for patterning the regenerating limb.[113,224-232] Consequently, it is believed that fibroblasts are involved in the respecification of positional identity during regeneration. What is the information in fibroblasts that conveys patterning?

One transcription factor family that has been shown to confer positional identity in regenerating limbs is Meis. Meis1 and Meis2 are expressed in the proximal region of the developing and regenerating limbs.[233-235] When overexpressed in the regenerating limb, Meis-positive cells fail to contribute to distal limb tissues and instead show a preference for proximal structures.[233] This result is similar to the proximalizing activity of Prod1.[155] Both Prod1 and Meis genes are activated by retinoic acid,[154,236] indicating that these genes convey the proximalization activity of retinoic acid signaling. Future work needs to address whether Prod1 and Meis are acting in the same signal transduction mechanism or as separate pathways to instruct cells to assume a proximal fate.

Developmental studies have shown that Hox genes are important for patterning along several body axes,[237] because mutations of Hox genes result in the transformation of segmented structures along the anterior-posterior or proximal-distal axis.[238] Expression data show that Hox genes are re-expressed in regenerating appendages,[239,240] and that Hox genes are required for their regeneration.[241] Furthermore, Meis is a transcriptional cofactor of Hox genes,[242] suggesting that Meis regulates patterning through its interaction with Hox genes. What needs to be clarified is whether patterning information is only involved in the patterning process during regeneration, or whether patterning genes like Meis and Hox activity promote the regeneration process.

Intriguingly, expression analyses of human fibroblasts from both embryonic and adult tissues indicate that fibroblasts in the body have position-dependent Hox gene transcription profiles.[243] Fibroblasts removed from different regions of the body and subsequently cultured for several generations in standard culture media maintain specific Hox gene expression profiles that correlated to specific regions of the body.[244] These results indicate that patterning information is already in place before injury, which brings up two intriguing questions: (1) if human cells maintain regional identity, then can this information be used for the regeneration of organs and appendages, or (2) conversely, is the maintenance of such regional specific information in human fibroblasts (the inability to reprogram this positional information) an impediment to regeneration? It still needs to be determined how flexible the region-specific expression profiles are and whether maintaining specific Hox gene profiles prevents the regeneration process.

MicroRNAs

Within the last few years, microRNAs have been shown to have important roles in the formation and patterning of tissues and recent experiments in regenerating tissues have also shown the involvement of these transcripts in regenerating tissues of fish and salamanders.[245-247] The microRNA-133 is expressed in the uninjured adult fin but is down regulated in the regenerating fin.[248] Loss of miRNA-133 restores regeneration capacity in zebrafish fins even when the regenerates are deprived of the required Fgf signaling.[248] miRNA-133 was shown to be involved in suppressing cell proliferation by inhibiting the cell cycle regulator *mps1*,[248] a gene required for cell proliferation in the fish fin blastema.[249] miRNA-133 has also been shown to antagonize muscle differentiation.[250] Its role in antagonizing cell proliferation in the regenerating fin while it antagonizes inhibitors of cell proliferation in the developing muscle indicates that even one miRNA has diverse activities that may depend on tissue specificity or other regulatory components. Currently, it is unclear how cell proliferation and differentiation status of these blastema cells are mechanistically linked. Therefore, it will be worthwhile to determine if miRNA-133 is part of a molecular mechanism that coordinates cell proliferation with cell differentiation.

The microRNA miR-196 is also expressed in the blastema as well as the dorsal and lateral cells of the adjacent spinal cord just proximal to the blastema of the regenerating salamander tail.[251] Inhibition of miR196 causes abnormally shortened tails with spinal cord defects, while overexpression of a mimic miR-196 increased the length of the regenerate.[251] Further exploration of the detailed mechanisms through which miR-196 and other miRNAs regulate tissue regeneration should show additional layers of regulation that determine how differentiated cells become blastema cells and how blastema cells regenerate new tissues.

CONCLUSION

It is an exciting time in the field of regeneration biology, because the molecular biology underlying a centuries worth of observations of the cell biology is being defined. Although still at an early stage, the current development of new molecular tools in fish, amphibians and other animal models for regeneration is providing the technical advances necessary to investigate how progenitor cells contribute to the reconstruction of damaged or missing compound structures. Such studies are defining how terminally differentiated cells obtain the appropriate plasticity and what mechanisms control the proliferation and differentiation status of these cells to recreate well-organized structures to their near original form. This information together with the advancements in mammalian stem cell biology will allow more detailed comparisons between mammalian and regeneration animal models. These comparisons may not only open up new avenues for reprogramming adult cells into stem cells but also provide crucial information for instructing reprogrammed cells to regenerate functional organs and appendages.

REFERENCES

1. Lewitzky M, Yamanaka S. Reprogramming somatic cells torwards pluripotency by defined factors. Curr Opin Biotechnol 2007; 18:467-473.
2. Graf T, Enver T. Forcing cells to change lineages. Nature 2009; 462:587-594.
3. Yamanaka S. A fresh look at ips cells. Cell 2009; 137:13-17.
4. Okada TS. Transdifferentiation: flexibility in cell differentiation. Oxford: Oxford University Press; 1991.
5. Graw J. Genetic aspects of embryonic eye development in vertebrates. Dev Genet 1996; 18:181-197.
6. Reyer RW, Wooliftt RA, Withersty LT. Stimulation of lens regeneration from the newt dorsal iris when implanted into the blastema of the regenerating limb. Dev Biol 1973; 32:258-281.
7. Hyaga M, Kodama R, Eguchi G. Basic fibroblast growth factor as one of the essential factors regulating lens transdifferentiation of pigment epithelial cells. Int J Dev Biol 1993; 37:319-326.
8. Hayashi T, Mizuno N, Ueda Y et al. Fgf2 triggers iris-derived lens regeneration in newt eye. Mech Dev 2004; 121:519-526.
9. Hayashi T, Mizuno N, Owaribe K et al. Regulated lens regeneration from isolated pigment epithelial cells of newt iris in culture in response to fgf2/4. Differentiation 2002; 70:101-108.
10. Grogg MW, Call MK, Okamoto M et al. Bmp-inhibition-driven regulation of six-3 underlies induction of newt lens regeneration. Nature 2005; 438:858-862.
11. Halder G, Callaerts P, Gehring WJ. Induction of ectopic eyes by targeted expression of the eyeless gene in drosophila. Science 1995; 267:1788-1792.
12. Altmann CR, Chow RL, Lang RA et al. Lens induction by pax-6 in Xenopus laevis. Dev Biol 1997; 185:119-123.
13. Rio-Tsonis KD, Washabaugh CH, Tsonis PA. Expression of pax-6 during urodele eye development and lens regeneration. Proc Natl Acad Sci USA 1995; 92:5092-5096.
14. Madhavan M, Haynes TL, Frisch NC et al. The role of pax-6 in lens regeneration. Proc Natl Acad Sci USA 2006; 103:14848-14853.
15. Agata K, Kobayashi H, Itoh Y et al. Genetic charaterization of the multipotent dedifferentiated state of pigmented epithelial cells in vitro. Development 1993; 118:1993.
16. Maki N, Suetsugu-Maki R, Tarui H et al. Expression of stem cell pluripotency factors during regeneration in newts. Dev Dyn 2009; 238:1613-1616.
17. Maki N, Takechi K, Sano S et al. Rapid accumulation of nucleostemin in nucleolus during newt regeneration. Dev Dyn 2007; 236:941-950.
18. Freeman G. Lens regeneration from cornea in Xenopus laevis. J Exp Zool 963; 154:39-66.

19. Filoni S, Bosco L, Cioni C. The role of the neural retina in lens regeneration from cornea in larval Xenopus leavis. Acta Embryol Morphol Exp 1982; 3:15-28.
20. Filoni S, Bernardini S, Cannata SM. Experimental analysis of lens-forming capacity in Xenopus borealis larvae. J Exp Zool2006; 305A:538-550.
21. Bosco L, Filoni S. Relationships between presence of the eye cup and maintenance of lens-forming capacity in larval xenopus laevis. Dev Growth Differ 1992; 34:619-625.
22. Grainger RM, Henry JJ, Henderson R. Reinvestigation of the role of the optic vesicle in embryonic lens induction. Development 1988; 102:517-526.
23. Gargioli C, Giambra V, Santoni S et al. The lens-regenerating competence in the outer cornea and epidermis of larval Xenopus laevis is related to pax6 expression. J Anat 2008; 212:612-620.
24. Stone LS. The role of retinal pigment cells in regenerating neural retinae of adult salamander eyes. J Exp Zoo 1950; 113:9-31.
25. Bonnet C. Sur les Reproductions des Salamandres. Oeuvres d'Histoire naturelle et de Philosophie. Neutchatel 1781; 5:356.
26. Sakami S, Histomi O, Sakakibara S et al. Downregulation of otx2 in the dedifferentiated RPE cells of regenerating newt retina. Developmental Brain Research 2005; 155:49-59.
27. De Leeuw AM, Gaur VP, Saari JC et al. Immunolocalization of cellular retinol-, retinaldhyde- and retinoic acid-binding proteins in rat retina during pre- and postnatal development. J Neurocytol 1990; 19:253-264.
28. Park CM, Hollenberg MJ. Basic fibroblast growth factor induces retinal regeneration in vivo. Dev Biol 1989; 134:201-205.
29. Zhao S, Thornquist SC, Barnstable CJ. In vitro transdifferentiation of embryonic rat retinal pigment epithelium to neural retina. Brain Res 1995; 677:300-310.
30. Stroeva OG, Mitashov VI. Retinal pigment epithelium: proliferation and differentiation during development and regeneration. Int Rev Cytol 1983; 83:211-293.
31. Park CM, Hollenberg MJ. Induction of retinal regeneration in vivo by growth factors. Dev Biol 1991; 148:322-333.
32. Reh TA, Levine EM. Multipotent stem cells and progenitors in the vertebrate retina.J Neurobiol1998; 80:206-220.
33. Sakami S, Etter P, Reh TA. Activin signaling limits the competence for retinal regeneration from the pigment epithelium. Mech Dev 2008; 125:106-116.
34. Fischer AJ, Reh TA. Müller glia are a potential source of neural regeneration in the postnatal chicken retina. Nat Neurosci 2001; 4:247-252.
35. Thummel R, Kassen SC, Enright JM et al. Characterization of Müeller glia and neuronal progenitors during adult zebrafish retinal regeneration. Exp Eye Res 2008; 87:433-444.
36. Bernardos RL, Barthel LK, Meyers JR et al. Late-stage neuronal progenitors in the retina are radial Müeller glia that function as retinal stem cells. J Neurosci 2007; 27:7028-7040.
37. Raymond PA, Barthel LK, Bernardos RL et al. Molecular characterization of retinal stem cells and their niches in adult zebrafish. BMC Dev Biol 2006; 6:36-53.
38. Braisted JE, Essman TF, Raymond PA. Selective regeneration of photoreceptors in goldfish retina. Development 1994; 120:2409-2419.
39. Braisted JE, Raymond PA. Regeneration of dopaminergic neurons in goldfish retina. Development 1992; 114:913-919.
40. Raymond PA. The unique origin of rod photoreceptors in the teleost retina. Trends Neurosci 1985; 8:12-17.
41. Raymond PA, Reifler MJ, Rivlin PK. Regeneration of goldfish retina: rod precursors are a likely source of regenerated cells.J Neurobiol1988; 19:431-463.
42. Kirsche W, Kirsche K. Experimentelle untersuchungen zur frage der regeneration und funktion des tectums opticum von carassius carassuis. Z Mikrosk-Anat Forsch 1961; 67:140-182.
43. Parish CL, Beljajeva A, Arenas E et al. Midbrain dopaminergic neurogenesis and behavioural recovery in a salamander lesion-induced regeneration model. Development 2007; 134:2881-2887.
44. Endo T, Yoshino J, Kado K et al. Brain regeneration in anuran amphibians. Dev Growth Differ 2007; 49:121-129.
45. Okamoto M, Ohsawa H, Hayashi T et al. Regeneration of retinotectal projections after optic tectum removal in adult newts. Mol Vis 2007; 13:2112-2118.
46. Kirsche K, Kirsche W. Regenerative vorgänge im telencephalon von Ambystoma mexicanum. Journal für Hirnforschung 1963/1964; 6:421-436.
47. Richter W, Kranz D. Autoradiographische Untersuchungen der postnatalen Proliferationsaktivität in den Matrixzonen des Telencephalons und des Diencephalons beim Axolotl (Ambystoma mexicanum), unter Berücksichtigung der Proliferation im olfactorischen Organ. Zeitschrift Mikrosk-Anat Forschung 1981; 95:883-904.

48. Kirsche W. The significance of matrix zones for brain regeneration and brain transplantation with special consideration of lower vertebrates. In: Wallace RB, Das GD, eds. Neural Tissue Transplantation Research. New York: Springer-Verlag, 1983:65-104.
49. Stevenson JA, Yoon MG. Regeneration of optic nerve fibers enhances cell proliferation in the goldfish optic tectum. Brain Res 1978; 153:345-351.
50. Stevenson JA, Yoon MG. Kenetics of cell proliferation in the halved tectum of adult goldfish. Brain Res 1980; 184:11-22.
51. Kaslin J, Ganz J, Geffarth M et al. Stem cells in the adult zebrafish cerebellum: initiation and maintenance of a novel stem cell niche. J Neurosci 2009; 29:6142-6153.
52. Grandel H, Kaslin J, Ganz J et al. Neural stem cells and neurogenesis in the adult zebrafish brain: origin, proliferation dynamics, migration and cell fate. Dev Biol 2006; 295:263-277.
53. Zupanc GKH, Ott R. Cell proliferation after lesions in the cerebellum of adult teleost fish: time course, origin and type of new cells produced. Exp Neurol 1999; 160:78-87.
54. Zupanc GKH, Hinsch K, Gage FH. Proliferation, migration, differentiation and long-term survival of new cells in the adult zebrafish brain. J Comp Neurol 2005; 488:290-319.
55. Ekström P, Johnsson C-M, Ohlin L-M. Ventricular proliferation zones in the brain of an adult teleost fish and their relation to neuromeres and migration (secondary matrix) zones. J Comp Neurol 2001; 436:92-110.
56. Becker CG, Becker T. Gradients of ephrin-a2 and ephrin-a5b mrna during retinotopic regeneration of the optic projection is adult zebrafish. J Comp Neurol 2000; 427:469-483.
57. Springer AD. Normal and abnormal retinal projections following the crush of one optic nerve in goldfish (Carrassius auratus). J Comp Neurol 1981; 199:87-95.
58. Hatten ME. Central nervous system neuronal migration. Annu Rev Neurosci 1999; 22:511-539.
59. Campbell K, Goetz M. Radial glia: multi-purpose cells for vertebrate brain development. Trends Neurosci 2002; 25:235-238.
60. Noctor SC, Flint AC, Weissman TA et al. Neurons derived from radial glial cells establish radial units in neocortex. Nature 2001; 409:714-720.
61. Holder N, Clarke JDW, Kamalati T, et al. Heterogeneity in spinal radial glia demonstrated by intermediate filament expression and hrp labelling. J Neurocytol 1990; 19:915-928.
62. O'Hara CM, Egar MW, Chernoff EAG. Reorganization of the ependyma during axolotl spinal cord regeneration: changes in intermediate filament and fibronectin expression. Dev Dyn 1992; 193:103-115.
63. Mchedlishvili L, Epperlein HH, Telzerow A et al. A clonal analysis of neural progenitors during axolotl spinal cord regeneration reveals evidence for both spatially restricted and multipotent progenitors. Development 2007; 134:2083-2093.
64. Butler EG, Ward MB. Reconstitution of the spinal cord after ablation in adult triturus. Dev Biol 1967; 15:454-486.
65. Egar MW, Singer M. The role of ependyma in spinal cord after ablation in adult triturus. Exp Neurol 1972; 37:422-430.
66. Nordlander RH, Singer M. The role of ependyma in regeneration of the spinal cord in the urodele amphibian tail. J Comp Neurol 1978; 180:349-374.
67. Echeverri K, Tanaka EM. Ectoderm to mesoderm lineage switching during axolotl tail regeneration. Science 2002; 298:1993-1996.
68. Schnapp E, Kragl M, Rubin L et al. Hedgehog signaling controls dorsoventral patterning, blastema cell proliferation and cartilage induction during axolotl tail regeneration. Development 2005; 132:3243-3253.
69. Beck CW, Christen B, Slack JMW. Molecular pathways needed for regeneration of spinal cord and muscle in a vertebrate. Dev Cell 2003; 5:429-439.
70. Sugiura T, Tazaki A, Ueno N et al. Xenopus wnt-5a induces an ectopic larval tail at injured site, suggesting a crucial role for noncanonical wnt signal in tail regeneration. Mech Dev 2009; 126:56-67.
71. Becker T, Wullimann MF, Becker CG et al. Axonal regrowth after spinal cord transection in adult zebrafish. J Comp Neurol 1997; 337:577-595.
72. Bernstein JJ. Relation of spinal cord regeneration to age in adult goldfish. Exp Neurol 1964; 9:161-174.
73. Reimer MM, Soerensen I, Kuscha V et al. Motor neuron regeneration in adult zebrafish. J Neurosci 2008; 28:8510-8516.
74. Becker T, Lieberoth BC, Becker CB et al. Differences in the regenerative response of neuronal cell populations and indications for plasticity in intraspinal neurons after spinal cord transection in adult zebrafish. Mol Cell Neurosci 2005; 30:265-278.
75. Anderson MJ, Waxman SG, Laufer M. Fine structure of regenerated ependyma and spinal cord in Sternarchus albifrons. Anat Rec 1983; 205:73-83.
76. Anderson MJ, Rossettl DL, Lorenz LA. Neuronal differentiation in vitro from precursor cells of regenerating spinal cord of the adult teleost Apteronotus albifrons. Cell Tissue Res 1994; 278:243-248.
77. Anderson MJ, Waxman SG, Fong HL. Explant cultures of teleost spinal cord: source of neurite outgrowth. Dev Biol 1987; 119:601-604.

78. Reimer MM, Kuscha V, Wyatt C et al. Sonic hedgehog is a polarized signal for motor neuron regeneration in adult zebrafish. J Neurosci 2009; 29(48):15073-15082.
79. Bernstein JJ, Gelderd JB. Regeneration of the long spinal tracts in the goldfish. Brain Res 1970; 20:33-38.
80. Bhatt DH, Otto SJ, Depoister B et al. Cyclic amp-induced repair of zebrafish spinal circuits. Science 2004; 305:254-258.
81. Oberpriller JO, Oberpriller JC. Response of the adult newt ventricle to injury. J Exp Zoo 1974; 187:249-260.
82. Poss KD, Wilson LG, Keating MT. Heart regeneration in zebrafish. Science 2002; 298:2188-2190.
83. Bader D, Oberpriller JO. Repair and reorganization of minced cardiac muscle in the adult newt (Notophthalmus viridescens. J Morphol 1978; 155:349-358.
84. Betencourt-Dias M, Mittnacht S, Brockes JP. Heterogeneous proliferative potential in regenerative adult newt cardiomyocytes. J Cell Biol 2003; 116:4001-4009.
85. Flink IL. Cell cycle reentry of ventricular and atrial cardiomyocytes and cells within the epicardium following amputation of the ventricular apex in the axolotl, Amblystoma mexicanum. Anat Embryol 2002; 205:235-244.
86. Laube F, Heister M, Scholz C et al. Re-programming of newt cardiomyocytes is induced by tissue regeneration. J of Cell Science 2006; 119:4719-4729.
87. Kikuchi K, Holdaway JE, Werdich AA et al. Primary contribution to zebrafish heart regeneration by gata4+ cardiomyocytes. Nature 2010; 464:601-605.
87a. Joplin C, Sleep E, Raya M et al. Zebrafish heart regeneration occurs by cardiomyocyte dedifferentiation and proliferation. Nature 2010; 464:606-609.
88. Lepilina A, Coon AN, Kikuchi K et al. A dynamic epicardial injury response supports progenitor cell activity during zebrafish heart regeneration. Cell 2006; 127:607-619.
89. Raya A, Koth CM, Buescher D et al. Activation of notch signaling pathway precedes heart regeneration in zebrafish. Proc Natl Acad Sci USA 2003; 100:11889-11895.
90. Lien CL, Schebesta M, Makino S et al. Gene expression analysis of zebrafish heart regeneration. PLoS Biology 2006; 4:1386-1396.
91. Lavine KJ, Yu K, White AC et al. Endocardial and epicardial derived fgf signals regulate myocardial proliferation and differentiation in vivo. Dev Cell 2005; 8:85-95.
92. Brockes JP, Kumar A. Appendage regeneration in adult vertebrates and implications for regenerative medicine. Science 2005; 310:1919-1923.
93. Poss KD, Keating MT, Nechiporuk A. Tales of regeneration in zebrafish. Dev Dyn 2003; 226:202-210.
94. Straube WL, Tanaka EM. Reversibility of the differentiated state: regeneration in amphibians. Artif Organs 2006; 30:743-755.
95. Muneoka K, Fox WF, Bryant S. Cellular contribution from dermis and cartilage to the regenerating llimb blastema in axolotls. Dev Biol 1986; 116:256-260.
96. Gardiner DM, Muneoka K, Bryant SV. The migration of dermal cells during blastema formation in axolotls. Dev Biol 1986; 118:488-493.
97. Bryant SV, Endo T, Gardiner DM. Vertebrate limb regeneration and the origin of the limb stem cells. Int J Dev Biol 2002; 46:887-896.
98. Hay ED. Electron microscopic observations of muscle dedifferentiation in regenerating Amblystoma limbs. Dev Biol 1959; 1:555-585.
99. Hay ED. The fine structure of differentiating muscle in the salamander tail. Zeitschrift für Zellforschung 1963; 59:6-34.
100. Mauro A. Satellite cell of skeletal muscle fibers. J Biophys Biochem Cytol 1961; 9:493-495.
101. Collins CA, Olsen I, Zammit PS et al. Stem cell function, self-renewal and behavioral heterogeneity of cells from the adult muscle satellite cell niche. Cell 2005; 122:289-301.
102. Montarras D, Morgan J, Collins C et al. Direct isolation of satellite cells for skeletal muscle regeneration. Science 2005; 309:2064-2067.
103. Morrison JI, Loeoef S, He P et al. Salamander limb regeneration involves the activation of a multipotent skeletal muscle satellite cell population. J Cell Biol 2006; 172:433-440.
104. Thornton CS. The histogenesis of muscle in the regenerating fore limb of larval Amblyostoma punctatum. J Morphol 1938; 62:17-47.
105. Kintner CR, Brockes JP. Monoclonal antibodies identify blastemal cells derived from dedifferentiating muscle in newt limb regeneration. Nature 1984; 308:67-69.
106. Lo DC, Allen F, Brockes JP. Reversal of muscle differentiation during urodele limb regeneration. Proc Natl Acad Sci USA 1993; 90:7230-7234.
107. Kumar A, Velloso CP, Imokawa Y et al. Plasticity of retrovirus-labelled myotubes in the newt limb regeneration blastema. Dev Biol 2000; 218:125-136.
108. Echeverri K, Clarke JDW, Tanaka EM. In vivo imaging indicates muscle fiber dedifferentiation is a major contributor to the regenerating tail blastema. Dev Biol 2001; 236:151-164.

109. Tanaka EM, Drechsel DN, Brockes JP. Thrombin regulates S-phase re-entry by cultured newt myotubes. Current Biology 1999; 9:792-799.
110. Rollman-Dinsmore C, Bryant SV. The distrubution of marked dermal cells from small localized implants in limb regenerates. Dev Biol 1984; 106:275-281.
111. Muneoka K, Fox WF, Bryant SV. Cellular contibution from dermis and cartilage to the regenerating limb blastema in axololts. Dev Biol 1986; 116:256-260.
112. Chalkey DT. A quantitative histological analysis of forelimb regeneration in Triturus viridescens. J Morphol 1954; 94:21-70.
113. Kragl M, Knapp D, Nacu E et al. Cells keep a memory fo their tissue origin during axolotl limb regeneration. Nature 2009; 460:60-65.
114. Namenwirth M. The inheritance of cell differentiation during limb regeneration in the axolotl. Dev Biol 1974; 41:42-56.
115. Steen TP. Stability of chondrocyte differentiation and contribution of muscle to cartilage during limb regeneration in the axolotl (Siredon mexicanum). J Exp Zoo 1968; 167:49-77.
116. Holtzer H, Avery G, Holtzer S. Some properties of the regenerating limb blastema cells of salamanders. Biological Bulletin 1954; 107:313.
117. Pietsch P. Differentiation in regeneration: i. The development of muscle and cartilage following deplantation of regenerating limb blastemata of amblystoma larvae. Dev Biol 1961; 3:255-264.
118. Pietsch P. The effect of heterotropic musculature on myogenesis during limb regeneration in Amblystoma larvae. The Anat Rec 1961; 141:295-303.
119. Steen TP. Stability of chondrocyte differentiation and contribution of muscle to cartilage during limb regeneration in the axolotl (Siredon mexicanum). J Exp Zoo 1968; 167:49-78.
120. Morrison JI, Borg P, Simon A. Plasticity and recovery of skeletal muscle satellite cells during limb regeneration. FASEB Journal 2009; 24:Nov 6. [Epub ahead of print].
121. Odelberg SJ, Kollhoff A, Keating MT. Dedifferentiation of mammalian myotubes induced by msx1. Cell 2000; 103:1099-1109.
122. Mescher AL. Effects of adult newt limb regeneration of partial and complete skin flaps over the amputation surface. J Exp Zoo 1976; 195:117-128.
123. Thornton CS. The effect of apical cap removal on limb regeneration in amblystoma larvae. J Exp Zoo 1957; 134:357-381.
124. Tassava RA, Garling DJ. Regenerative responses in larval axolotl limbs with skin grafts over the amputation surface. J Exp Zoo 1979; 208:97-110.
125. Polezhaev LV, Faworina WN. Über die Rolle des Epithels in den Anfänglichen Entwicklungstadien einer Regenerationsanlage der Extremität beim Axolotl. Wilhelm Roux' Archiv Entwicklungsmech Org 1935; 133:701-727.
126. Tank PW. Skin of non-limb origin blocks regeneration of the newt forlimb. Prog Clin Biol Res 1983; 110:565-575.
127. Riddiford LM. Autoradiographic studies of tritiated thymidine infused into the blastema of the early regenerate in the adult newt, triturus. J Exp Zoo 1960; 144:25-31.
128. Hay ED, Fischman DA. Origin of the blastema in regenerating limbs of the newt triturus viridescens. Dev Biol 1961; 3:26-59.
129. Saunders JW, Gasseling MT, Errick JE. Inductive activity and enduring cellular constitution of a supernumerary apical ectodermal ridge grafted to the limb bud of the chick embryo. Dev Biol 1976; 50:16-25.
130. Saunders JW. The proximo-distal sequence of origin of the parts of the chick wing and the role of the ectoderm. J Exp Zoo 1948; 108:363-403.
131. Carlson BM. Morphogenetic interactions between rotated skin duffs and underlying stump tissues in regenerating axolotl forelimbs. Dev Biol 1974; 39:263-285.
132. Bryant SV, Iten LE. Supernumerary limbs in amphibians: experimental production in Notophthalmus viridescens and a new interpretation of their formation. Dev Biol 1976; 50:212-234.
133. Thornton CS. Influence of an eccetric epidermal cap on limb regeneration in Ambystoma larvae. Dev Biol 1960; 2:551-569.
134. Thornton CS, Thornton MT. The regeneration of accessory limb parts following epidermal cap transplantation in urodeles. Experimentia 1965; 21:146-148.
135. Lee Y, Hami D, Val SD et al. Maintenance of blastemal proliferation by functionally diverse epidermis in regenerating zebrafish fins. Dev Biol 2009; 331:270-280.
136. Viviano CM, Horton CE, Maden M et al. Synthesis and release of 9-cis retinoic acid by the urodele wound epidermis. Development 1995; 121:3753-3762.
137. Stoick-Cooper CL, Weidinger G, Riehle KJ et al. Distinct wnt signaling pathways have opposing roles in appendage regeneration. Development 2007; 134:479-489.
138. Laforest L, Brown CW, Poleo G et al. Involvement of the sonic hedgehog, patched1 and bmp2 genes in patterning of the zebrafish dermal fin rays. Development 1998; 125:4175-4184.

139. Maden M. Vitamin a and pattern formation in the regenerating limb. Nature 1982; 295:672-675.
140. White JA, Boffa MB, Jones B et al. A zebrafish retinoic acid receptor expressed in the regenerating caudal fin. Development 1994; 120:1861-1872.
141. Saxena S, Niazi IA. Effect of vitamin A excess on hind limb regeneration in tadpoles of the toad, bufo andersonii (boulenger). Indian J Exp Biol 1977; 15:435-439.
142. Ghosh S, Roy S, Seguin C et al. Analysis of the expression and function of wnt-5a and wnt-5b in developing and regenerating axolotl (Ambystoma mexicanum) limbs. Dev Growth Differ 2000; 50:289-297.
143. Christensen RN, Weinstein M, Tassava RA. Expression of fibroblast growth factors 4, 8 and 10 in limbs, flanks and blastemas of ambystoma. Dev Dyn 2002; 223:193-203.
144. Geraudie J, Singer M. Necessity of an adequate nerve supply for regeneration of the amputated pectoral fin in the teleost fundulus. J Exp Zoo 1985; 234:367-374.
145. Singer M. The influence of the nerve in regeneration of the amphibian extremity. Q Rev Biol 1952; 27:169-200.
146. Singer M. The nervous system and regeneration of the forelimb of adult triturus. J Exp Zoo 1942; 90:377-399.
147. Goss RJ, Stagg MW. The regeneration of fins and fin rays in Fundulus heteroclitus. J Exp Zoo 1957; 136:487-507.
148. Endo T, Bryant SV, Gardiner DM. A stepwise model system for limb regeneration. Dev Biol 2004; 270:135-145.
149. Satoh A, Gardiner DM, Bryant SV et al. Nerve-induced ectopic limb blastemas in the axolotl are equivalent to amputation-induced blastemas. Dev Biol 2007; 312:231-244.
150. Satoh A, Graham GMC, Bryant SV et al. Neurotrophic regulation of epidermal dedifferentiation during wound healing and limb regeneration in the axolotl (Ambystoma mexicanum). Dev Biol 2008; 319:321-335.
151. Singer M, Inoue S. The nerve and the epidermal apical cap in regeneration of the forelimb of adult triturus. J Exp Zoo 964; 115:105-116.
152. Kumar A, Gates PB, Brockes JP. Positional indentity of adult stem cells in salamander limb regeneration. Comptes Rendus Biologies 2007; 330:485-490.
153. Kumar A, Godwin JW, Gates PB et al. Molecular basis for the nerve dependence of limb regeneration in an adult vertebrate. Science 2007; 318:772-777.
154. Silva SMd, Gates PB, Brockes JP. The newt orthology of CD59 is implicated in proximodistal identity during amphibian limb regeneration. Dev Cell 2002; 3:547-555.
155. Echeverri K, Tanaka EM. Proximodistal patterning during limb regeneration. Dev Biol 2005; 279:391-401.
156. Brockes JP, Kintner CR. Glial growth factor and nerve-dependent proliferation in the regeneration blastema of urodele amphibians. Cell 1986; 45:301-306.
157. Albert P, Boilly B, Courty J et al. Stimulation in cell culture of mesenchymal cells of newt limb blastemas by edgfi and ii (basic or acidic fgf). Cell Differ 1987; 21:63-68.
158. Wang L, Marchionni MA, Tassava RA. Cloning and neuronal expression of a type iii newt neuregulin and rescue of denerated, nerve-dependent newt limb blastemas. J Neurobiol2000; 43:150-158.
159. Mullen LM, Bryant SV, Torok MA et al. Nerve dependency of regeneration: the role of distal-less and fgf signaling in amphibian limb regeneration. Development 1996; 122:3487-3497.
160. Kuehn LC, Schulman HM, Ponka P. Iron-transferrin requirements and transferrin receptor expression in proliferating cells. In: Ponka P, Schulman HM, Woodworth RC, eds. Iron Transport and Storage. Boca Raton: CRC Press, 1990:149-191.
161. Mescher AL, Connell E, Hsu C et al. Transferrin is necessary and sufficient for the neural effect on growth in amphibian limb regeneration blastemas. Dev Growth Differ 1997; 39:677-684.
162. Whitehead GG, Makino S, Lien CL et al. Fgf20 is essential for initiating zebrafish fin regeneration. Science 2005; 310:1957-1960.
163. Poss KD, Shen J, Nechiporuk A et al. Roles for fgf signaling during zebrafish fin regeneration. Dev Biol 2000; 222:347-358.
164. Kubo F, Takeichi M, Nakagawa S. Wnt2b controls retinal cell differentiation at the ciliary marginal zone. Development 2003; 130:587-598.
165. Kubo F, Takeichi M, Nakagawa S. Wnt2b inhibits differentiation of retinal progenitor cells in the absence of notch activity by downregulating the expression of proneural genes. Development 2005; 132:2759-2770.
166. Kawakami Y, Esteban CR, Raya M et al. Wnt/β-catenin signaling regulates vertebrate limb regeneration. Genes and Development 2006; 20:3232-3237.
167. Yokoyama H, Ogino H, Stoick-Cooper CL et al. Wnt/β-catenin signaling has an essential role in the initiation of limb regeneration. Dev Biol 2007; 306:170-178.
168. Lin G, Slack JMW. Requirement for wnt and fgf signaling in Xenopus tadpole tail regeneration. Dev Biol 2008; 316:323-335.

169. Levesque M, Gatien S, Finnson K et al. Transforming growth factor: β signaling is essential for limb regeneration in axolotls. PLoS One 2007; 11:2-14.
170. Ho DM, Whitman M. Tgf-B signaling is required for multiple processes during Xenopus tail regeneration. Dev Biol 2008; 315:203-216.
171. Inman GJ, Nicolas FJ, Callahan JF et al. Sb-431542 is a potent and specific inhibitor of transforming growth factor-B superfamily type I activin receptor-like kinase (alk) receptors alk4, alk5 and alk7. Molecular Pharmacology 2002; 62:65-74.
172. Jazwinska A, Badakov R, Keating MT. Activin-βa signaling is required for zebrafish fin regeneration. Current Biology 2007; 17:1390-1395.
173. Smith A, Avaron F, Guay D et al. Inhibition of bmp signaling during zebrafish fin regeneration disrupts fin growth and scleroblast differentiation and function. Dev Biol 2006; 299:438-454.
174. Roy S, Gardiner DM. Cyclopamine induces digit loss in regenerating axolotl limbs. J Exp Zool2002; 293:186-190.
175. Quint E, Smith A, Avaron F et al. Bone patterning is altered in the regenerating zebrafish caudal fin after ectopic expression of sonic hedgehog and bmp2b or exposure to cyclopamine. Proc Natl Acad Sci USA 2002; 99:8713-8718.
176. Roy S, Gardiner DM, Bryant SB. Vaccinia as a tool for functional analysis in regenerating limbs: ectopic expression of shh. Dev Biol 2000; 218:199-205.
177. Endo T, Tamura K, Ide H. Analysis of gene expression during Xenopus forelimb regeneration. Dev Biol 2000; 220:296-306.
178. Yakushiji N, Suzuki M, Satoh A et al. Correlation between shh expression and DNA methylation status of the limb-specific shh enhancer region during limb regenertion in amphibians. Dev Biol 2007; 312:171-182.
179. Yakushiji N, Suzuki M, Satoh A et al. Effects of activation of hedgehog signaling on patterning, growth and differentiation in Xenopus froglet limb regeneration. Dev Dyn 2009; 238:1887-1896.
180. Dent JN. Limb regeneration is larvae and metamorphosing individuals of the south african clawed toad. J Morphol 1962; 110:61-77.
181. Endo T, Yokoyama H, Tamura K et al. Shh expression in developing and regenerating limb buds of xenopus laevis. Dev Dyn 1997; 209:227-232.
182. Moneoka K, Holler-Dinsmore G, Bryant SV. Intrinsic control of regenerative loss in Xenopus laevis limb. J Exp Zoo 986; 240:47-54.
183. Lee Y, Grill S, Sanchez A et al. Fgf signaling instructs position-dependent growth rate during zebrafish fin regeneration. Development 2005; 132:5173-5183.
184. Kizil C, Otto GW, Geisler R et al. Simplet controls cell proliferation and gene transcription during zebrafish caudal fin regeneration. Dev Biol 2009; 325:329-340.
185. Thermes V, Candal E, Alunni A et al. Medaka simplet (fam53b) belongs to a family of novel vertebrate genes controlling cell proliferation. Development 2006; 133:1881-1890.
186. Vogelstein B, Lane D, Levine AJ. Surfing the p53 network. Nature 2000; 408:307-310.
187. Villiard E, Brinkmann H, Moiseeva O et al. Urodele p53 tolerates amino acid changes found in p53 varients linked to human cancer. BMC Evolutionary Biology 2007; 7:180-194.
188. Ferretti P, Brockes JP. Culture of newt cells from different tissues and their expression of a regeneration-associated antigen. J Exp Zoo 1988; 247:77-91.
189. Tanaka EM, Gann AAF, Gates PB et al. Newt myotubes reenter the cell cycle by phosphorylation of the retinoblastoma protein. J Cell Biol 1997; 136:155-165.
190. Florini JR, Ewton DZ, Magri KA. Hormones, growth factors and myogenic differentiation. Annual Review Physiology 1991; 53:201-216.
191. Olson EN. Proto-oncogenes in the regulatory circuit for myogenesis. Seminars in Cell Biology 1992; 3:127-136.
192. Schneider JW, Gu W, Zhu L et al. Reversal of terminal differentiation mediates by p107 in Rb-/- muscle cells. Science 1994; 264(1467-1471).
193. Davidson DR, Crawley A, Hill RE et al. Position-dependent expression of two related homeobox genes in developing vertebrate limbs. Nature 1991; 352:429-431.
194. Summerbell D, Lewis JH, Wolpert L. Positional information in chick limb morphogenesis. Nature 1973; 244:492-496.
195. Robert B, Sassoon D, Jacq B et al. Hox-7, a mouse homeobox gene with a novel pattern of expression during embryogenesis. EMBO Journal 1989; 8:91-100.
196. Song K, Wang Y, Sassoon D. Expression of hox-7.1 in myoblasts inhibits terminal differentiation and induces cell transformation. Nature 1992; 360:477-481.
197. Woloshin P, Song K, Degnin C et al. Msx1 inhibits myod expression in fibroblast x 10t1/2 cell hybrids. Cell 1995; 82:611-620.
198. Akimenko MA, Johnson SL, Westerfield M et al. Differential induction of four msx homeobox genes during fin development and regeneration in zebrafish. Development 1995; 121:347-357.

199. Carlson MRJ, Bryant SV, Gardiner DM. Expression of msx-2 during development, regeneration and wound healing in axolotl limbs. J Exp Zoo 1998; 282:715-723.
200. Crews L, Gates PB, Brown R et al. Expression and activity of the newt msx-1 gene in relation to limb regeneration. Proc R Soc Lond B 1995; 259:161-171.
201. Kumar A, Velloso CP, Imokawa Y et al. The regenerative plasticity of isolated urodele myofibers and its dependence on msx1. PLoS Biology 2004; 2:1168-1176.
202. Hu G, Lee H, Price SM et al. Msx homeobox genes inhibit differentiation through upregulation of cyclin D1. Development 2001; 128:2373-2384.
203. Thummel R, Bai S, Michael P et al. Inhibition of zebrafish fin regeneration using in vivo electroporation of morpholinos against fgfr1 and msxb. Dev Dyn 2006; 235:335-346.
204. Barnes RM, Firulli AB. A twist of insight-the role of twist-family bHLH factors in development. Int J Dev Biol 2009; 53:909-924.
205. Nüsslein-Volhard C, Wieschaus E, Kluding H. Mutations affecting the pattern of the larval cuticle in drosophila melanogaster. Roux's Archives of Dev Biol 1984; 193:267-282.
206. Simpson P. Maternal-zygotic gene interactions during formation of the dorsoventral pattern in Drosophila embryos. Genetics 1983; 105:615-632.
207. Bate M, Rushton E, Currie DA. Cells with persistant twist expression are the embryonic precursors of adult muscles in Drosophila. Development 1991; 113:79-89.
208. Currie DA, Bate M. The development of adult abdominal muscles in drosophila: myoblasts express twist and are associated with nerves. Development 1991; 113:19-102.
209. Hjiantoniou E, Anayasa M, Nicolaou P et al. Twist induces reversal of myotubes formation. Differentiation 2007; 76:182-192.
210. Satoh A, Bryant SV, Gardiner DM. Regulation of dermal fibroblast dedifferentiation and redifferentiation during wound healining and limb regeneration in the axolotl. Dev Growth Differ 2008; 50:743-754.
211. Scaal M, Fürchtbauer EM, Brand-Saberi B. cDermo-1 expression indicates a role in avian skin development. Anat Embryol 2001; 203:1-7.
212. Li L, Cserjesi P, Olson EN. Dermo-1: a novel twist-related bHLH protein expressed in the developing dermis. Dev Biol 1995; 172:280-292.
213. Seale P, Sabourin LA, Girgis-Gabardo A et al. Pax7 is required for the specification of myogenic satellite cells. Cell 2000; 102:777-786.
214. Seale P, Polesskaya A, Rudnicki MA. Adult stem cell specification by wnt signaling in muscle regeneration. Cell Cycle 2003; 2:418-419.
215. Maekawa M, Takashima N, Arai Y et al. Pax6 is required for production and maintenance of progenitor cells in postnatal hippocampal neurogenesis. Genes Cells 2005; 10:1001-1014.
216. Lang D, LM M, Huang L et al. Pax3 functions at a nodal point in melanocyte stem cell differentiation. Nature 2005; 433:884-887.
217. Lagha M, Sato T, Bajard L et al. Regulation of skeletal muscle stem cell behavior by pax3 and pax7. Cold Spring Harbor Symposia of Quantitative Biology 2008; 73:307-315.
218. Seale P, Ishibashi J, Scime A et al. Pax7 is necessary and sufficient for the myogenic specification of cd45+:sca1+ stem cells from injured muscle. PLoS Biology 2004; 2:E130.
219. Asakura A, Seale P, Girgis-Gabardo A et al. Myogenic specification of side population cells in skeletal muscle. J Cell Biol 2002; 159:123-134.
220. Jackson KA, Mi T, Goodell MA. Hematopoeitic potential of stem cells isolated from murine skeletal muscle. Proc Natl Acad Sci USA 1999; 96:14482-14486.
221. Gussoni E, Soneoka Y, Strickland CD et al. Dystrophin expression in the mdx mouse restored by stem cell transplantation. Nature 1999; 401:390-394.
222. Chen Y, Lin G, Slack JMW. Control of muscle regeneration in the Xenopus tadpole tail by pax7. Development 2006; 133:2303-2313.
223. McKinnell IW, Ishibashi J, Grand FL et al. Pax7 activates myogenic genes by recruitment of a histone methyltransferase complex. Nat Cell Biol 2008; 10:77-84.
224. Slack JMW. Morphogenetic properties of the skin in axolotl limb regeneration. J Embryol Exp Morphol 1980; 58:265-288.
225. Slack JMW. Positional information in the forelimb of the axolotl: properties of the posterior skin. J Embryol Exp Morphol 1983; 73:233-247.
226. Rollman-Dinsmore C, Bryant SV. Pattern regulation between hind- and forelimbs after blastema exchanges and skin grafts in Notophthalmus viridescens. J Exp Zoo 1982; 223:51-56.
227. Tank PW. The ability of localized implants of whole or minced dermis to disrupt pattern formation in the regeneration forelimb of the axolotl. Am J Anat 1981; 162:315-326.
228. Dunis DA, Namenwirth M. The role of grafted skin in the regeneration of x-irradiated axolotl limbs. Dev Biol 1977; 56:97-109.

229. Holder N. Organization of connective tissue patterns by dermal fibroblasts in the regenerating axolotl limb. Development 1989; 105:585-593.
230. Mescher AL. The cellular basis of limb regeneration in urodeles. Int J Dev Biol 1996; 40:785-795.
231. Bryant SV, Gardiner DM. Limb development and regeneration. Am Zool 1987; 27:675-696.
232. Carlson BM. Multiple regeneration from axolotl limb stumps bearing cross-transplanted minced muscle regenerates. Dev Biol 1975; 45:203-208.
233. Mercader N, Tanaka EM, Torres M. Proximodistal identity during vertebrate limb regeneration is regulated by meis homeodomain proteins. Development 2005; 132:4131-4142.
234. Mercader N, Selleri L, Criado LM et al. Ectopic meis1 expression in the mouse limb bud alters P-Dpatterning in a pbx1-independent manner. Int J Dev Biol 2008; 53:1483-1494.
235. Capdevila J, Tsukui T, Esteban CR et al. Control of vertebrate limb outgrowth by the proximal factor meis2 and distal antagonism of bmp by gremlin. Molecular Cell 1999; 4:839-849.
236. Mercader N, Leonardo E, Piedra ME et al. Opposing RA and fgf signals control proximodistal vertebrate limb development through regulation of meis genes. Development 2000; 127:3961-3970.
237. Izpisua-Belmonte JC, Tickle C, Dolle P et al. Expression of the homeobox hox-4 genes and the specification of position in chick wing development. Nature 1991; 350:585-589.
238. Krumlauf R. Hox genes in vertebrate development. Cell 1994; 78:191-201.
239. Gardiner DM, Blumberg B, Komine Y et al. Regulation of hoxa expression in developing and regenerating axolotl limbs. Development 1995; 121:1731-1741.
240. Geraudie J, Birraux VB. Posterior hoxa genes expression during zebrafish bony ray development and regeneration suggests their involvement in scleroblast differentiation. Dev Genes Evol 2003; 213:182-186.
241. Thummel R, Ju M, Michael P et al. Both hoxc13 orthologs are functinally important for zebrafish tail fin regeneration. Dev Genes Evol 2007; 217:413-420.
242. Moens CB, Selleri L. Hox cofactors in vertebrate development. Dev Biol 2006; 291:193-206.
243. Rinn JL, Bondre C, Gladstone HB et al. Anatomic demarcation by positional variation in fibroblast gene expression programs. PLoS Genetics 2006; 2:10841096.
244. Chang HY, Chi JT, Dudoit S et al. Diversity, topographic differentiation and positional memory in human fibroblasts. Proc Natl Acad Sci USA 2002; 99:12877-12882.
245. Makarev E, Spence JR, Rio-Tsonis KD et al. Identification of microRNAs and other small RNAs from the adult newt eye. Mol Vis 2006; 12:1386-1391.
246. Tsonis PA, Call MK, Grogg MW et al. MicroRNAs and regeneration: let-7 members as potential regulators of dedifferentiation in lens and inner ear hair cell regeneration of the adult newt. Biochem Biophys Res Commun 2007; 362:940-945.
247. Thatcher EJ, Paydar I, Anderson KK et al. Regulation of zebrafish fin regeneration by microRNAs. Proc Natl Acad Sci USA 2008; 105:18384-18389.
248. Yin VP, Thomson JM, Thummel R et al. Fgf-dependent depletion of microRNA-133 promotes appendage regeneration in zebrafish. Genes Dev 2008; 22:728-733.
249. Poss KD, Nechiporuk A, Hillam AM et al. Msp1 defines a proximal blastemal proliferative compartment essential for zebrafish fin regeneration. Development 2002; 129:5141-5149.
250. Chen JF, Mandel EM, Thomson JM et al. The role of microRNA-1 and microRNA-133 in skeletal muscle proliferation and differentiation. Nat Genet 2006; 38:228-233.
251. Sehm T, Sachse C, Frenzel C et al. Mir-196 is an essential early-stage regulator of tail regeneration, upstream of key spinal cord patterning events. Dev Biol 2009; 334:468-480.

CHAPTER 14

REPROGRAMMING OF SOMATIC CELLS TO PLURIPOTENCY

Masato Nakagawa* and Shinya Yamanaka

Abstract: Reprogramming of somatic cells into pluripotent stem cells has been achieved by introducing four transcription factors, Sox2, Oct3/4, Klf4 and c-Myc, in 2006. These induced pluripotent stem (iPS) cells have raised hopes for a new era of regenerative medicine because they can avoid the ethical problems and innate immune rejection associated with embryonic stem cells. However, the underlying molecular mechanism of reprogramming still remains unclear. In this chapter, we look back at the history of reprogramming research ranging from amphibian to mammalian cells and discuss our recent understanding of the molecular mechanisms of reprogramming and the possibility of utilizing reprogrammed cells for regenerative medicine.

INTRODUCTION

Numerous attempts have been made by many researchers to resolve worthwhile and difficult questions regarding "reprogramming". During development, different stem cell populations differentiate into several lineages of somatic cells. The environment (niche) around stem cells, such as secreted factors, cell-cell/cell-matrix communications, mechanical pressures and other stimulations, coordinates the orderly events that underlie the process of differentiation. Although data about differentiation are increasing, studies of the reverse process, de-differentiation, are lagging behind. Somatic cell reprogramming (nuclear reprogramming) has been first accomplished in frog. This discovery later led not only to the cloned sheep Dolly but also to the generation of induced pluripotent stem (iPS) cells. Thus, differentiation, once considered to proceed in only one direction, can be rewound experimentally.

*Corresponding Author: Masato Nakagawa— Center for iPS Research and Application, Kyoto University, 53 Kawahara-cho, Shogoin, Sakyo-ku, Kyoto 606-8507, Japan.
Email: nakagawa@cira.kyoto-u.ac.jp

The Cell Biology of Stem Cells, edited by Eran Meshorer and Kathrin Plath.
©2010 Landes Bioscience and Springer Science+Business Media.

SOMATIC NUCLEAR REPROGRAMMING IN FROG

The nuclear reprogramming of somatic cells and the production of fertile adults from reprogrammed cells have succeeded in Xenopus laevis by somatic cell nuclear transfer (SCNT).[1] SCNT has been the most conventional method for reprogramming. Both differentiated and pluripotent cells of the same organism contain the same DNA sequence, however the readout of the DNA sequence varies greatly among different cell types, leading to different gene expression profiles, for example. Expression is regulated by several chromatin modifications, such as DNA methylation and histone modifications. During SCNT, the nuclei are removed from somatic cells and inserted into enucleated oocytes. After the transfer, the nuclei are reprogrammed through de-modification of DNA and chromatin by the new environment. In the pre-implantation embryos containing the reprogrammed somatic nuclei, the developmental stages proceed as in normal development, albeit at lower efficiency. In a key experiments, the nuclei of fully differentiated intestinal epithelial cells were isolated and transplanted into oocytes, which resulted in reprogramming of the injected nuclei.[1] The authors managed to obtain fertile male and female frogs. This and other studies[2-4] revealed that the differentiation process of somatic cells is reversible and that oocytes have the ability of reprogramming somatic nuclei. In other words, the reprogramming factors exist in the eggs.

BIRTH OF A CLONED ANIMAL, DOLLY

About thirty years after the reprogramming of somatic cell nuclei in frogs, mammalian cells were also successfully reprogrammed.[5] They used three kinds of somatic cells for donor nuclei, including mammary epithelial cells, fetal fibroblasts and embryo-derived cells from lambs. Using single nuclei from these cells, nuclear transfer was performed and offsprings were successfully obtained. This was the first report of cloned mammalian animals. Although several cloned lambs were born upon nuclear reprogramming, the efficiency seemed to depend on the donor cell type. The number of pregnancies using the nuclei from fetal fibroblasts or embryo-derived cells was more than that of nuclei from mammary epithelial cells. This result suggested that the mammalian nuclei from early developmental stages are readily reprogrammed.

CHANGING CELL FATE BY DEFINED FACTORS, MyoD

In the reprogramming experiments of frogs or sheep, they used oocytes that contain many proteins and other factors. Since many mechanisms might be involved in the reprogramming of somatic cells, it is difficult to explain the mechanisms of reprogramming by SCNT. Treatment with 5-azacytidine (5-aza), a DNA demethylation reagent, alone converted mouse embryonic cells, C3H/10T1/2CL8 and Swiss 3T3 cells to myogenic, adipogenic and chondrogenic cells in cell culture system (this type of change is not considered "reprogramming" but "conversion"). [6] However, it is unclear whether treatment with 5-aza caused a complete conversion. Another study showed that a single gene could achieve the conversion of some cells to other types. The gene is MyoD, a basic-helix-loop-helix transcription factor,[7] a master gene for regulation of myogenesis.[8] Introduction of MyoD alone converted mouse fibroblasts, adipoblasts, or monkey kidney cells to myoblasts. This study suggested that cells could be converted to another lineage of cells only by ectopic expression of inducing factors. There are other similar reports about cell fate conversion.[9] Monocytic precursors were converted into erythroid-megakaryocytic cells by GATA1,[10] B cells into macrophages by C/EBPα,[11] B cells into T-cells by Pax5 ablation[12] and fibroblasts into macrophage-like cells by PU.1 and C/EBPα/β.[13] These studies demonstrated that single factors can have a large impact on cell fate and specification.

REPROGRAMMING OF SOMATIC CELLS BY CELL FUSION

Reprogramming of somatic cell nuclei has also been achieved by in vitro cell fusion experiments. Nuclear reprogramming of mouse adult thymocytes was reproducibly achieved by fusion with ES cells, confirmed by the reprogramming by V-(D)-J DNA rearrangement of the T-cell receptor.[14] The Oct3/4-GFP reporter gene that is inactivated in thymocytes, is activated forty-eight hours after cell fusion. Contribution of this thymocyte-ES hybrid cells to early mouse development demonstrated its pluripotency. This study implied existence of reprogramming factors in ES cells. Very recently, it has been reported that demethylation of DNA by AID, activation-induced cytidine deaminase, is required for the reprogramming by the cell fusion experiments.[15] In this study, mouse ES cells and human fibroblasts were fused to generate interspecies heterokaryons. Knockdown of AID inhibited the demethylation of the Oct4 promoter region and resulted in the suppression of Oct4 and other pluripotent genes at the initiation of reprogramming. These results suggest that demethylation of DNA is important for the reprogramming process.

GENERATION OF INDUCED PLURIPOTENT STEM CELLS BY Sox2, Oct3/4, Klf4 AND c-Myc

All cells of our body were constructed with cells differentiated from the epiblast progenitors of the mammalian blastocysts, which are pluripotent. Embryonic stem cells (ES cells) were generated by their derivation from these cells in vitro.[16-18] ES cells also show pluripotency similar to epiblast progenitor cells of the blastocyst and infinite growth ability in vitro. ES cells are therefore considered to be an extremely useful tool for regenerative medicine.

Although human ES cells could be used in cell transplantation therapy, their clinical application faces ethical objections against utilizing human embryos. One solution is to generate pluripotent cells directly from somatic cells without the use of pre-implantation embryos. Studies of SCNT and cell fusion revealed that somatic cells could be reprogrammed into pluripotent state, even though the molecular mechanism is not clear. However, the efficiency is low and reproducibility is not consistent and thus these techniques are not conducted routinely. To resolve these problems, direct somatic nuclear reprogramming by defined factors was studied.

At first, our laboratory hypothesized that the factors that play important roles in the maintenance of ES cell pluripotency could function as reprogramming factors. We searched the genes that are highly and specifically expressed in ES cells, but not in somatic cells, using in silico expression data analysis. There were many candidates for reprogramming factors and we named them ECATs (ES cell associated transcripts).[19]

Nanog, one of the ECATs, was found to be a very important gene for the maintenance of ES cell pluripotency.[19] Nanog-knockout (KO) ES cells lost their pluripotency and differentiated into extraembryonic endoderm lineage. Another group was able to establish Nanog-KO ES cells.[20] These Nanog-KO ES cells contributed to chimeric mice but not to germ cells. The activity of self-renewal was lower in Nanog-KO ES cells than in wild-type ES cells. Moreover, Nanog-KO ES cells easily differentiated into primitive endoderm-like cells. These results indicated that Nanog is not necessary for self-renewal of ES cells but functions in inhibiting the differentiation, namely in maintaining the ES cell state. ECATs also contained the genes that have been shown to be important for ES cell pluripotency, such as Sox2 and Oct3/4.[21,22] Finally, we focused on twenty-four genes as potential reprogramming factors.[23]

These factors were classified into three groups. First group contained Nanog, Sox2 and Oct3/4, which play important roles in the maintenance of ES cell pluripotency. The tumor-associated genes, Tcl1, Stat3 and c-Myc, were classified into the second group. ECAT1, Esg1 and Klf4, originally found in our laboratory and which might specifically function in ES cells, were placed in the third group. All twenty-four genes were transduced into mouse embryonic fibroblasts by retroviral infection induced ES-like cells and we obtained ES cell-like colonies.[23] We refined the twenty-four factors into four factors, including Sox2, Oct3/4, Klf4 and c-Myc (Fig. 1) by omitting one of the factors at a time. The morphologically ES-like cells showed pluripotency similar to

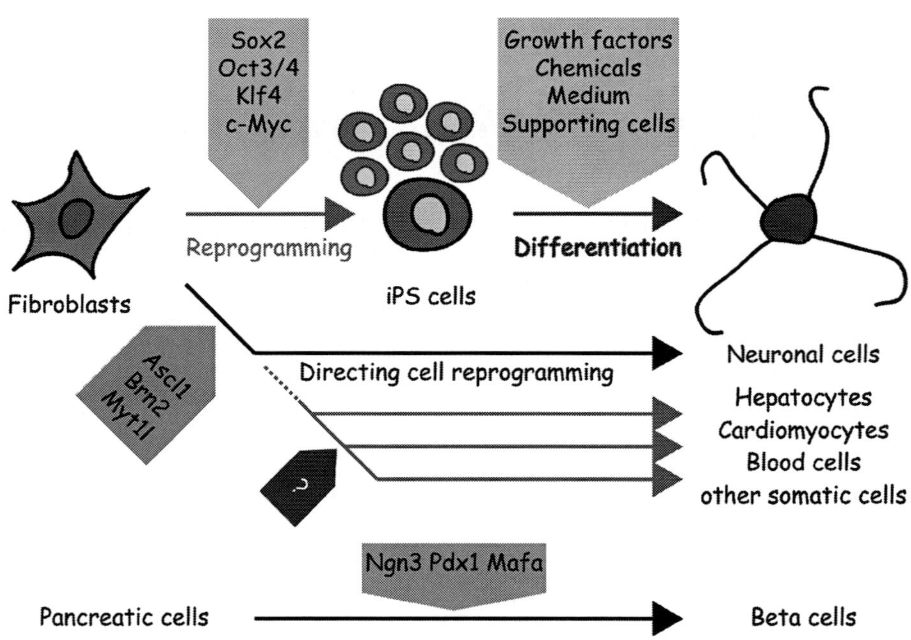

Figure 1. Reprogramming pathway. Fibroblasts are reprogrammed into iPS cells by defined factors (Sox2, Oct3/4, Klf4 and c-Myc [other combinations are also available]). iPS cells differentiate into several kinds of somatic cells by defined conditions. Neuronal cells are directly generated from fibroblasts. Beta-cells are also directly generated from pancreatic exocrine cells in vivo. Gray lines show unidentified reprogramming pathways.

ES cells and infinite cell growth similar to cancer cells. We named these cells induced pluripotent stem cells, or iPS cells. For the first-generation iPS cells, they were selected by G418 resistance derived from a Fbx15-reporter (Fbx-iPSC). Fbx15 is an ES cell specific gene that is, accordingly, not expressed in fibroblasts, but highly expressed in ES cells. Fbx-iPSCs could generate chimeric fetus but failed in adult chimeric mice, which indicated that Fbx-iPSCs did not share the exactly the same properties with ES cells and were incompletely reprogrammed cells.

To produce iPS cells which could make adult chimeric mice, we and others constructed a Nanog-reporter followed by puromycin resistance for selection of high quality iPSCs.[24] Adult chimeric mice were obtained using second-generation iPSCs (Nanog-iPSCs), which were competent for germline transmission.[24-26] In addition to the reporter gene, the timing of drug selection was also important for the generation of iPS cells that were competent for germline transmission. Fully differentiated cells including tail-derived fibroblasts,[23] hepatocytes[27] and gastric epithelial cells,[27] were also successfully reprogrammed into iPS cells by the same four factors. These results indicate that somatic cells can be reprogrammed by defined factors. Many laboratories have by now reported the generation of mouse iPS cells from a variety of cell sources.

Human iPS cells were also successfully generated by the same four factors (SOX2, OCT3/4, KLF4 and c-MYC) with some modifications.[28] At the same time, human iPS cells were generated by other factors including SOX2, OCT3/4, NANOG and LIN28.[29] Human iPS cells seem to be indistinguishable from human ES cells in morphology, growth rate and differentiation activity.

METHODS FOR iPS CELL INDUCTION

At first, iPS cells were generated from fibroblasts by retroviral transduction of four transcription factors. This method was highly efficient and reproducible for iPS cell induction. Using retroviruses, the DNA of the genes used for reprogramming, together with small portions of retroviral vectors, integrate into the host genome, increasing the risk of tumor formation in iPS cell-derived chimeric mice. We detected reactivation of retroviral c-Myc in these tumors,[24] suggesting that usage of retroviral c-Myc should be avoided for iPS cell generation. Because of this result, we tried to induce iPS cells without retroviral c-Myc for the production of safer iPS cells and succeeded in it (3F-iPSC),[30,31] although the efficiency was quite low. The morphology, growth rate and differentiation activity were similar to those of four factor-induced iPS cells (4F-iPSC). In chimeric mice derived from 3F-iPSCs, the tumorigenicity dramatically decreased compared to 4F-iPSC chimeric mice. These results indicated that c-Myc is dispensable for iPS cell induction and retroviral c-Myc should be omitted for generation of safer iPS cells. According to our recent study, the efficiency of germline transmission for 3F-iPSC is lower than that for 4F-iPSC (Aoi et al, unpublished results). It may imply that reprogramming is not fully achieved by three factors only. Identification of new factors is required to take the place of c-Myc for generation of safe and high quality iPS cells. Recent report has reported that Tbx3 improves the germline competency of mouse iPS cells instead of c-Myc.[32] Three factors and Tbx3 could increase the efficiency of iPSC generation but not the numbers of iPSC colonies. Although the frequency of germline transmission by Tbx3 increased, the tumorigenicity of chimera mice was not determined. Further observations will be required.

For the generation of safer iPS cells, several methods introducing reprogramming factors into fibroblasts have been reported. We previously reported that transfection of plasmid vectors by general transfection reagents could generate mouse iPS cells and these cells were competent for germline transmission.[33] Generation of iPS cells has also been reported by other groups using alternative methods.[34-48] Current progress of iPS cells has been well summarized in another review.[49] It is currently still unclear which of these methods is best.

MOLECULAR MECHANISM FOR iPS CELL GENERATION

Although transduction of four transcription factors into fibroblasts can change somatic cells into pluripotent cells, the molecular mechanism is unclear. Sox2 and Oct3/4 are known to be important factors in maintaining ES cell pluripotency and functions by similar mechanisms during iPS cell induction. c-Myc is famous as an oncogene and highly expressed in many tumors. The function of c-Myc is thought to remodel the chromatin structure and activate several genes. There are about thirty thousand DNA-binding regions recognized by c-Myc with or without adaptor proteins. Klf4 has a dual function acting both as an oncogene and a tumor suppressor gene.[50,51] During iPS cell induction, Klf4 might suppress the c-Myc function that negatively regulates iPS cell generation.[52,53] Although the function of each factor has been well reported, it is still unknown why transduction of defined three or four factors reprograms somatic cells to pluripotent cells.

DIRECTED CELL REPROGRAMMING: β-CELLS FROM PANCREATIC CELLS

For regenerative medicine, specific healthy somatic cells are required. Using iPS cell techniques, we can make some several different kinds of somatic cells by in vitro differentiation from established iPS cells. However, the efficiency, activity and purity of the differentiated cells from iPS cells are still too low to be used for regenerative medicine, especially for transplantation therapy. At the same time when iPS cell technology is robustly developing, it has been reported that pancreatic exocrine cells were changed to β-cells in vivo[54] (Fig. 1). This in vivo reprogramming was achieved by transduction of defined factors directly into pancreatic cells using an adenoviral system. This study suggested that most cells might have the ability of directed cell reprogramming into other cell types without going through iPS cells.

DIRECTED CELL REPROGRAMMING: NEURONAL CELLS FROM FIBROBLASTS

The possibility has recently emerged that somatic cells could be reprogrammed to pluripotent stem cells in vitro or directly reprogrammed to other types of somatic cells in vivo. After the report of in vivo directing cell reprogramming, induced somatic cells were generated in vitro.[55] The research group expressed defined factors in mouse fibroblasts and changed them into neuronal cells. The defined factors were only three neuronal-lineage-specific transcription factors. These cells were named as induced neuronal (iN) cells (Fig. 1). In iN cells, neuron-specific genes were expressed and functional synapses were also generated. This technique is very useful for production of disease-specific abnormal cells in terms of periods for generation.

DISEASE iPS CELLS FOR CLINICAL APPLICATIONS

Human iPS cells may advance regenerative medicine (Fig. 2). For cell transplantation therapy, ES cells were thought to be good sources. However, it is almost impossible for most patients to have their own ES cells. In contrast, iPS cells are derived from patients' own fibroblasts (or other cells), which can conquer the rejection problem of ES cells. Cell transplantation therapy has the potential to develop quickly. Importantly, human iPS cells now can be produced from patient somatic cells that contain genetic mutations for diseases (disease-iPSC) and the derived disease-iPSC also possess the genetic mutations. Using disease-iPS cells, it is potentially possible to reproduce the disease of patients by differentiation into abnormal cells observed in them. These cells are useful for analysis of pathology, drug screening, toxicology and study for side effects of drugs. These kinds of approaches were difficult to do before the birth of iPS cells.

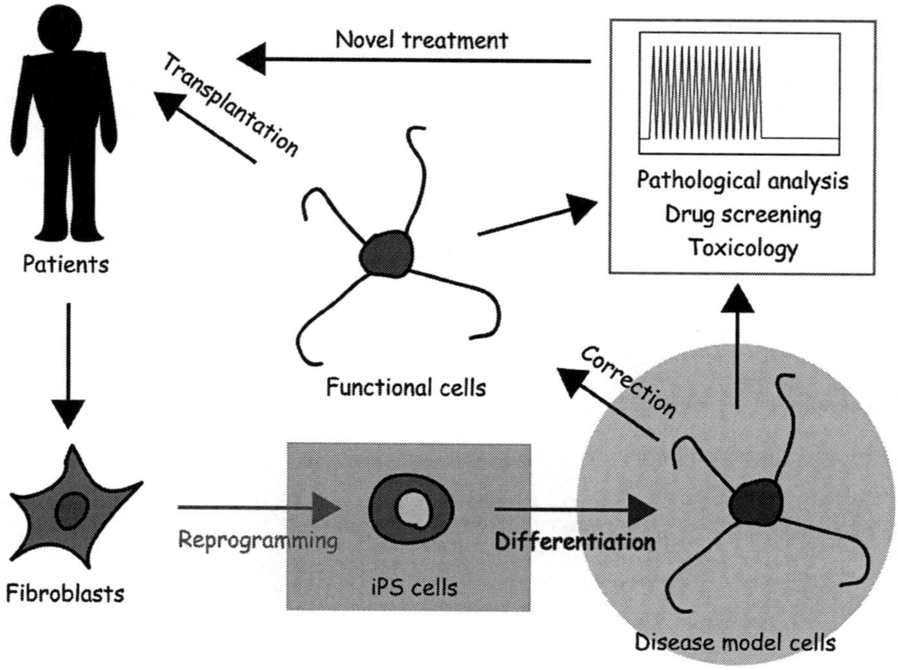

Figure 2. Application of iPS cells for regenerative medicine. Fibroblasts (or other cells) are collected from patients. The cells are reprogrammed into iPS cells and disease model cells are obtained by in vitro differentiation. For transplantation therapy, disease cells are corrected to functional cells and normal cells from patients are transplanted into their body. Disease model cells are also useful for pathological analysis, drug screening and toxicology. From these studies, novel treatment or drugs might be identified and applied for patients.

CONCLUSION

Reprogramming of somatic cells was just a dream to many researchers for a long time. Reprogramming experiments started in frogs, yielding cloned animals and eventually produced iPS cells by direct reprogramming. Reprogramming or conversion might commonly occur in differentiated cells by activation of key factors which are repressed in those cells by some exquisite mechanisms, such as methylation of genomic DNA or chromatin modifications. Once the reprogramming factors are activated, the cell properties changes and cells are reprogrammed to other cell types. Reprogramming might be easier than previously thought. But the exact combination of reprogramming factors is important for the generation of the exact cell type. Researchers will identify more combinations for generation of various cell types in the future (Fig. 1). Directed cell reprogramming is also a promising technique for obtaining disease-specific abnormal cells. It is possible that disease model cells for neurological disorders are generated from patients' skin fibroblasts directly in the short-term. Further studies are required for the application of these cells.

The iPS cell technology has been well established by many researchers. Basic research, especially on mouse and human ES cells, has helped the success for iPS cell generation. Human iPS cells might improve the current levels of regenerative medicine. The quality of iPS cells seems to be high enough for application, however iPS cell technology is not fully matured and there are several problems, such as selection and characterization of iPS cell clones, development of generation methods (viruses, plasmids, proteins, or chemicals), optimization of culture conditions and validation methods for the determination of safe iPS cells.

REFERENCES

1. Gurdon JB, Uehlinger V. "Fertile" intestine nuclei. Nature 1966; 210(5042):1240-1241.
2. Briggs R, King TJ. Transplantation of Living Nuclei From Blastula Cells into Enucleated Frogs' Eggs. Proc Natl Acad Sci USA 1952; 38(5):455-463.
3. King TJ, Briggs R. Changes in the Nuclei of Differentiating Gastrula Cells, as Demonstrated by Nuclear Transplantation. Proc Natl Acad Sci USA 1955; 41(5):321-325.
4. Gurdon JB. The developmental capacity of nuclei taken from intestinal epithelium cells of feeding tadpoles. J Embryol Exp Morphol 1962; 10:622-640.
5. Wilmut I, Schnieke AE, McWhir J et al. Viable offspring derived from fetal and adult mammalian cells. Nature 1997; 385(6619):810-813.
6. Taylor SM, Jones PA. Multiple new phenotypes induced in 10T1/2 and 3T3 cells treated with 5-azacytidine. Cell 1979; 17(4):771-779.
7. Tapscott SJ. The circuitry of a master switch: Myod and the regulation of skeletal muscle gene transcription. Development 2005; 132(12):2685-2695.
8. Davis RL, Weintraub H, Lassar AB. Expression of a single transfected cDNA converts fibroblasts to myoblasts. Cell 1987; 51(6):987-1000.
9. Graf T, Enver T. Forcing cells to change lineages. Nature 2009; 462(7273):587-594.
10. Kulessa H, Frampton J, Graf T. GATA-1 reprograms avian myelomonocytic cell lines into eosinophils, thromboblasts and erythroblasts. Genes Dev 1995; 9(10):1250-1262.
11. Xie H, Ye M, Feng R et al reprogramming of B cells into macrophages. Cell 2004; 117(5):663-676.
12. Nutt SL, Heavey B, Rolink AG et al. Commitment to the B-lymphoid lineage depends on the transcription factor Pax5. Nature 1999; 401(6753):556-562.
13. Feng R, Desbordes SC, Xie H et al. PU.1 and C/EBPalpha/beta convert fibroblasts into macrophage-like cells. Proc Natl Acad Sci USA 2008; 105(16):6057-6062.
14. Tada M, Takahama Y, Abe K et al. Nuclear reprogramming of somatic cells by in vitro hybridization with ES cells. Curr Biol 2001; 11(19):1553-1558.

15. Bhutani N, Brady JJ, Damian M et al. Reprogramming towards pluripotency requires AID-dependent DNA demethylation. Nature 2010; 463(7284):1042-7.
16. Evans MJ, Kaufman MH. Establishment in culture of pluripotential cells from mouse embryos. Nature 1981; 292(5819):154-156.
17. Martin GR. Isolation of a pluripotent cell line from early mouse embryos cultured in medium conditioned by teratocarcinoma stem cells. Proc Natl Acad Sci USA 1981; 78(12):7634-7638.
18. Thomson JA, Itskovitz-Eldor J, Shapiro SS et al. Embryonic stem cell lines derived from human blastocysts. Science 1998; 282(5391):1145-1147.
19. Mitsui K, Tokuzawa Y, Itoh H et al. The Homeoprotein Nanog Is Required for Maintenance of Pluripotency in Mouse Epiblast and ES Cells. Cell 2003; 113(5):631-642.
20. Chambers I, Silva J, Colby D et al. Nanog safeguards pluripotency and mediates germline development. Nature 2007; 450(7173):1230-1234.
21. Boyer LA, Lee TI, Cole MF et al. Core Transcriptional Regulatory Circuitry in Human Embryonic Stem Cells. Cell 2005; 122:947-956.
22. Wang J, Rao S, Chu J et al. A protein interaction network for pluripotency of embryonic stem cells. Nature 2006; 444(7117):364-368.
23. Takahashi K, Yamanaka S. Induction of pluripotent stem cells from mouse embryonic and adult fibroblast cultures by defined factors. Cell 2006; 126(4):663-676.
24. Okita K, Ichisaka T, Yamanaka S. Generation of germ-line competent induced pluripotent stem cells. Nature 2007; 448:313-317.
25. Maherali N, Sridharan R, Xie W et al. Directly reprogrammed fibroblasts show global epigenetic remodelling and widespread tissue contribution. Cell Stem Cell 2007; 1(1):55-70.
26. Wernig M, Meissner A, Foreman R et al. In vitro reprogramming of fibroblasts into a pluripotent ES cell-like state. Nature 2007; 448:318-324.
27. Aoi T, Yae K, Nakagawa M et al. Generation of Pluripotent Stem Cells from Adult Mouse Liver and Stomach Cells Science 2008; 321:699-702.
28. Takahashi K, Tanabe K, Ohnuki M et al. Induction of pluripotent stem cells from adult human fibroblasts by defined factors. Cell 2007; 131(5):861-872.
29. Yu J, Vodyanik MA, Smuga-Otto K et al. Induced pluripotent stem cell lines derived from human somatic cells. Science 2007; 318(5858):1917-1920.
30. Nakagawa M, Koyanagi M, Tanabe K et al. Generation of induced pluripotent stem cells without Myc from mouse and human fibroblasts. Nat Biotechnol 2008; 26(1):101-106.
31. Wernig M, Meissner A, Cassady JP et al. c-Myc is dispensable for direct reprogramming of mouse fibroblasts. Cell Stem Cell 2008; 2(1):10-12.
32. Han J, Yuan P, Yang H et al. Tbx3 improves the germ-line competency of induced pluripotent stem cells. Nature 463(7284):1096-1100.
33. Okita K, Nakagawa M, Hyenjong H et al. Generation of Mouse Induced Pluripotent Stem Cells Without Viral Vectors. Science 2008; 322(5903):949-953.
34. Carey BW, Markoulaki S, Hanna J et al. Reprogramming of murine and human somatic cells using a single polycistronic vector. Proc Natl Acad Sci USA 2008.
35. Sommer CA, Stadtfeld M, Murphy GJ et al. iPS Cell Generation Using a Single Lentiviral Stem Cell Cassette. Stem Cells. 2009; 27(3):543-9.
36. Stadtfeld M, Nagaya M, Utikal J et al. Induced Pluripotent Stem Cells Generated Without Viral Integration. Science 2008; 322(5903):945-949.
37. Gonzalez F, Barragan Monasterio M, Tiscornia G et al. Generation of mouse-induced pluripotent stem cells by transient expression of a single nonviral polycistronic vector. Proc Natl Acad Sci USA 2009.
38. Hotta A, Cheung AYL, Farra N et al. Isolation of human iPS cells using EOS lentiviral vectors to select for pluripotency. Nature Methods. 2009; advance online publication.
39. Kaji K, Norrby K, Paca A et al. Virus-free induction of pluripotency and subsequent excision of reprogramming factors. Nature 2009.
40. Lyssiotis CA, Foreman RK, Staerk J et al. Reprogramming of murine fibroblasts to induced pluripotent stem cells with chemical complementation of Klf4. Proc Natl Acad Sci USA 2009.
41. Shao L, Feng W, Sun Y et al. Generation of iPS cells using defined factors linked via the self-cleaving 2A sequences in a single open reading frame. Cell Research. 2009; advance online publication.
42. Soldner F, Hockemeyer D, Beard C et al. Parkinson's Disease Patient-Derived Induced Pluripotent Stem Cells Free of Viral Reprogramming Factors. Cell 2009; 136:964-977.
43. Sommer CA, Stadtfeld M, Murphy GJ et al. Induced pluripotent stem cell generation using a single lentiviral stem cell cassette. Stem Cells 2009; 27(3):543-549.
44. Woltjen K, Michael IP, Mohseni P et al. piggyBac transposition reprograms fibroblasts to induced pluripotent stem cells. Nature 2009.

45. Zhou W, Freed CR. Adenoviral Gene Delivery Can Reprogram Human Fibroblasts to Induced Pluripotent Stem Cells. Stem Cells 2009.
46. Kim D, Kim CH, Moon JI et al. Generation of human induced pluripotent stem cells by direct delivery of reprogramming proteins. Cell Stem Cell 2009; 4(6):472-476.
47. Zhou H, Wu S, Joo JY et al. Generation of induced pluripotent stem cells using recombinant proteins. Cell Stem Cell 2009; 4(5):381-384.
48. Jia F, Wilson KD, Sun N et al. A nonviral minicircle vector for deriving human iPS cells. Nat Methods; 7(3):197-199.
49. Chen L, Liu L. Current progress and prospects of induced pluripotent stem cells. Sci China C Life Sci 2009; 52(7):622-636.
50. Zhao W, Hisamuddin IM, Nandan MO et al. Identification of Kruppel-like factor 4 as a potential tumor suppressor gene in colorectal cancer. Oncogene 2004; 23(2):395-402.
51. Rowland BD, Bernards R, Peeper DS. The KLF4 tumour suppressor is a transcriptional repressor of p53 that acts as a context-dependent oncogene. Nat Cell Biol 2005; 7(11):1074-1082.
52. Cartwright P, McLean C, Sheppard A et al. LIF/STAT3 controls ES cell self-renewal and pluripotency by a Myc-dependent mechanism. Development 2005; 132(5):885-896.
53. Sumi T, Tsuneyoshi N, Nakatsuji N et al. Apoptosis and differentiation of human embryonic stem cells induced by sustained activation of c-Myc. Oncogene 2007; 26(38):5564-5576.
54. Zhou Q, Brown J, Kanarek A et al. In vivo reprogramming of adult pancreatic exocrine cells to beta-cells. Nature 2008; 455(7213):627-632.
55. Vierbuchen T, Ostermeier A, Pang ZP et al. Direct conversion of fibroblasts to functional neurons by defined factors. Nature 2010; 463(7284):1035-41

INDEX

P

R

S

T

X

Y